VOLUME SIXTY NINE

ADVANCES IN
QUANTUM CHEMISTRY
Energetic Materials

EDITORIAL BOARD

VOLUME SIXTY NINE

ADVANCES IN QUANTUM CHEMISTRY

Energetic Materials

Edited by

JOHN R. SABIN

Quantum Theory Project
University of Florida
Gainesville, Florida

Series Editors

JOHN R. SABIN
and ERKKI BRÄNDAS

ELSEVIER

AMSTERDAM • BOSTON • HEIDELBERG • LONDON
NEW YORK • OXFORD • PARIS • SAN DIEGO
SAN FRANCISCO • SINGAPORE • SYDNEY • TOKYO
Academic Press is an imprint of Elsevier

Academic Press is an imprint of Elsevier
The Boulevard, Langford Lane, Kidlington, Oxford, OX5 1GB, UK
32 Jamestown Road, London NW1 7BY, UK
Radarweg 29, PO Box 211, 1000 AE Amsterdam, The Netherlands
225 Wyman Street, Waltham, MA 02451, USA
525 B Street, Suite 1800, San Diego, CA 92101-4495, USA

First edition 2014

ISBN: 978-0-12-800345-9
ISSN: 0065-3276

14 15 16 17 12 11 10 9 8 7 6 5 4 3 2 1

Printed and bound by CPI Group (UK) Ltd, Croydon, CR0 4YY

CONTENTS

PREFACE

It is with great pleasure that we bring you a thematic volume of *Advances in Quantum Chemistry* focusing on energy-rich materials. On initiation, such materials decompose and release large amounts of energy of various types, for example, heat, light, and kinetic energy of the fragments. Consequently, they are frequently used as explosives, incendiary devices, propellants, and pyrotechnics. More recently, there has also been discussion of the use of energy-rich materials as a mechanism for energy storage. The detailed understanding of the electronic structure of these systems and the mechanism of their decomposition is necessary for design and application of such materials. Thus, in this volume, we present eight chapters by experts in the field who discuss precisely these points.

This volume follows the two-volume contribution[1] to the series *Theoretical and Computational Chemistry* of a decade ago edited by Peter Politzer and Jane S. Murray and includes some of the material learned since that time.

John R. Sabin
Editor

[1] Energetic Materials: Part 1. Decomposition, Crystal and Molecular Properties, and Part 2. Detonation, Combustion, Volumes 12 and 13 of Theoretical and Computational Chemistry, edited by Peter Politzer and Jane S. Murray, Elsevier, Amsterdam, 2003.

CONTRIBUTORS

Rodney J. Bartlett
Quantum Theory Project, University of Florida, Gainesville, Florida, USA

Elliot R. Bernstein
Department of Chemistry, Colorado State University, Fort Collins, Colorado, USA

Laurence E. Fried
Energetic Materials Center, Lawrence Livermore National Laboratory, Livermore, California, USA

Marta Gałyńska
Theoretical Chemistry Division, Chemistry Department, Lund University, Box 124, Lund, Sweden

Maija M. Kuklja
MSE Department, University of Maryland, College Park, Maryland, and National Science Foundation, Arlington, Virginia, USA

Ming-Chang Lin
Center for Interdisciplinary Molecular Science, Department of Applied Chemistry, National Chiao Tung University, Hsinchu, Taiwan

M. Riad Manaa
Energetic Materials Center, Lawrence Livermore National Laboratory, Livermore, California, USA

Jane S. Murray
Department of Chemistry, University of New Orleans, New Orleans, Louisiana, USA

N.T. Nghia
Center for Interdisciplinary Molecular Science, Department of Applied Chemistry, National Chiao Tung University, Hsinchu, Taiwan, and Department of Physical Chemistry, Hanoi University of Science and Technology, Hanoi, Vietnam

Ajith Perera
Quantum Theory Project, University of Florida, Gainesville, Florida, USA

Petter Persson
Theoretical Chemistry Division, Chemistry Department, Lund University, Box 124, Lund, Sweden

Peter Politzer
Department of Chemistry, University of New Orleans, New Orleans, Louisiana, USA

Putikam Raghunath
Center for Interdisciplinary Molecular Science, Department of Applied Chemistry, National Chiao Tung University, Hsinchu, Taiwan

Betsy M. Rice
Weapons and Materials Research Directorate, Army Research Laboratory, Aberdeen Proving Ground, Aberdeen, Maryland, USA

DeCarlos E. Taylor
Weapons and Materials Research Directorate, Army Research Laboratory, Aberdeen Proving Ground, Aberdeen, Maryland, USA

CHAPTER ONE

Detonation Performance and Sensitivity: A Quest for Balance

Peter Politzer[1], Jane S. Murray
Department of Chemistry, University of New Orleans, New Orleans, Louisiana, USA
[1]Corresponding author: e-mail address: ppolitze@uno.edu

Contents

Abstract

In designing new explosives, a key goal is to achieve an optimum balance between two inherently contradictory objectives: a high level of detonation performance and low sensitivity to accidental initiation of detonation. There appears to be a reasonably good understanding of the determinants of detonation performance, and there are procedures available for quantitative predictions of satisfactory reliability. The situation regarding sensitivity is more complex. However, progress has been made in identifying at least some of the factors that are involved. It is accordingly possible to propose a series of structural features that can be guidelines in the quest for excellent detonation performance combined with low sensitivity.

1. AN UNEASY COEXISTENCE

An explosive compound is intrinsically metastable. The input of a relatively small amount of energy can initiate very rapid self-propagating

Advances in Quantum Chemistry, Volume 69
ISSN 0065-3276
http://dx.doi.org/10.1016/B978-0-12-800345-9.00001-5

decomposition producing gaseous products at high temperatures and pressures (detonation).[1–5] The term "sensitivity" refers to the ease with which this initiation can occur; more specifically, how vulnerable is the compound to accidental, undesirable detonation caused by an unintended stimulus? Such a stimulus could be impact, shock, heat, friction, sparks, etc.

A long-standing goal in the explosives area is to combine a high level of detonation performance with low sensitivity. The challenge is that these objectives are inherently somewhat contradictory, and factors promoting one of them are often in conflict with the other. Realistically, one generally seeks an optimum balance between detonation performance and sensitivity. There are, of course, other important considerations that are also involved in evaluating potential energetic compounds: difficulty and cost of synthesis, processability, etc. These will not be addressed in this chapter.

Our focus will be upon trying to design new explosives that combine high detonation performance and low sensitivity to a greater extent than is currently available. Ideally, this would require (a) a good understanding of the factors that govern performance and sensitivity and (b) some means of assessing how well a proposed compound or formulation would achieve the desired levels of each. The latter information should be available prior to attempts to prepare the proposed explosive. As will be seen, this is possible to a greater extent with respect to detonation performance than for sensitivity.

2. PREDICTING DETONATION PERFORMANCE

Two key properties that are used to evaluate explosives are the detonation velocity D and the detonation pressure P. The former is the stable velocity of the shock front that characterizes detonation, while the latter is the stable pressure that is developed behind the front.[1–5] High values of both D and P are desirable.

The detonation velocity and pressure can be determined experimentally.[6–10] However, our present interest is in proposed explosives that have not yet been prepared. For such, D and P can be predicted with parameterized computer codes, for example, RUBY,[11] TIGER,[12] BKW,[13] EXPLO5,[14] CHEETAH,[15] and ZMWNI,[16] or they can be calculated by various semiempirical procedures.[17–21] Particularly useful in contemplating the design of new explosive compounds are the Kamlet–Jacobs equations for D and P.[17] They have been shown to have good accuracy and reliability in comparisons with experimental values,[17,18,21,22] and, what is very important, they explicitly identify four primary determinants of D and P.

The Kamlet–Jacobs equations, Eqs. (1.1) and (1.2), apply to C, H, N, O-containing compounds.[17] They express detonation velocity and detonation pressure as being governed by (a) the number of moles N of gaseous detonation products per gram of the explosive compound, (b) their average molecular mass M_{ave} in g/mol, (c) the heat release $Q=-\Delta H$, where ΔH is the enthalpy change in the detonation reaction and Q is in calories per gram of compound, and (d) the loading density d of the compound, in g/cm^3:

$$D(\text{km/s}) = 1.01\left[N^{0.5}M_{ave}^{0.25}Q^{0.25}(1+1.30d)\right] \tag{1.1}$$

$$P(\text{kbar}) = 15.58\left[NM_{ave}^{0.5}Q^{0.5}d^2\right] \tag{1.2}$$

It is sometimes convenient to define a parameter φ,

$$\varphi = NM_{ave}^{0.5}Q^{0.5} \tag{1.3}$$

in terms of which Eqs. (1.1) and (1.2) become

$$D(\text{km/s}) = 1.01\varphi^{0.5}(1+1.30d) \tag{1.4}$$

$$P(\text{kbar}) = 15.58\varphi d^2 \tag{1.5}$$

The density d appears to a higher power in Eqs. (1.1) and (1.2) than do any of the other variables, attesting to its importance as a determinant of D and P. However, it is not as dominant, on a relative basis, as is sometimes believed. For example, RDX (**1**) has $D=8.754$ km/s and $P=347$ kbar at a loading density of 1.80 g/cm^3, while TATB (**2**) has lower D and P, 7.860 km/s and 315 kbar, at a greater loading density of 1.895 g/cm^3.[13] The reason is the much higher value of Q for RDX.[22]

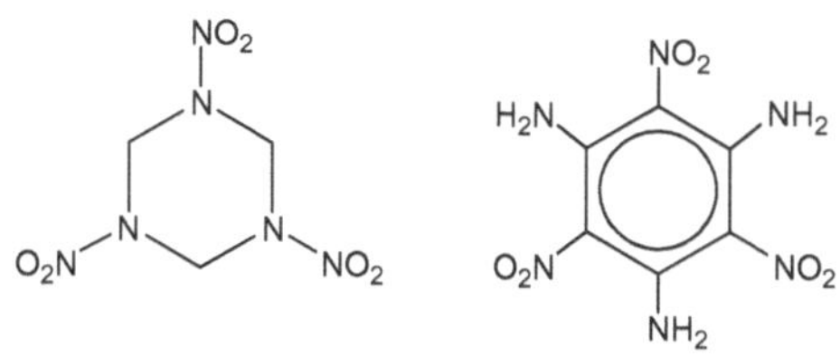

1, RDX **2**, TATB

A computational approach has been developed for predicting the crystal densities of C, H, N, O compounds.[23,24] The resulting values are commonly used for d in Eqs. (1.1) and (1.2), even though actual loading densities are typically less than those for the pure crystals.

The quantities N and M_{ave} in Eqs. (1.1) and (1.2) can be determined if it is known how much of each type of gaseous product results from the detonation reaction. In order to find Q, one needs ΔH for the detonation reaction, which can be obtained as the difference between the enthalpies of formation of the products and the explosive compound, taking into account the respective physical states (solid, liquid, or gaseous). (Using the gas phase value for the explosive leads to overestimating both D and P, by roughly 2% and 4%, respectively.[22]) A procedure for calculating solid phase enthalpies of formation of C, H, N, O compounds is given by Byrd and Rice.[25] For the detonation products (see below), experimental values are available.[26] Once ΔH is known, then $Q = -\Delta H$.

Evaluating N, M_{ave}, and Q clearly requires knowing—or assuming—what is the composition of the detonation products. In principle, there can be a wide variety of these. Even if we consider only the most common types of explosives—composed of just carbon, hydrogen, nitrogen, and oxygen atoms—the detonation products can include H_2O, H_2, O_2, CO_2, CO, NH_3, CH_4, NO, N_2, and solid carbon.[13] Furthermore, the relative amounts of these change, for a given explosive, depending upon its loading density. In practice, however, the products are normally found to be primarily H_2O, N_2, CO_2, and/or CO.[13,17] Furthermore, the main consequence of decreasing the loading density is to reduce the CO_2/CO ratio.

There have accordingly been proposed several sets of "rules" for predicting the detonation products of C, H, N, O explosives. Kamlet and Jacobs suggested that for loading densities near the pure crystal value, the products can be assumed to be N_2, H_2O, and CO_2, with oxygens being used for H_2O before CO_2 and any remaining carbon going to the solid state.[17] This is supported by data generated by the RUBY[17] and the BKW[13] codes. In contrast, the Kistiakowsky–Wilson, the modified Kistiakowsky–Wilson, and the Springall–Roberts rules put greater emphasis upon CO although not excluding CO_2.[3,27] Thus, for RDX (**1**), the Kamlet–Jacobs protocol would write the detonation reaction as

$$C_3H_6O_6N_6(s) \rightarrow 3N_2(g) + 3H_2O(g) + 1.5CO_2(g) + 1.5C(s) \qquad (1.6)$$

whereas by the Kistiakowsky–Wilson,

$$C_3H_6O_6N_6(s) \rightarrow 3N_2(g) + 3H_2O(g) + 3CO(g) \qquad (1.7)$$

If one's interest is in the number of moles of gaseous products, then Eqs. (1.6) and (1.7) differ significantly. However, for calculating the detonation properties via Eqs. (1.1) and (1.2), the two sets of products depicted in Eqs. (1.6) and (1.7) are found to yield similar results. For a loading density of 1.80 g/cm^3,[13] and using an RDX enthalpy of formation of 18.9 kcal/mol,[25]

$$\text{Eq.}(1.6): \quad D = 8.81\,\text{km/s}, \quad P = 345\,\text{kbar}$$
$$\text{Eq.}(1.7): \quad D = 8.95\,\text{km/s}, \quad P = 355\,\text{kbar}$$

For comparison, the experimental D and P for RDX at this loading density are $D = 8.754$ km/s and $P = 347$ kbar.[13] Thus, the detonation properties obtained with the two different sets of products are in quite good agreement with each other *and* with the experimental values.

The products in Eq. (1.6) include 1.5 mol of CO_2(g) and 1.5 C(s). In Eq. (1.7), these are replaced by 3CO(g). Why then are D and P predicted by Eqs. (1.1) and (1.2) to be approximately the same for the two reactions? Since the density is unchanged, the answer must lie in the product $NM_{\text{ave}}^{0.5}Q^{0.5}$, that is, the quantity φ. This has been discussed earlier.[22] Equation (1.7) has more moles of gaseous products, making N larger; however, M_{ave} is smaller because CO has a lower molecular mass than CO_2. Q is also smaller for Eq. (1.7) because the enthalpy of formation of CO is much less negative (−26.42 kcal/mol) than that of CO_2 (−94.05 kcal/mol)[26]; for C(s) it is zero by definition. Since N appears to a higher power in Eqs. (1.1) and (1.2) than do M_{ave} and Q, the changes in these quantities largely balance; φ is approximately the same for the two reactions, as are therefore D and P.

It has indeed been demonstrated that all four of the sets of detonation product rules mentioned above yield, for a given compound, similar values of D and P via Eqs. (1.1) and (1.2).[28] They are also all close to the experimental values, since Eqs. (1.1) and (1.2) in conjunction with the Kamlet–Jacobs rules have been shown to produce good agreement with experiment.[17,18,21,22]

The near-constancy of φ *for a given explosive* combined with Eqs. (1.4) and (1.5) suggests that

$$D \sim d \tag{1.8}$$

$$P \sim d^2 \tag{1.9}$$

Equation (1.8) explains why experimental D have been found to correlate linearly with d alone for various compounds and ranges of d.[13,29]

In summary, Eqs. (1.1) and (1.2) identify four factors upon which the detonation velocity and detonation pressure depend, and they provide a means for obtaining reasonable estimates of D and P for compounds that have not yet been prepared or characterized. We have recently used these equations in evaluating several proposed new explosives.[30,31,31a]

3. PREDICTING SENSITIVITY

The issue of sensitivity is a particularly challenging one. There is no equivalent of the Kamlet–Jacobs equations which—for diverse categories of explosive compounds—provide insight into what determines detonation performance and also allow its prediction quantitatively with satisfactory accuracy.

The first point to recognize is that different stimuli can initiate accidental, unwanted detonation: impact, shock, heat, electric sparks, friction, etc. A variety of techniques are used to measure an explosive's sensitivity to each of these.[1,32–35] Do sensitivities to the various types of stimuli correlate with each other? Sometimes yes, but not in all cases. Storm et al. did obtain satisfactory relationships, for various types of compounds, between measured shock and impact sensitivities and between measured shock sensitivities and the calculated critical temperatures at which thermal decompositions become self-sustaining.[35] On the other hand, the compound **3**, for example, is reported to be quite sensitive to impact but not to friction or sparks[36]; compound **4** is insensitive to electrostatic discharge, moderately sensitive to friction, and quite sensitive to impact.[37]

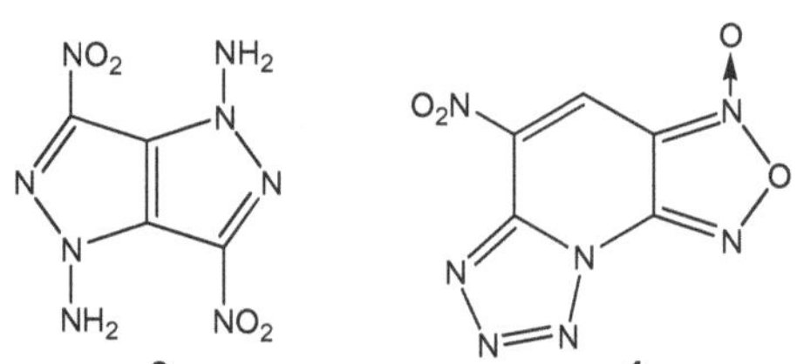

3 **4**

The most widely cited type of sensitivity is that toward impact, which we will accordingly emphasize, although much of the discussion will apply as well to shock sensitivity. Impact sensitivity is commonly measured by dropping a given mass upon a sample of the explosive material and determining the height from which 50% of the drops initiate reaction. This height is labeled h_{50}; in the United States, its value in centimeters is widely used as a measure of sensitivity: the smaller is h_{50}, the more sensitive is the explosive.

However, the magnitude of h_{50} also depends upon the mass being used, and unfortunately this is often not specified. For this reason, many researchers prefer to convert h_{50} to the impact energy, which takes the mass into account:

$$\text{Impact energy} = mgh_{50} \tag{1.10}$$

In Eq. (1.10), m is the mass being used and g is the acceleration of gravity. The less is the impact energy required to initiate reaction, the more sensitive is the explosive. For a mass of 2.5 kg, which is frequently used, $h_{50}=100$ cm corresponds to impact energy = 24.5 J = 24.5 Nm. In practice, the sensitivities of most C, H, N, O explosives range from about $h_{50}=10$ cm (impact energy = 2.5 J), which is very sensitive, to about $h_{50}=300$ cm (impact energy = 74 J), which is relatively insensitive.[35] More detailed descriptions of the drop-height technique can be found elsewhere.[1,32,34,35]

While this procedure is straightforward in principle, it can be quite challenging to obtain reproducible results.[32,34] A major problem is that impact sensitivity is very dependent upon the physical state of the sample—the sizes, shapes, and hardness of the crystals, the roughness of their surfaces, their purity, lattice defects, etc.[38–41] For instance, impact sensitivity can be considerably decreased by reducing particle size. The detailed manner of preparation of the sample of explosive—crystallization, purification, grinding, etc.—can make a big difference. For example, sensitivity may vary significantly from one polymorphic form of a solid to another,[42,43] and hence from one crystallization technique to another.[44]

The strong dependence of impact sensitivity upon physical factors has given rise to some skepticism as to whether meaningful relationships between measured sensitivities and intrinsic molecular/crystalline properties can be found. This reaction seems unduly pessimistic. After all, many properties in chemistry depend upon physical conditions; this is why the concept of standard states was introduced. With regard to sensitivity measurements, great care must be taken to ensure that highly specific and consistent preparation and testing procedures are used under fixed atmospheric conditions.[34,45,46]

Notwithstanding the pessimism that is sometimes expressed, there have in fact been reported correlations between impact or shock sensitivity and quite a variety of molecular or crystalline properties: bond energies,[47,48] bond lengths,[42,49] bond polarities,[50] band gaps,[51,52] electrostatic potentials,[53,54] atomic charges,[33,55–57] heats of detonation,[33,58]

NMR chemical shifts,[33] rates of vibrational energy transfer,[59–61] heats of fusion,[33] oxygen balance,[45,46] etc. These correlations are generally limited to compounds of a specific type, for example, nitramines, nitroaromatics, etc.

Ironically, such correlations have also occasionally provoked a negative response,[2,62] in part because they have been established with so many seemingly disparate properties. While those relationships that are computationally based provide a useful predictive capability in evaluating proposed compounds, it can be argued that the overall range of properties is too large to afford any insight into factors governing sensitivity.

It must be kept in mind, however, that correlations are frequently symptomatic, that is, they *reflect* causation but do not directly reveal it. We suggest that many of the relationships cited above simply reflect the importance of a "trigger linkage."[45,63] This term refers to the concept that certain types of bonds are particularly likely to be broken by the input of energy from an external stimulus and their ruptures then "trigger" the further exothermic and self-sustaining decomposition that characterizes detonation. Commonly proposed trigger linkages include $C—NO_2$ in nitroaromatics, nitroaliphatics, and nitroheterocycles; $N—NO_2$ in nitramines; $O—NO_2$ in nitrate esters; and $N—N_2$ in organic azides.[33,62–64] We emphasize, however, that rupture of one of the aforementioned bonds is not the only mechanism that can lead to detonation initiation; there are certainly other possibilities,[33,45,46,62–65] such as molecular rearrangements. For molecules that follow such alternative pathways, relationships such as those mentioned above may not predict the proper impact sensitivity.

We will consider in detail how the relationships cited may reflect the ruptures of trigger linkages. Correlations of sensitivity with $C—NO_2$ and $N—NO_2$ dissociation energies[47,48] and bond lengths[42,49] clearly are focusing upon how readily the respective trigger linkages can be broken. It has further been demonstrated that the strengths of $C—NO_2$ and $N—NO_2$ bonds vary inversely with the magnitudes of the positive electrostatic potentials in the C—N and N—N internuclear regions[65–68] that are caused by the movement of electronic charge to the NO_2 oxygens. This explains correlations of sensitivity not only with electrostatic potentials[53,54] but also with the atomic charges on NO_2 groups,[33,55–57] bond polarities,[50] and the NMR chemical shifts of NO_2 nitrogens,[33] all of which reflect electronic charge distributions in C—N and N—N bond regions. Diminished band gaps have been linked to electronic excitations that can break $N—NO_2$ bonds[69]; this accounts for correlations of sensitivities with band gaps.[51,52] Detonation initiation has been linked to lattice vibrational energy (from an external stimulus) being transferred to molecular modes and resulting in the breaking of $C—NO_2$ and $N—NO_2$ bonds[2,70]—hence the relationship of rate of vibrational energy transfer to sensitivity.[59–61] Zeman has invoked the Evans–Polanyi–Semenov analysis to relate the heat of detonation to the activation energy of the decomposition,[71] which—for trigger linkage rupture—corresponds to the bond dissociation energy; this can explain sensitivity correlating with detonation heat release.[33,58] Finally, a higher oxygen balance, and therefore oxygen content, means that more CO_2 can be produced in the detonation, resulting in a higher heat release, as pointed out in Section 2 and in earlier work.[22]

Thus, nearly all (and perhaps all) of the sensitivity correlations cited can be rationalized as reflecting the roles of trigger linkages in detonation initiation, despite the quite different properties that are involved. However, we stress again that these correlations can only be invoked for the specific types of compounds for which they were developed, and only if breaking one of the presumed trigger linkages is the key to the particular initiation process.[56,68] The latter need not be the case; initiation can be via another mechanism,[33,45,46,62–65] for example, molecular rearrangement, and/or it may be affected by crystalline factors, as will be discussed.

An interesting feature of the types of correlations that have been mentioned is that some of them can be improved by scaling the property being related to sensitivity, using a global feature of the entire molecule as the scaling factor. For instance, trigger bond energy may be divided by (a) total molecular energy,[72,73] (b) detonation heat density,[4] (c) detonation heat per covalent bond,[74] and (d) detonation heat per atom.[75] Other examples include dividing the electrostatic potential by the molecular surface area[67] and inverse bond lengths by molecular mass.[49]

Scaling by some variation of the heat of detonation has been justified by invoking Arrhenius kinetics, rate $\sim\exp[-E_{act}/kT]$, where E_{act} is the activation energy, T the absolute temperature, and k the Boltzmann constant. It is argued that replacing E_{act} by the trigger bond energy and assuming that the heat release governs the local temperature suggests that the logarithm of the sensitivity should vary with the ratio of bond energy to heat release.[4,74,75] This neglects the role of the heat capacity, which can vary from one compound to another and also changes with the temperature.

These scaled relationships have had some success, as have the others cited earlier, although the variety of scaling factors again raises the question: what do they really reflect? The fact that good results are sometimes obtained without such factors may simply mean that the magnitudes of the latter are very similar for the particular series of compounds.[67]

4. EXAMINATION OF SOME CONCEPTS RELATING TO SENSITIVITY

4.1. Trigger linkages

The various correlations that were discussed in Section 3 emphasize the trigger linkage concept. However, there are ambiguities associated with this. First is the fact that rupture of a presumed trigger linkage may not be the key initiating event. There are numerous other possibilities,[62,76] including nitro/aci tautomerization,[77–80] furazan/furoxan formation,[81,82] nitro/nitrite isomerization,[83] release of N_2 from a heterocyclic ring,[84,85] etc.

Even when breaking a trigger linkage does play a role in the initiation, there may be additional issues. Thus, Kamlet compared two series of nitroaliphatics, one having $C(NO_2)_3$ groups and the other $N{-}NO_2$.[45] For compounds with similar oxygen balances, the former were less sensitive. This was initially attributed to the fact that $C{-}NO_2$ bonds are in general stronger than $N{-}NO_2$.[4,65,68,75] However, the first-order decomposition activation energies are very similar for both series of compounds. The apparent explanation is that while the decomposition of the $C{-}NO_2$ compounds does proceed via first-order $C{-}NO_2$ scission, for the $N{-}NO_2$ this is true only at the beginning, followed shortly by autocatalysis by the initially generated NO_2 product.[45] Thus, a comparison of first-order activation energies is meaningless, because the nitramine decompositions quickly adopt different mechanisms, which greatly increase their rates and hence the sensitivities of the compounds.

If a particular detonation initiation process is believed to be governed by rupture of a specific trigger linkage, then the activation energy for the decomposition is often taken to be the dissociation energy of that bond. There are at least two problems with this. One is that bond energies can differ depending upon whether the molecule is in the bulk solid, at a void or surface, or in the gas phase.[86–88] A second problem is that at high pressures and temperatures, the Arrhenius activation energy can differ significantly from the "intrinsic" value, by an amount proportional to the volume change in the activation process.[89]

Notwithstanding the various issues that have been raised, the trigger linkage concept continues to be qualitatively extremely important, and it has been used, explicitly or implicitly, in developing some perhaps surprisingly effective sensitivity correlations. However, it should be invoked cautiously, with an awareness of possible complications.

4.2. Molecular electrostatic potentials

We have already pointed out that impact sensitivity has been correlated quite well with the positive electrostatic potentials in the internuclear regions of trigger linkages,[53,54] and that these potentials, in turn, correlate (inversely) with the respective bond energies.[65–68] We wish now to look at the relationship between electrostatic potential and sensitivity from a more conceptual standpoint.

The electrostatic potential $V(\mathbf{r})$ that the nuclei and electrons of a molecule create at a point $\mathbf{r}$ in the surrounding space is given by

$$V(\mathbf{r}) = \sum_A \frac{Z_A}{|\mathbf{R}_A - \mathbf{r}|} - \int \frac{\rho(\mathbf{r}')d\mathbf{r}'}{|\mathbf{r}' - \mathbf{r}|} \quad (1.11)$$

where Z_A is the charge on nucleus A, located at $\mathbf{R}_A$, and $\rho(\mathbf{r})$ is the electronic density of the molecule. The sign of $V(\mathbf{r})$ in any region depends upon whether the positive contribution of the nuclei or the negative one of the electrons is dominant there.

It is important to emphasize that the electrostatic potential is a real physical property; it should not be confused with atomic charges, which have no rigorous basis and have been defined in many different ways that can give quite contradictory results.[90] $V(\mathbf{r})$ can be determined experimentally, by diffraction methods,[91–93] as well as computationally. It is a property of fundamental significance,[90,94,95] which can serve as the basic determinant of all molecular properties.[96] Atomic and molecular energies, for instance, can be expressed rigorously in terms of the electrostatic potentials at their nuclei.[94,97]

When $V(\mathbf{r})$ is used to interpret and predict molecular interactive behavior, it is commonly computed on the "surface" of the molecule and labeled $V_S(\mathbf{r})$. The surface is usually taken to be the 0.001 au (electrons/bohr^3) contour of the molecule's electronic density $\rho(\mathbf{r})$, as suggested by Bader et al.[98] $V_S(\mathbf{r})$ has been shown to be very useful for analyzing noncovalent interactions.[90,94,99] It has in fact been demonstrated that a variety of condensed phase physical properties that depend upon noncovalent interactions can be approximated analytically as functions of certain features of $V_S(\mathbf{r})$, for example, its locally most positive and most negative values (the $V_{S,max}$

and $V_{S,min}$), its positive and negative averages and variances, etc. What is particularly significant is that properties pertaining to condensed phases can be represented empirically in terms of quantities computed for a single (i.e., gas phase) molecule. $V_S(\mathbf{r})$ is apparently able to generally account satisfactorily for the molecule's interactions in solid, liquid, and solution phases.

An example that is very relevant in the context of this chapter is crystal density. The densities of C, H, N, O molecular and ionic solids can be estimated with overall reasonable accuracy using the molecular and ionic volumes encompassed by the 0.001 au surfaces plus electrostatic interaction terms based upon the molecular or ionic $V_S(\mathbf{r})$.[23,24,100]

In general, the surface electrostatic potentials of C, H, N, O explosives differ markedly from those of typical organic molecules.[54,90,94] For most organic molecules, the regions of negative $V_S(\mathbf{r})$—which are due primarily to lone pairs and π electrons—are likely to be stronger than the positive, although the latter may be larger in area. An example is shown in Fig. 1.1, which depicts the $V_S(\mathbf{r})$ of *m*-hydroxypyridine, **5**. The negative region is above the nitrogen and oxygen lone pairs and the π electrons of the ring, the most negative values (the $V_{S,min}$) being −37, −21, and

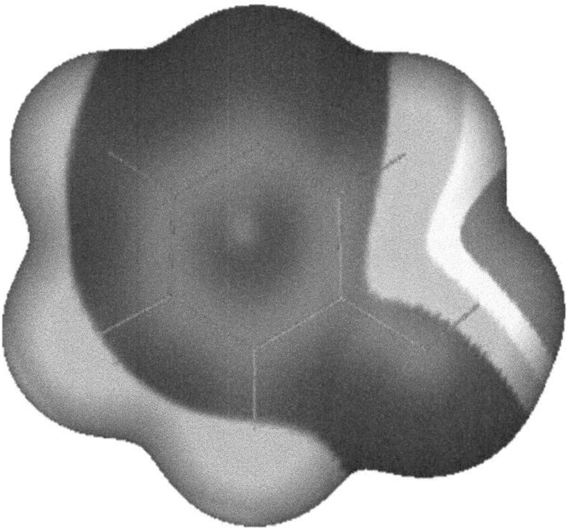

Figure 1.1 Electrostatic potential on the 0.001 au molecular surface of *meta*-hydroxypyridine. The nitrogen is at the top, the hydroxyl group is at the lower right. Color ranges, in kcal/mol, are: red, more positive than 30; yellow, 15–30; green, 0–15; blue, negative. There is a $V_{S,max}$ of 58 kcal/mol (hydroxyl hydrogen) and $V_{S,min}$ of −37 kcal/mol (nitrogen), −21 kcal/mol (oxygen), and −11 kcal/mol (ring). The calculations were at the B3PW91/6-31G(d,p) level. (For interpretation of the references to color in this figure legend, the reader is referred to the online version of this chapter.)

−11 kcal/mol, respectively. The strongest positive potential ($V_{S,max}$) is associated with the hydroxyl hydrogen, 58 kcal/mol.

C, H, N, O explosives, however, commonly have highly electron-withdrawing components on their peripheries—NO_2 groups, aza nitrogens, etc.—which deplete the electronic charge in the central portions of the molecules. This leaves strongly positive potentials in these regions and above C—NO_2 and N—NO_2 bonds. There are weakly negative potentials on the outsides of the molecules, associated with the lone pairs of the NO_2 oxygens and aza nitrogens. Overall, the $V_S(\mathbf{r})$ of most C, H, N, O explosives are dominated by the positive potentials, as can be seen in Fig. 1.2 for the explosive compound 2,4,6-trinitropyridine (**6**). $V_S(\mathbf{r})$ is positive over most of the molecule, particularly above the ring ($V_{S,max}$=38 kcal/mol) and the C—NO_2 bonds (*ortho* $V_{S,max}$=35 kcal/mol, *para* $V_{S,max}$=39 kcal/mol). The only negative regions are due to the lone pairs of the NO_2 oxygens ($V_{S,min}$=−15 to −21 kcal/mol) and the ring nitrogen ($V_{S,min}$=−28 kcal/mol). Other examples have been presented on numerous occasions.[53,64,65,68,101]

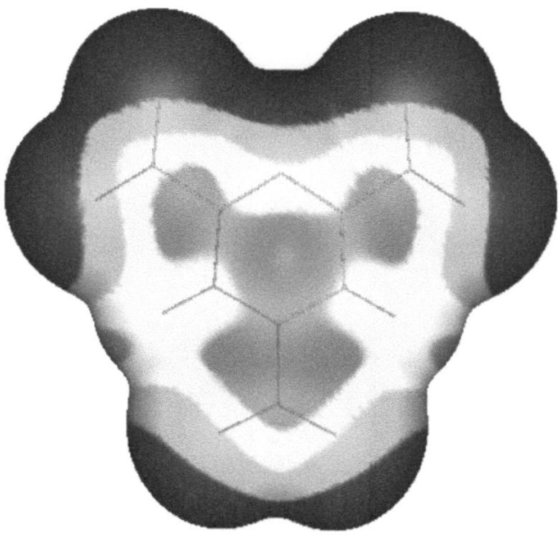

Figure 1.2 Electrostatic potential on the 0.001 au molecular surface of 2,4,6-trinitropyridine. The ring nitrogen is at the top. Color ranges, in kcal/mol, are: red, more positive than 30; yellow, 15–30; green, 0–15; blue, negative. There are $V_{S,max}$ of 39 and 35 kcal/mol (above the *para* and *ortho* C—NO_2 bonds, respectively) and 38 kcal/mol (above the center of the ring). There are $V_{S,min}$ of −28 kcal/mol (ring nitrogen) and −15 to −21 kcal/mol (NO_2 oxygens). The calculations were at the B3PW91/6-31G (d,p) level. (For interpretation of the references to color in this figure legend, the reader is referred to the online version of this chapter.)

The molecular surface potentials of C, H, N, O explosives therefore tend to be characterized by an anomalous imbalance and separation between strongly positive central regions and weakly negative peripheral ones. These features distinguish most C, H, N, O explosives from other organic molecules. Can these features also be related, at least symptomatically, to the explosives' metastabilities and sensitivities? Pauling's electroneutrality postulate offers some encouragement; it implies a tendency for the electrostatic potential of a molecule to be as uniform and close to neutral as possible, with deviations from this promoting instability.[102]

Qualitatively, and for a given type of compound (e.g., nitroaromatics), it has been found that sensitivity does generally increase as $V_S(\mathbf{r})$ over the inner portion of the molecule becomes more positive.[64,65,68,101,103–105] For instance, compare the surface electrostatic potentials of 2,4,6-trinitropyridine, **6**, and 2,6-diamino-3,5-dinitropyrazine-1-oxide, **7** (Figs. 1.2 and 1.3). Both are known explosives, but **7** is considerably less sensitive (h_{50} = 117 cm, impact energy = 29 J[106]) than **6** (h_{50} = 18–27 cm, impact energy = 4.5–6.5 J[107]). This is what would be predicted from the much more positive surface potential in the central portion of **6** (Fig. 1.2).

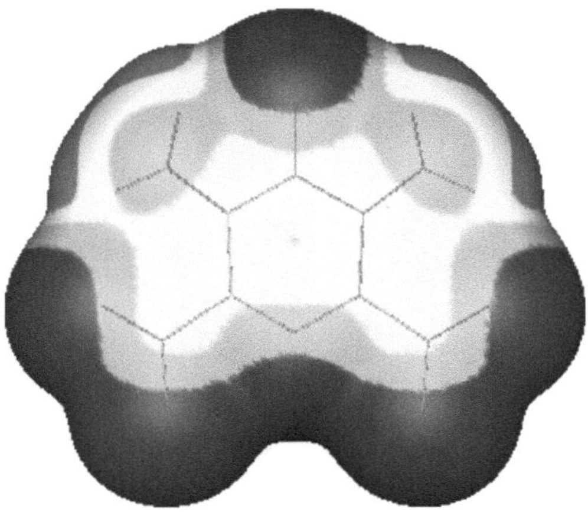

Figure 1.3 Electrostatic potential on the 0.001 au molecular surface of 2,6-diamino-3,5-dinitropyrazine-1-oxide. The N→O linkage is at the top, with the NH_2 groups on both sides of it. Color ranges, in kcal/mol, are: red, more positive than 30; yellow, 15–30; green, 0–15; blue, negative. There are $V_{S,max}$ of 46 kcal/mol (amino hydrogens), 30 kcal/mol (above the center of the ring), and 24 kcal/mol (above the C—NO_2 bonds). The $V_{S,min}$ are −38 kcal/mol (unsubstituted ring nitrogen), −25 to −27 kcal/mol (nitro oxygens), and −16 kcal/mol (N-oxide oxygen). The calculations were at the B3PW91/6-31G(d,p) level. (For interpretation of the references to color in this figure legend, the reader is referred to the online version of this chapter.)

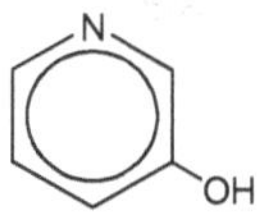

5

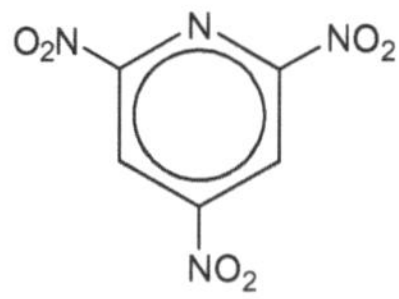

6

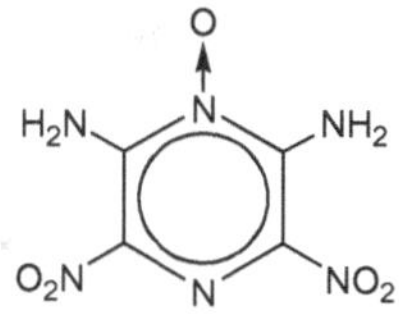

7

To some extent, it has also been possible to put the relationship between impact sensitivity and surface potential imbalance on a quantitative basis; this was mentioned already in Section 3. Different categories of compounds—nitroaromatics, nitramines, nitroheterocycles, etc.—are again treated separately. Initial correlations were between sensitivities and the most positive values of $V_S(\mathbf{r})$ above C—NO_2 bonds.[53,67] These were eventually expanded to give a more detailed picture of the potential imbalance, in terms of the positive and negative averages and variances of $V_S(\mathbf{r})$ in addition to its local maxima,[54] and a relationship was also developed for nitramines.

Why should the anomalous surface potential imbalance be linked to sensitivity? One proposed reason is that it is symptomatic of the weakening of trigger linkages.[65,68] It has long been known that there are local positive $V_S(\mathbf{r})$ maxima above C—NO_2 and N—NO_2 bonds.[65–68,108] This is true even if there is only a single NO_2 group present. In a molecule of an explosive, there are typically several strongly electron-withdrawing components, such as NO_2 groups and aza nitrogens, which cause a shifting of electronic charge from the inner portion of the molecule to its periphery. This gives rise to the strongly positive central regions characteristic of explosive molecules. As part of this, the $V_S(\mathbf{r})$ maxima above the C—NO_2 and N—NO_2 bonds also become more positive, suggesting a depletion of charge in these bonds that weakens them. It has indeed been shown that C—NO_2 and N—NO_2 dissociation energies decrease as the potential maxima in the bond regions become more positive.[65–68] Accordingly, if the rupture of one of these bonds is a key initial or early step in the initiation of detonation, then the anomalous surface potential imbalance—which parallels the weakening of the bond—should be at least symptomatic of sensitivity.

The surface potential imbalance may also be related to sensitivity through the strain that results when the crystal lattice is deformed by an external force, such as impact. An example is the shear strain that accompanies one portion of the lattice shifting ("slipping") past another.[5,109–111] The repulsion between strongly positive surface potentials on neighboring molecules that are moving past each other increases the energy needed for the deformation and thus localizes some of the externally introduced energy. This may be sufficient to create a "hot spot" (Section 4.3) and initiate detonation. The well-known insensitivity of TATB (**2**) has been interpreted in terms of its crystal structure, which consists of planar molecules in parallel two-dimensional layers; it has been shown that these layers can slip past each other with relatively little shear strain and energy requirement.[109–111] This is consistent with the fact that the molecular surface potential of TATB is relatively balanced,[101] not having the dominant positive central region that characterizes most explosives.

There are as well other ways in which the strongly positive surface potential of an explosive may promote detonation initiation. For example, in a nitro/nitrite isomerization, C—NO_2→C—ONO,[83,112] the interaction of the nitro oxygen with the carbon should be favored by the very positive electrostatic potentials around many C—NO_2 carbons (see Fig. 1.2). The extremely high sensitivity of Si-PETN, **8**, a silicon analogue of the well-known explosive PETN (pentaerythritol tetranitrate, **9**), has been attributed to a rearrangement in which one of the linking oxygens interacts with the silicon.[113] This rearrangement is encouraged by a particularly strong positive potential (a σ-hole) on the silicon.[114]

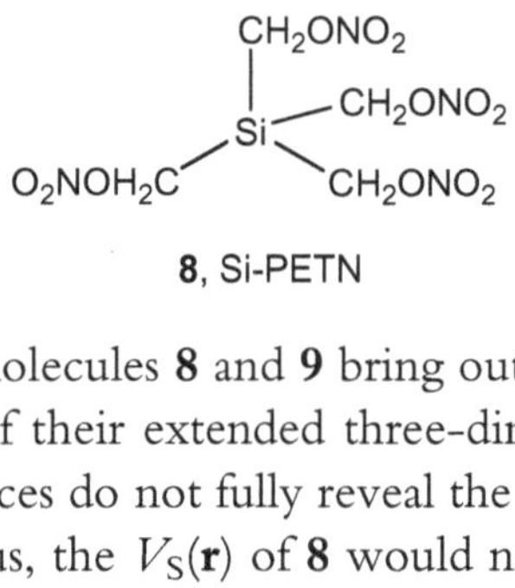

8, Si-PETN

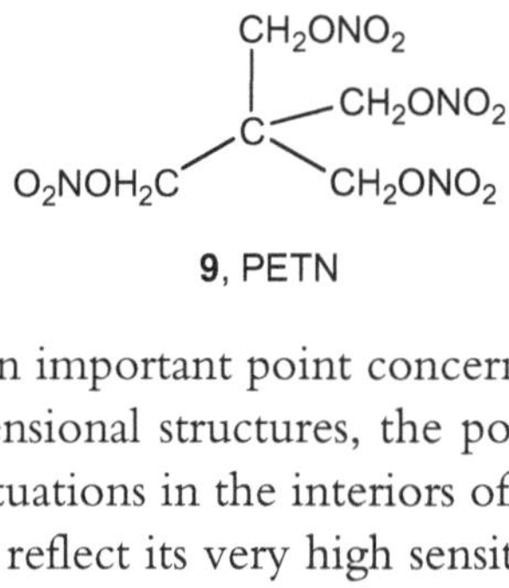

9, PETN

The molecules **8** and **9** bring out an important point concerning $V_S(\mathbf{r})$. Because of their extended three-dimensional structures, the potentials on their surfaces do not fully reveal the situations in the interiors of the molecules. Thus, the $V_S(\mathbf{r})$ of **8** would not reflect its very high sensitivity.

4.3. Free space in crystal lattice

Our emphasis so far has been upon the relationship of sensitivity to molecular properties. However, it is well established that crystal factors are also extremely important. One sign of this is that different polymorphs of a solid explosive can have markedly different sensitivities.[42–44]

Lattice defects—vacancies, voids, dislocations, misalignments, cracks, etc.—are believed to promote sensitivity, at least in part because they are associated with the formation of "hot spots."[2,5,38,63,64,115–118] These are small regions of the crystal lattice in which is localized some portion of the energy coming from the external stimulus, for example, impact or shock. If the hot spot energy is sufficient, it can be used to initiate the molecular processes (such as bond breaking and rearrangements) that lead to self-sustaining exothermal chemical decomposition, with the release of energy and gaseous products. A high pressure shock wave may be created that propagates through the compound at supersonic velocity (detonation).[2,5,13,119]

The input of mechanical energy into an explosive compound via shock or impact leads to compression whereby some of that energy, combined with strain energy due to lattice defects, produces hot spots.[2,115,120] The larger is the change in volume due to compression, the greater is the resulting thermal energy. Rice et al. concluded that "the first step in the reaction sequence leading to detonation is the compression of the energetic solid to a critical density."[121] Since it is the thermal energy that is produced is believed to initiate detonation, it seems reasonable to infer that among the factors governing the sensitivity of the explosive compound should be the amount of free space in its crystal lattice and its compressibility.

Support for this hypothesis comes from the work of Dick et al.,[123–125] who observed that PETN (**9**) is much more sensitive to shock parallel to the [110] and [001] crystallographic directions than parallel to the [101] and [100]. Kunz subsequently showed that the compressibility of PETN is greater in the [001] direction than in the [100].[126] The insensitive explosive TATB (**2**) has a very low compressibility [111]. Also consistent with a possible link between sensitivity and free space in the lattice is the finding that the dissociation energies of the N—NO_2 bonds in RDX (**1**) and in HMX (**10**)[86–88] and the C—NO_2 in H_3C—NO_2[122] are less when the molecules are at the crystal surfaces or by lattice voids (i.e., free space) than when they are in the bulk crystals.

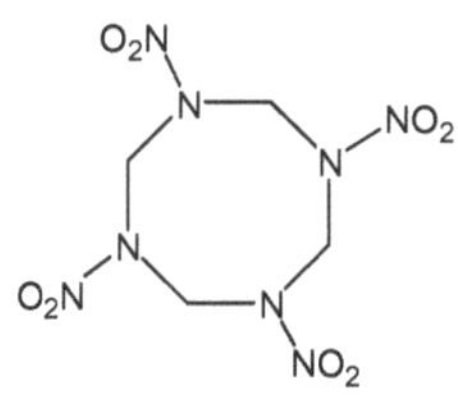

10, HMX

We have accordingly investigated the possibility of there being at least a rough relationship between impact sensitivity and free space in the crystal lattice.[127–129] The free space per molecule in the unit cell, ΔV, can be obtained by

$$\Delta V = V_{eff} - V_{int} \tag{1.12}$$

V_{eff} is the effective volume per molecule that would completely fill the unit cell:

$$V_{eff} = M/d \tag{1.13}$$

where M is the molecular mass and d is the crystal density. V_{int} is the intrinsic gas phase molecular volume. It is not obvious how to define V_{int}, and we considered several possibilities.[127,128] Taking V_{int} to be the space enclosed by the 0.001 au contour of the electronic density, which is such a useful molecular surface for many purposes (Section 4.2), is not acceptable because $M/V(0.001)$ is actually an approximation of the crystal density[23,24] and therefore, by Eq. (1.13), is similar to V_{eff}.

In identifying an appropriate V_{int}, we were guided by the work of Eckhardt and Gavezzotti.[130] They surveyed the crystal structures of 38 C, H, N, O energetic compounds and found that the packing coefficients vary between 0.71 and 0.83, with an average of 0.77. (This is higher than for typical organic compounds, which average about 0.70[130]; it reflects the greater densities of energetic compounds.) We found that defining V_{int} to be the volume within the 0.003 au contour of the electronic density yields packing coefficients for explosives (given by V_{int}/V_{eff}) that reproduce well the range and average value of Eckhardt and Gavezzotti's energetic compounds. Consequently, we settled upon $V_{int} = V(0.003)$.[128,129]

We have calculated the free lattice space, ΔV, for a diverse group of 25 explosives of different chemical types.[129] Their known crystal densities were used to determine V_{eff}, Eq. (1.13), and the V_{int} were computed. Comparing the trends in ΔV and in measured impact sensitivities, there was observed to be a general tendency for compounds having more free space in the crystal lattice (larger ΔV) to be more sensitive (smaller h_{50} and impact energy).

The database of 25 explosives that was used contains eight nitramines. Most of them are quite sensitive, with $h_{50} < 35$ cm (impact energy < 8.6 J)[129]; the least sensitive has $h_{50} = 55$ cm (impact energy $= 13$ J). While these nitramine sensitivities do show an overall indication of increasing with ΔV, their relative lack of variation despite a wide range of ΔV values indicates that the dependence of nitramine sensitivities upon ΔV is rather weak.

For the other types of compounds included in the database, however, the free space per molecule in the crystal lattice does appear to be a factor related to sensitivity to detonation initiation, as suggested by the earlier discussion of the role of compression. Determining ΔV can therefore permit at least a rough estimate of the impact sensitivity of a proposed target compound; procedures are available for computationally predicting the crystal density[23,24] and for finding V_{int},[131] and thus obtaining ΔV via Eqs. (1.12) and (1.13). This has been done recently for several proposed explosives.[30,31,31a]

5. THE QUEST FOR BALANCE

In designing new explosives, a central issue is attaining the desired balance between a high level of performance and low sensitivity. We will now consider some features that can promote either or both of these objectives.

5.1. N/C ratio in molecular framework

For at least two reasons, it is desirable to have a high ratio of nitrogen atoms to carbon atoms in the molecular framework of an explosive compound:

1. Nitrogens increase the enthalpy of formation of the compound and hence the heat release Q upon detonation.[132–134] This is illustrated, for example, by the trend in the enthalpies of formation of the polyazines in going from one to two to three nitrogens.[134] The basis for this nitrogen effect is that the enthalpy of formation reflects the breaking of the very strong N≡N bond in N_2 versus creating the weaker C—N, C=N, N—N, and/or N=N bonds.
2. The presence of a nitrogen instead of a C—H unit results in a greater crystal density,[132–137] attributable at least in part to the nitrogen atom having a larger mass and smaller volume than the C—H.[135] For instance, replacing one C—H by N in going from 1,3,5-trinitrobenzene (**11**) to 2,4,6-trinitropyridine (**6**) increases the density from 1.68 to 1.77 g/cm^3.[107]

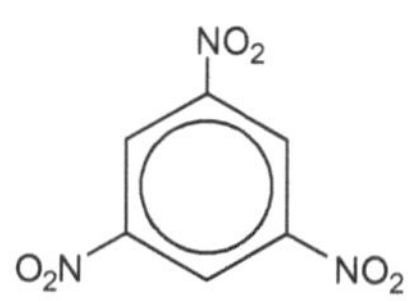

11

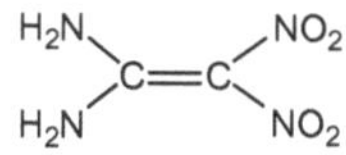

12, FOX-7

Since high enthalpies of formation and high densities both enhance detonation performance (Section 2), there is considerable interest in potential explosives with large N/C ratios,[30,31,31a,36,37,103–107,132,133,136–139] including molecules in which several nitrogens are linked together. The latter feature is called nitrogen catenation, and it has the drawback that it is frequently destabilizing.[140,141] This may be due to repulsion between the lone pairs on neighboring nitrogens,[134,142,143] the weakness of any N—N single bonds that may be present and the possibility of decomposition by losing the very stable N_2.[141] However, this destabilization can be diminished by converting some of the nitrogens to N-oxides ($N^+ \rightarrow O^-$)[134,139,142]—which also may produce a higher density and enthalpy of formation.[36,133,144,145] N-oxides are accordingly attracting a great deal of attention.[31,31a,36,37,132–134,137–139,144,145]

5.2. Planar molecules

As was mentioned in Section 4.2, crystal lattices composed of parallel or near-parallel layers are often less vulnerable to sensitizing shear strain caused by impact or shock, because the layers may be able to slip or slide past each other relatively easily.[109–111] This minimizes the strain-induced localization of externally introduced energy, that is, the formation of hot spots. This factor has been invoked to explain the low sensitivities of some explosives with layered crystal structures.[109–111,146,147]

The likelihood of a lattice having parallel or near-parallel layers is enhanced when the molecules are planar, particularly if there is the possibility of significant intermolecular noncovalent interactions within the layers, such as hydrogen bonding. Planar molecules also pack better in the crystal lattice,[148] resulting in desirably higher densities as well as less free space, which is often associated with reduced sensitivity (Section 4.3). Note however that a high density does not necessarily imply efficient packing; for instance, CL-20 (hexanitrohexa-azaisowurtzitane) has a high density and also a large ΔV.[129]

In the present context, it is interesting to compare TATB (**2**) and FOX-7 (1,1-diamino-2,2-dinitroethylene, **12**). Both are essentially planar in their crystal lattices and are involved in considerable intermolecular hydrogen bonding between the NH_2 and the NO_2 groups of neighboring molecules.[109–111] These apparent similarities led initially to the expectation that FOX-7 might be as insensitive as TATB. However, this is not the case[101]—while FOX-7 is indeed relatively insensitive ($h_{50}=126$ cm, impact energy = 31 J), it is not nearly to the extent of TATB ($h_{50}=490$ cm, impact energy = 120 J). An important difference between the two molecules is that TATB is symmetrical and FOX-7 is asymmetrical. This gives rise to markedly different molecular surface electrostatic potentials[101]; the $V_S(\mathbf{r})$ of

TATB is much more balanced and has positive and negative regions of generally moderate magnitudes, while FOX-7 shows a gradient in $V_S(\mathbf{r})$ from strongly positive in the NH_2 region to quite negative in the NO_2.

Both TATB and FOX-7 have layered lattice structures, but the TATB layers are planar whereas those of FOX-7 are zigzag.[109,111] As a result, FOX-7 is more susceptible to shear strain. A significant portion of the externally introduced energy is thereby localized in the C—NO_2 bonds of FOX-7, considerably reducing the barriers to breaking them. Kuklja and Rashkeev showed that whereas the energies required for C—NO_2 rupture are very similar for TATB and for FOX-7 in the isolated molecules, they can be quite different in shear-strained crystals, that for TATB being the higher.[109]

5.3. Amino substituents

It is well established that amino groups tend to diminish sensitivity.[35,36,65,101,149,150] A commonly cited example is the decreasing sensitivity in going from 1,3,5-trinitrobenzene (**11**) to its mono-, di-, and triamino derivatives.[35,62,65,101] The NH_2 group is an electron donor through resonance; in molecules in which charge delocalization is possible, it can therefore partially counteract the surface potential imbalance that is caused by electron-withdrawing components and is linked to sensitivity (Section 4.2). The electronic charge provided by NH_2 groups can also strengthen possible trigger linkages, for example, C—NO_2 and N—NO_2.

The presence of NH_2 substituents can give rise to inter- and intramolecular hydrogen bonding, which stabilizes the system. Furthermore, intermolecular hydrogen bonding may increase thermal conductivity, thus promoting the diffusion and dissipation of hot spot energy.[116] Note that TATB, known for its very low sensitivity, has been reported to have "the highest thermal conductivity of the common organic explosive molecules."[120]

Any or all of the above factors may help to explain the reduced sensitivity that tends to accompany amino substituents. The drawback to the stabilizing influence of NH_2 groups upon explosive compounds is that there is likely to be an accompanying decrease in the enthalpy of formation, and consequently the heat release Q upon detonation. For instance, 1,3,5-trinitrobenzene (**11**) has a solid phase enthalpy of formation of −10.4 kcal/mol; the introduction of just one amino group lowers it to −20.1 kcal/mol.[119]

Intuitively, it may seem that strong intermolecular hydrogen bonding should also offer the benefits of higher density and better packing. However, an extensive survey of crystallographic data revealed no relationship between hydrogen bonding and density.[148] This may seem surprising until one thinks of ice, in which the strong intermolecular hydrogen bonds create an open structure that is famously less dense than the liquid.

5.4. Some final comments

We will conclude by summarizing the structural features that can serve as guidelines in the quest for compounds with a high level of detonation performance combined with low sensitivity.

1. The molecules should be planar, for better packing and increased densities and for enhancing the likelihood of a crystal lattice composed of parallel layers that are less vulnerable to sensitizing shear strain.
2. The molecular frameworks should have large N/C ratios, for higher enthalpies of formation and densities.
3. There should be some amino substituents. This promotes stabilizing hydrogen bonding, which may also increase thermal conductivity. If charge delocalization is feasible, it may diminish surface electrostatic potential imbalance and strengthen possible trigger linkages.
4. If there are fewer carbons in the molecular framework, then $N^+ \rightarrow O^-$ linkages may provide sufficient oxygens to oxidize the carbons and hydrogens that are present and to simultaneously stabilize any nitrogen catenation.

In the context of oxygen sources and trigger linkages, it should be noted that $N{-}NO_2$ and $O{-}NO_2$ bonds are generally weak[4,65,68,75,151,152] and that compounds containing them are usually quite sensitive.[35,101] Organic azides also tend to be sensitive.[104,153]

Recent computational studies along the lines discussed in this chapter provide encouragement that compounds with the features listed above can combine very good performance with relatively low sensitivity.[31,31a] However, these predictions and estimates, while encouraging, must be followed by equally important assessments of synthetic feasibility, cost, and processability. This chapter is only the beginning of the story.

ACKNOWLEDGMENT

We greatly appreciate the support of this work by the Office of Naval Research, contract number N00014-12-1-0535, Program Officer Dr. Clifford Bedford.

REFERENCES

1. Iyer, S.; Slagg, N. Molecular Aspects in Energetic Materials. In *Structure and Reactivity*; Liebman, J. F.; Greenberg, A. Eds.; VCH Publishers: New York, 1988; pp 255–285 (Chapter 7).
2. Dlott, D. D. Fast Molecular Processes in Energetic Materials. In *Energetic Materials. Part 2. Detonation, Combustion*; Politzer, P.; Murray, J. S. Eds.; Elsevier: Amsterdam, 2003; pp 125–191 (Chapter 6).
3. Akhavan, J. *The Chemistry of Explosives*; 2nd ed.; Royal Society of Chemistry: Cambridge, UK, 2004.
4. Fried, L. E.; Manaa, M. R.; Pagoria, P. F.; Simpson, R. L. Design and Synthesis of Energetic Materials. *Annu. Rev. Mater. Res.* **2001**, *31*, 291–321.
5. Shackelford, S. A. Role of Thermochemical Decomposition in Energetic Material Initiation Sensitivity and Explosive Performance. *Cent. Eur. J. Energ. Mater.* **2008**, *5*, 75–101.
6. Honrbegr, H.; Volk, F. The Cylinder Test in the Context of Physical Detonation Measurement Methods. *Propell. Explos. Pyrotech.* **1989**, *14*, 198–211.
7. Licht, H.-H. Performance and Sensitivity of Explosives. *Propell. Explos. Pyrotech.* **2000**, *25*, 126–132.
8. Prinse, W. C. Development of Fiber Optic Sensors at TNO for Explosion and Shock Wave Measurements. *Proc. SPIE* **2000**, *4183*, 748–758.
9. Trzciński, W. A.; Cudziło, S.; Szymańczyk, L. Determination of the Detonation Pressure from a Water Test. *Eng. Trans.* **2001**, *49*, 443–458.
10. Trzciński, W. A.; Cudziło, S.; Chyłek, Z.; Szymańczyk, L. Detonation Properties of 1,1-Diamino-2,2-Dinitroethene (DADNE). *J. Hazard. Mater.* **2008**, *157*, 605–612.
11. Levine, H. B.; Sharples, R. E. *Operator's Manual for RUBY: Lawrence Livermore Laboratory Report UCRL-6815*; Lawrence Livermore Laboratory: Livermore, CA, 1962.
12. Cowperthwaite, M.; Zwister, M. W. H. *TIGER Computer Program Documentation*; SRI Publication No. 2106; Stanford Research Institute, 1973.
13. Mader, C. L. *Numerical Modeling of Explosives and Propellants*; 2nd ed.; CRC Press: Boca Raton, FL, 1998.
14. Sućeska, M. Calculation of Detonation Properties in EXPLO5 Computer Program. *Mater. Sci. Forum* **2004**, *465–466*, 325–330.
15. Bastea, S.; Fried, L. E.; Glaesemann, K. R.; Howard, W. M.; Sovers, P. C.; Vitello, P. A. *CHEETAH 5.0, User's Manual*; Lawrence Livermore National Laboratory: Livermore, CA, 2006.
16. Grys, S.; Trzciński, W. A. Calculation of Combustion, Explosion and Detonation Characteristics of Energetic Materials. *Cent. Eur. J. Energ. Mater.* **2010**, 7, 97–113.
17. Kamlet, M. J.; Jacobs, S. J. Chemistry of Detonation. I. A Simple Method for Calculating Detonation Properties of C,H,N,O Explosives. *J. Chem. Phys.* **1968**, *48*, 23–55.
18. Urbański, T. *Chemistry and Technology of Explosives* Pergamon Press: Oxford, UK, 1984.
19. Stine, J. R. On Predicting Properties of Explosives—Detonation Velocity. *J. Energ. Mater.* **1990**, *8*, 41–73.
20. Urtiew, P. A.; Hayes, B. Empirical Estimate of Detonation Properties in Condensed Explosives. *J. Energ. Mater.* **1991**, *9*, 297–318.
21. Shekhar, H. Studies on Empirical Approaches for Estimation of Detonation Velocity of High Explosives. *Cent. Eur. J. Energ. Mater.* **2012**, *9*, 39–48.
22. Politzer, P.; Murray, J. S. Some Perspectives on Estimating Detonation Properties of C,H,N,O Compounds. *Cent. Eur. J. Energ. Mater.* **2011**, *8*, 209–220.
23. Politzer, P.; Martínez, J.; Murray, J. S.; Concha, M. C.; Toro-Labbé, A. An Electrostatic Interaction Correction for Improved Crystal Density Predictions. *Mol. Phys.* **2009**, *107*, 2095–2101.
24. Rice, B. M.; Byrd, E. F. C. Evaluation of Electrostatic Descriptors for Predicting Crystalline Density. *J. Comput. Chem.* **2013**, *34*, 2146–2151.

25. Byrd, E. F. C.; Rice, B. M. Improved Prediction of Heats of Formation of Energetic Materials. *J. Phys. Chem. A* **2006**, *110*, 1005–1013, erratum: *J. Phys. Chem. A* **2009**, *113*, 5813.
26. Lias, S. G.; Bartmess, J. E.; Liebman, J. F.; Holmes, J. L.; Levin, R. D.; Mallard, W. G. Gas-Phase Ion and Neutral Thermochemistry. *J. Phys. Chem. Ref. Data* **1988**, *17*(Suppl. 1).
27. Muthurajan, H.; How Ghee, A. Software Development for the Detonation Product Analysis of High Energetic Materials. *Cent. Eur. J. Energ. Mater.* **2008**, *5*(3–4), 19–35.
28. Politzer, P.; Murray, J. S. Detonation Product Composition and Detonation Properties. *Cent. Eur. J. Energ. Mater.* **2014**, in press.
29. Gibbs, T. R.; Popolato, A. In *LASL Explosive Property Data*, University of California Press: Berkeley, CA, 1980.
30. Politzer, P.; Lane, P.; Murray, J. S. Computational Characterization of a Potential Energetic Compound: 1,3,5,7-Tetranitro-2,4,6,8-Tetraazacubane. *Cent. Eur. J. Energ. Mater.* **2011**, *8*, 39–52.
31. Politzer, P.; Lane, P.; Murray, J. S. Computational Characterization of Two Di-1,2,3,4-Tetrazine Tetraoxides, DTTO and Iso-DTTO, as Potential Energetic Compounds. *Cent. Eur. J. Energ. Mater.* **2013**, *10*, 37–52.
31a. Politzer, P.; Lane, P.; Murray, J. S. Tricyclic Polyazine N-Oxides as Proposed Energetic Compounds. *Cent. Eur. J. Energ. Mater.* **2013**, *10*, 305–323.
32. Sućeska, M. *Test Methods for Explosives*; Springer-Verlag: New York, 1995.
33. Zeman, S. Sensitivities of High Energy Compounds. *Struct. Bond.* **2007**, *125*, 195–271.
34. Doherty, R. M.; Watt, D. S. Relationship Between RDX Properties and Sensitivity. *Propell. Explos. Pyrotech.* **2008**, *33*, 4–13.
35. Storm, C. B.; Stine, J. R.; Kramer, J. F. Sensitivity Relationships in Energetic Materials. In *Chemistry and Physics of Energetic Materials*; Bulusu, S. N. Ed.; Kluwer: Dordrecht, The Netherlands, 1990; pp 605–639 (Chapter 27).
36. Pagoria, P. F.; Lee, G. S.; Mitchell, A. R.; Schmidt, R. D. A Review of Energetic Materials Synthesis. *Thermochim. Acta* **2002**, *384*, 187–204.
37. Huynh, M. H. V.; Hiskey, M. A.; Chavez, D. E.; Gilardi, R. D. Preparation, Characterization, and Properties of 7-Nitrotetrazolo[1,5-*f*]Furazano[4,5-*b*]Pyridine 1-Oxide. *J. Energ. Mater.* **2005**, *23*, 99–106.
38. Armstrong, R. W.; Coffey, C. S.; DeVost, V. F.; Elban, W. L. Crystal Size Dependence for Impact Sensitivities of Cyclotrimethylenetrinitramine. *J. Appl. Phys.* **1990**, *68*, 979–984.
39. Herrmann, M.; Engel, W.; Eisenreich, N. Thermal Expansion, Transitions, Sensitivities and Burning Rates of HMX. *Propell. Explos. Pyrotech.* **1992**, *17*, 190–195.
40. Elbeih, A.; Husarova, A.; Zeman, S. Path to ε-HNIW with Reduced Impact Sensitivity. *Cent. Eur. J. Energ. Mater.* **2011**, *8*, 173–182.
41. Wang, Y.; Jiang, W.; Song, X.; Deng, G.; Li, F. Insensitive HMX (Octahydro-1,3,5,7-Tetranitro-1,3,5,7-Tetrazocine) Nanocrystals Fabricated by High-Yield, Low-Cost Mechanical Milling. *Cent. Eur. J. Energ. Mater.* **2013**, *10*, 277–287.
42. Kohno, Y.; Maekawa, K.; Tsuchioka, T.; Hashizume, T.; Imamura, A. A Relationship Between the Impact Sensitivity and the Electronic Structures for the Unique N-N Bond in the HMX Polymorphs. *Combust. Flame* **1994**, *96*, 343–350.
43. Simpson, R. L.; Urtiew, P. A.; Ornellas, D. L.; Moody, G. L.; Scribner, K. J.; Hoffman, D. M. Cl-20 Performance Exceeds That of HMX and Its Sensitivity Is Moderate. *Propell. Explos. Pyrotech.* **1997**, *22*, 249–255.
44. McCrone, W. C. Letter to the Editor. *Chemical & Engineering News* July 5, 1999, p 2.
45. Kamlet, M. J. In *Proceedings of the 6th Symposium (International) on Detonation, San Diego, CA; Report No. ACR 221*, Office of Naval Research: Arlington, VA, 1976; pp 312–322.

46. Kamlet, M. J.; Adolph, H. G. The Relationship of Impact Sensitivity with Structure of Organic High Explosives. II. Polynitroaromatic Explosives. *Propell. Explos.* **1979**, *4*, 30–34.
47. Owens, F. J. Calculation of Energy Barriers for Bond Rupture in Some Energetic Molecules. *J. Mol. Struct. (Theochem.)* **1996**, *370*, 11–16.
48. Rice, B. M.; Sahu, S.; Owens, F. J. Density Functional Calculations of Bond Dissociation Energies for NO_2 Scission in Some Nitroaromatic Molecules. *J. Mol. Struct. (Theochem.)* **2002**, *583*, 69–72.
49. Politzer, P.; Murray, J. S.; Lane, P.; Sjoberg, P.; Adolph, H. G. Shock Sensitivity Relationships for Nitramines and Nitroaliphatics. *Chem. Phys. Lett.* **1991**, *181*, 78–82.
50. Delpuech, A.; Cherville, J. Relation entre la Structure Electronique et la Sensibilité au Choc des Explosifs Secondaires Nitrè. Critère Moléculaire de Sensibilité. *Propell. Explos.* **1978**, *3*, 169–175.
51. Zhang, H.; Cheung, F.; Zhao, F.; Cheng, X.-L. Band Gaps and the Possible Effect on Impact Sensitivity for Some Nitroaromatic Explosive Materials. *Int. J. Quantum Chem.* **2009**, *109*, 1547–1552.
52. Zhu, W.; Xiao, H. First-Principles Band Gap Criterion for Impact Sensitivity of Energetic Crystals: A Review. *Struct. Chem.* **2010**, *21*, 657–665.
53. Murray, J. S.; Lane, P.; Politzer, P. Relationships Between Impact Sensitivities and Molecular Surface Electrostatic Potentials of Nitroaromatic and Nitroheterocyclic Molecules. *Mol. Phys.* **1995**, *85*, 1–8.
54. Murray, J. S.; Lane, P.; Politzer, P. Effects of Strongly Electron-Attracting Components on Molecular Surface Electrostatic Potentials: Application to Predicting Impact Sensitivities of Energetic Molecules. *Mol. Phys.* **1998**, *93*, 187–194.
55. Mullay, J. Relationships Between Impact Sensitivity and Molecular Electronic Structure. *Propell. Explos. Pyrotech.* **1987**, *12*, 121–124.
56. Zhang, C. Review of the Establishment of Nitro Group Charge Method and Its Applications. *J. Hazard. Mater.* **2009**, *161*, 21–28.
57. Anders, G.; Borges, I., Jr. Topological Analysis of the Molecular Charge Density and Impact Sensitivity Models of Energetic Materials. *J. Phys. Chem. A* **2011**, *115*, 9055–9068.
58. Pepekin, V. I.; Korsunskii, B. L.; Denisaev, A. A. Initiation of Solid Explosives by Mechanical Impact. *Combust. Explos. Shock Waves* **2008**, *44*, 586–590.
59. Fried, L. E.; Ruggiero, A. J. Energy Transfer Rates in Primary, Secondary and Insensitive Explosives. *J. Phys. Chem.* **1994**, *98*, 9786–9791.
60. McNesby, K. L.; Coffey, C. S. Spectroscopic Determination of Impact Sensitivities of Explosives. *J. Phys. Chem. B* **1997**, *101*, 3097–3104.
61. Ye, S.; Koshi, M. Theoretical Studies of Energy Transfer Rates of Secondary Explosives. *J. Phys. Chem. B* **2006**, *110*, 18515–18520.
62. Brill, T. B.; James, K. J. Kinetics and Mechanisms of Thermal Decomposition of Nitroaromatic Explosives. *Chem. Rev.* **1993**, *93*, 2667–2692.
63. Kamlet, M. J.; Adolph, H. G. Some Comments Regarding Sensitivities, Thermal Stabilities and Explosive Performance Characteristics of Fluoronitromethyl Compounds. In *Proceedings of the Seventh Symposium (International) on Detonation; Report No. NSWCMP-82-334*, Naval Surface Warfare Center Silver Springs, MD, 1981; pp 60–67.
64. Politzer, P.; Murray, J. S. Sensitivity Correlations. In *Energetic Materials. Part 2. Detonation, Combustion*; Politzer, P.; Murray, J. S. Eds.; Elsevier: Amsterdam, 2003; pp 5–23 (Chapter 1).
65. Politzer, P.; Murray, J. S. Some Perspectives on Sensitivity to Initiation of Detonation. In *Green Energetic Materials*; Brinck, T. Ed.; Wiley: Chichester, UK, 2013.

66. Politzer, P.; Murray, J. S. C-NO_2 Dissociation Energies and Surface Electrostatic Potential Maxima in Relation to the Impact Sensitivities of Some Nitroheterocyclic Molecules. *Mol. Phys.* **1995**, *86*, 251–255.
67. Politzer, P.; Murray, J. S. Relationships Between Dissociation Energies and Electrostatic Potentials of C-NO_2 Bonds: Applications to Impact Sensitivities. *J. Mol. Struct.* **1996**, *376*, 419–424.
68. Murray, J. S.; Concha, M. C.; Politzer, P. Links Between Surface Electrostatic Potentials of Energetic Molecules, Impact Sensitivities and C-NO_2/N-NO_2 Bond Dissociation Energies. *Mol. Phys.* **2009**, *107*, 89–97.
69. Kuklja, M. M.; Stefanovich, E. V.; Kunz, A. B. An Excitonic Mechanism of Detonation Initiation in Explosives. *J. Chem. Phys.* **2000**, *112*, 3417–3423.
70. Hong, X.; Chen, S.; Dlott, D. D. Ultrafast Mode-Specific Intermolecular Vibrational Energy Transfer for Liquid Nitromethane. *J. Phys. Chem.* **1995**, *99*, 9102–9109.
71. Zeman, S. New Aspects of Initiation Reactivities of Energetic Materials Demonstrated on Nitramines. *J. Hazard. Mater.* **2006**, *A132*, 155–164.
72. Song, X.-S.; Cheng, X.-L.; Yang, X.-D. Relationship Between the Bond Dissociation Energies and Impact Sensitivities of Some Nitro-Explosives. *Propell. Explos. Pyrotech.* **2006**, *31*, 306–310.
73. Atalar, T.; Jungová, M.; Zeman, S. A New View of Relationships of the N-N Bond Dissociation Energies of Cyclic Nitramines. Part II. Relationships with Impact Sensitivity. *J. Energ. Mater.* **2009**, *27*, 200–216.
74. Mathieu, D. Theoretical Shock Sensitivity Index for Explosives. *J. Phys. Chem. A* **2012**, *116*, 1794–1800.
75. Mathieu, D. Toward a Physically Based Quantitative Modeling of Impact Sensitivities. *J. Phys. Chem. A* **2013**, *117*, 2253–2259.
76. Zeman, S. A Study of Chemical Micro-Mechanisms of Initiation of Organic Polynitro Compounds. In *Energetic Materials. Part 2. Detonation, Combustion*; Politzer, P.; Murray, J. S. Eds.; Elsevier: Amsterdam, 2003; pp 25–52 (Chapter 2).
77. Engelke, R.; Earl, W. L.; Rohlfing, C. M. Microscopic Evidence That the Nitromethane Aci Ion Is a Rate Controlling Species in the Detonation of Liquid Nitromethane. *J. Chem. Phys.* **1986**, *84*, 142–146.
78. Politzer, P.; Seminario, J. M.; Bolduc, P. R. A Proposed Interpretation of the Destabilizing Effect of Hydroxyl Groups on Nitroaromatic Molecules. *Chem. Phys. Lett.* **1989**, *158*, 463–469.
79. Fan, J.; Gu, Z.; Xiao, H.; Dong, H. Theoretical Study on Pyrolysis and Sensitivity of Energetic Compounds. Part 4. Nitro Derivatives of Phenols. *J. Phys. Org. Chem.* **1998**, *11*, 177–184.
80. Murray, J. S.; Lane, P.; Göbel, M.; Klapötke, T. M.; Politzer, P. Reaction Force Analyses of Nitro-Aci Tautomerizations of Trinitromethane, the Elusive Trinitromethanol, Picric Acid and 2,4-Dinitro-1H-Imidazole. *Theor. Chem. Acc.* **2009**, *124*, 355–363.
81. Murray, J. S.; Lane, P.; Politzer, P.; Bolduc, P. R.; McKenney, R. L., Jr. A Computational Analysis of Some Possible Hydrogen Transfer and Intramolecular Ring Formation Reactions of o-Nitrotoluene and o-Nitroaniline. *J. Mol. Struct. (Theochem.)* **1990**, *209*, 349–359.
82. Zeman, S.; Shu, Y.; Wang, X. Study on Primary Step of Initiation Mechanisms of Two Polynitro Arenes. *Cent. Eur. J. Energ. Mater.* **2005**, *2*(4), 47–54.
83. Gindulyte, A.; Massa, L.; Huang, L.; Karle, J. Proposed Mechanism of 1,1-Diamino-Dinitroethylene Decomposition: A Density Functional Theory Study. *J. Phys. Chem. A* **1999**, *103*, 11045–11051.
84. Storm, C. B.; Ryan, R. R.; Ritchie, J. P.; Hall, J. N.; Bachrach, S. M. Structural Basis of the Sensitivities of 1-Picryl-1,2,3,-Triazole. *J. Phys. Chem.* **1989**, *93*, 1000–1007.

85. Politzer, P.; Grice, M. E.; Seminario, J. M. A Density Functional Analysis of the Decomposition of 4-Nitro-1,2,3-Triazole Through the Evolution of N_2. *Int. J. Quantum Chem.* **1997**, *61*, 389–392.
86. Kuklja, M. M. Thermal Decomposition of Solid Cyclotrimethylene Trinitramine. *J. Phys. Chem. B* **2001**, *105*, 10159–10162.
87. Tsiaousis, D.; Munn, R. W. Energy of Charged States in the RDX Crystal: Trapping of Charge-Transfer Pairs as a Possible Mechanism for Initiating Detonation. *J. Chem. Phys.* **2005**, *122*, 184708 (1–9).
88. Sharia, O.; Tsyshevsky, R.; Kuklja, M. M. Surface-Accelerated Decomposition of δ-HMX. *J. Phys. Chem. Lett.* **2013**, *4*, 730–734. http://dx.doi.org/10.1021/jz302166p, and references cited.
89. Bardo, R. D.; Hall, T. N.; Kamlet, M. J. Energies and Volumes of Activation for Condensed Detonating Explosives. *J. Chem. Phys.* **1982**, 77, 5858–5859.
90. Murray, J. S.; Politzer, P. The Electrostatic Potential: An Overview. *WIREs Comp. Mol. Sci.* **2011**, *1*, 153–163.
91. Stewart, R. F. On the Mapping of Electrostatic Properties from Bragg Diffraction Data. *Chem. Phys. Lett.* **1979**, *65*, 335–342.
92. Politzer, P.; Truhlar, D. G. Eds.; *Chemical Applications of Atomic and Molecular Electrostatic Potentials*; Plenum Press New York, 1981.
93. Klein, C. L.; Stevens, E. D. Charge Density Studies of Drug Molecules. In *Structure and Reactivity*; Liebman, J. F.; Greenberg, A. Eds.; VCH Publishers: New York, 1988; pp 25–64 (Chapter 2).
94. Politzer, P.; Murray, J. S. The Fundamental Nature and Role of the Electrostatic Potential in Atoms and Molecules. *Theor. Chem. Acc.* **2002**, *108*, 134–142.
95. Politzer, P.; Murray, J. S. Molecular Electrostatic Potentials: Some Observations. In *Concepts and Methods in Modern Theoretical Chemistry*; Ghosh, K.; Chattaraj, P. Eds.; Electronic Structure and Reactivity; Vol. 1, Taylor & Francis: Boca Raton, FL, 2013; (Chapter 9).
96. Ayers, P. W. Using Reactivity Indicators Instead of Electron Density to Describe Coulomb Systems. *Chem. Phys. Lett.* **2007**, *438*, 148–152.
97. Politzer, P. Atomic and Molecular Energies as Functional of the Electrostatic Potential. *Theor. Chem. Acc.* **2004**, *111*, 395–399.
98. Bader, R. F. W.; Carroll, M. T.; Cheeseman, J. R.; Chang, C. Properties of Atoms in Molecules: Atomic Volumes. *J. Am. Chem. Soc.* **1987**, *109*, 7968–7979.
99. Politzer, P.; Murray, J. S. Statistical Analysis of the Molecular Surface Electrostatic Potential: An Approach to Describing Noncovalent Interactions in Condensed Phases. *J. Mol. Struct. (Theochem.)* **1998**, *425*, 107–114.
100. Politzer, P.; Martínez, J.; Murray, J. S.; Concha, M. C. An Electrostatic Correction for Improved Crystal Density Predictions of Energetic Ionic Compounds. *Mol. Phys.* **2010**, *108*, 1391–1396.
101. Rice, B. M.; Hare, J. J. A Quantum Mechanical Investigation of the Relation Between Impact Sensitivity and the Charge Distribution in Energetic Molecules. *J. Phys. Chem. A* **2002**, *106*, 1770–1783.
102. Pauling, L. The Modern Theory of Valency. *J. Chem. Soc.* **1948**, 1461–1467.
103. Hammerl, A.; Klapötke, T. M.; Nöth, H.; Warchhold, M. Synthesis, Structure, Molecular Orbital and Valence Bond Calculations for Tetrazole Azide, CHN_7. *Propell. Explos. Pyrotech.* **2003**, *28*, 165–173.
104. Hammerl, A.; Klapötke, T. M.; Mayer, P.; Weigand, J. J. Synthesis, Structure, Molecular Orbital Calculations and Decomposition Mechanism for Tetrazolylazide CHN_7, Its Phenyl Derivative, $PhCN_7$ and Tetrazolylpentazole CHN_9. *Propell. Explos. Pyrotech.* **2005**, *30*, 17–26.

105. Klapötke, T. M.; Nordheiter, A.; Stierstorfer, J. Synthesis and Reactivity of an Unexpected Highly Sensitive 1-Carboxymethyl-3-Diazonio-5-Nitrimino-1,2,4-Triazole. *New J. Chem.* **2012**, *36*, 1463–1468.
106. Gilardi, R. D.; Butcher, R. J. 2,6-Diamino-3,5-Dinitro-1,4-Pyrazine 1-Oxide. *Acta Cryst. E* **2001**, *57*, 657–658.
107. Licht, H.-H.; Ritter, H. 2,4,6-Trinitropyridine and Related Compounds, Synthesis and Characterization. *Propell. Explos. Pyrotech.* **1988**, *13*, 25–29.
108. Politzer, P.; Laurence, P. R.; Abrahmsen, L.; Zilles, B. A.; Sjoberg, P. The Aromatic C-NO2 Bond as a Site for Nucleophilic Attack. *Chem. Phys. Lett.* **1984**, *111*, 75–78.
109. Kuklja, M. M.; Rashkeev, S. N. Shear-Strain-Induced Chemical Reactivity of Layered Molecular Crystals. *Appl. Phys. Lett.* **2007**, *90*, 151913 (1–3).
110. Zhang, C. Investigation of the Slide of the Single Layer of the 1,3,5-Triamino-2,4,6-trinitrobenzene Crystal: Sliding Potential and Orientation. *J. Phys. Chem. B* **2007**, *111*, 14295–14298.
111. Zhang, C.; Wang, X.; Huang, H. Π-Stacked Interactions in Explosive Crystals: Buffers Against External Mechanical Stimuli. *J. Am. Chem. Soc.* **2008**, *130*, 8359–8365.
112. Saxon, R. P.; Yoshimine, M. Theoretical Study of Nitro-Nitrite Rearrangement of Nitramide. *J. Phys. Chem.* **1989**, *93*, 3130–3135.
113. Liu, W.-G.; Zybin, S. V.; Dasgupta, S.; Klapötke, T. M.; Goddard, W. A., III Explanation of the Colossal Detonation Sensitivity of Silicon Pentaerythritol Tetranitrate (Si-PETN) Explosive. *J. Am. Chem. Soc.* **2009**, *131*, 7490–7491.
114. Murray, J. S.; Lane, P.; Nieder, A.; Klapötke, T. M.; Politzer, P. Enhanced Detonation Sensitivities of Silicon Analogs of PETN: Reaction Force Analysis and the Role of σ-Hole Interactions. *Theor. Chem. Acc.* **2010**, *127*, 345–354.
115. Tsai, D. H.; Armstrong, R. W. Defect-Enhanced Structural Relaxation Mechanism for the Evolution of Hot Spots in Rapidly Compressed Crystals. *J. Phys. Chem.* **1994**, *98*, 10997–11000.
116. Tarver, C. M.; Chidester, S. K.; Nichols, A. L., III Critical Conditions for Impact- and Shock-Induced Hot Spots in Solid Explosives. *J. Phys. Chem.* **1996**, *100*, 5794–5799.
117. White, C. T.; Barrett, J. J. C.; Mintmire, J. W.; Elert, M. L.; Robertson, D. H. Effects of Nanoscale Voids on the Sensitivity of Model Energetic Materials. *Mater. Res. Soc. Symp. Proc.* **1996**, *418*, 277.
118. Politzer, P.; Boyd, S. Molecular Dynamics Simulations of Energetic Solids. *Struct. Chem.* **2002**, *13*, 105–113, and references cited.
119. Meyer, R.; Köhler, J.; Homburg, A. *Explosives*; 6th ed.; Wiley-VCH: Weinheim, Germany, 2007.
120. Tarver, C. M.; Urtiew, P. A.; Tran, T. D. Sensitivity of 2,6-Diamino-3,5-Dinitropyrazine-1-Oxide. *J. Energ. Mater.* **2005**, *23*, 183–203.
121. Rice, B. M.; Mattson, W.; Trevino, S. F. Molecular-dynamics Investigation of the Desensitization of Detonable Material. *Phys. Rev. E* **1998**, *57*, 5106–5111.
122. Zhang, C. Stress-Induced Activation of Decomposition of Organic Explosives: A Simple Way to Understand. *J. Mol. Model.* **2013**, *19*, 477–483.
123. Dick, J. J. Effect of Crystal Orientation on Shock Initiation Sensitivity of Pentaerythritol Tetranitrate Explosive. *Appl. Phys. Lett.* **1984**, *44*, 859–861.
124. Dick, J. J.; Mulford, R. N.; Spencer, W. J.; Pettit, D. R.; Garcia, E.; Shaw, D. C. Shock Response of Pentaerythritol Tetranitrate Single Crystals. *J. Appl. Phys.* **1991**, *70*, 3572–3587.
125. Yoo, C. S.; Holmes, N. C.; Souers, P. C.; Wu, C. J.; Ree, F. H.; Dick, J. J. Anisotropic Shock Sensitivity and Detonation Temperature of Pentaerythritol Tetranitrate Single Crystal. *J. Appl. Phys.* **2000**, *88*, 70–75.
126. Kunz, A. B. An *Ab Initio* Investigation of Crystalline PETN. *Mater. Res. Soc. Symp. Proc.* **1996**, *418*, 287–292.

127. Pospíšil, M.; Vávra, P.; Concha, M. C.; Murray, J. S.; Politzer, P. Crystal Volume Factor in the Impact Sensitivities of Some Energetic Compounds. *J. Mol. Model.* **2010**, *16*, 895–901.
128. Pospíšil, M.; Vávra, P.; Concha, M. C.; Murray, J. S.; Politzer, P. Sensitivity and the Available Free Space per Molecule in the Unit Cell. *J. Mol. Model.* **2011**, *17*, 2569–2574.
129. Politzer, P.; Murray, J. S. Impact Sensitivity and Crystal Lattice Compressibility/Free Space. *J. Mol. Model.* **2014**, in press.
130. Eckhardt, C. J.; Gavezzotti, A. Computer Simulations and Analysis of Structural and Energetic Features of Some Crystalline Energetic Materials. *J. Phys. Chem. B* **2007**, *111*, 3430–3437.
131. Bulat, F. A.; Toro-Labbé, A.; Brinck, T.; Murray, J. S.; Politzer, P. Quantitative Analysis of Molecular Surfaces: Areas, Volumes, Electrostatic Potentials and Average Local Ionization Energies. *J. Mol. Model.* **2010**, *16*, 1679–1693.
132. Chavez, D. E.; Hiskey, M. A.; Gilardi, R. D. 3,3'-Azobis(6-Amino-1,2,4,5-Tetrazine): A Novel High-Nitrogen Energetic Material. *Angew. Chem. Int. Ed. Engl.* **2000**, *39*, 1791–1793.
133. Chavez, D. E.; Hiskey, M. A.; Naud, D. L. Tetrazine Explosives. *Propell. Explos. Pyrotech.* **2004**, *29*, 209–215.
134. Politzer, P.; Lane, P.; Murray, J. S. Computational Analysis of Relative Stabilities of Polyazine N-Oxides. *Struct. Chem.* **2013**, *24*, 1965–1974. http://dx.doi.org/10.1007/s11224-013-0277-2.
135. Stine, J. R. Molecular Structure and Performance of High Explosives. *Mater. Res. Soc. Symp. Proc.* **1993**, *296*, 3–12.
136. Murray, J. S.; Gilardi, R.; Grice, M. E.; Lane, P.; Politzer, P. Structures and Molecular Surface Electrostatic Potentials of High-Density C, N, H Systems. *Struct. Chem.* **1996**, 7, 273–280.
137. Wei, T.; Zhu, W.; Zhang, X.; Li, Y.-F.; Xiao, H. Molecular Design of 1,2,4,5-Tetrazine-Based High-Energy Density Materials. *J. Phys. Chem. A* **2009**, *113*, 9404–9412.
138. Licht, H.-H.; Ritter, H. New Energetic Materials from Triazoles and Tetrazines. *J. Energ. Mater.* **1994**, *12*, 223–235.
139. Churakov, A. M.; Tartakovsky, V. A. Progress in 1,2,3,4-Tetrazine Chemistry. *Chem. Rev.* **2004**, *104*, 2601–2616.
140. Bensen, F. R. *The High Nitrogen Compounds.* Wiley-Interscience: New York, 1984.
141. Fabian, J.; Lewars, E. Azabenzenes (Azines) the Nitrogen Derivatives of Benzene with One to Six N Atoms: Stability, Homodesmotic Stabilization Energy, Electron Distribution, and Magnetic Ring Current: A Computational Study. *Can. J. Chem.* **2004**, *82*, 50–69.
142. Wilson, K. J.; Perera, S. A.; Bartlett, R. J.; Watts, J. D. Stabilization of the Pseudo-Benzene N_6 Ring with Oxygen. *J. Phys. Chem. A* **2001**, *105*, 7693–7699.
143. Mandado, M.; Otero, N.; Mosquera, R. A. Local Aromaticity Study of Heterocycles Using *n*-Center Delocalization Indices: The Role of Aromaticity on the Relative Stability of Position Isomers. *Tetrahedron* **2006**, *62*, 12204–12210.
144. Li, J.; Huang, Y.; Dong, H. A Theoretical Study of Polynitropyridines and Their N-Oxides. *J. Energ. Mater.* **2005**, *23*, 133–149.
145. Lai, W.-P.; Lian, P.; Yu, T.; Chang, H.-B.; Xue, Y.-Q. Design and Density Functional Theoretical Study of Three Novel Pyrazine-Based High-Energy Density Compounds. *Comput. Theor. Chem.* **2011**, *963*, 221–226.
146. Kuklja, M. M.; Rashkeev, S. N. Molecular Mechanisms of Shear Strain Sensitivity of the Energetic Crystals DADNE and TATB. *J. Energ. Mater.* **2010**, *28*, 66–77.
147. Veauthier, J. M.; Chavez, D. E.; Tappan, B. C.; Parrish, D. A. Synthesis and Characterization of Furazan Energetics ADAAF and DOATF. *J. Energ. Mater.* **2010**, *28*, 229–249.

148. Dunitz, J. D.; Filippini, G.; Gavezzotti, A. A Statistical Study of Density and Packing Variations Among Crystalline Isomers. *Tetrahedron* **2000**, *56*, 6595–6601.
149. Agrawal, J. P. Past, Present and Future of Thermally-Stable Explosives. *Cent. Eur. J. Energ. Mater.* **2012**, *9*, 273–290.
150. Cao, X.; Wen, Y.; Xiang, B.; Long, X.; Zhang, C. Are Amino Groups Advantageous to Insensitive High Explosives (IHEs). *J. Mol. Model.* **2012**, *18*, 4729–4738.
151. Luo, Y.-R. *Comprehensive Handbook of Chemical Bond Energies*; CRC Press: Boca Raton, FL, 2007.
152. Fan, X.-W.; Ju, X.-H. Theoretical Studies on Four-Membered Ring Compounds with NF_2, ONO_2, N_3 and NO_2 Groups. *J. Comput. Chem.* **2008**, *29*, 505–513.
153. Spear, R. J.; Dagley, I. J. Synthetic Routes to Aliphatic C-Nitro Functionalities. In *Organic Energetic Compounds*; Marinkas, P. L. Ed.; Nova Science Publishers, Inc.: Commack, NY, 1994; pp 47–163 (Chapter 2).

CHAPTER TWO

On the Release of Stored Energy from Energetic Materials

Elliot R. Bernstein[1]
Department of Chemistry, Colorado State University, Fort Collins, Colorado, USA
[1]Corresponding author: e-mail address: erb@lamar.colostate.edu

Contents

Abstract

Grinding, shocking, rapidly crushing, or electrically stimulating organic solids generates a variety of excited electronic state species including atoms, ions, radicals, fragments, etc., that are reactive on an ultrafast (<100 fs) timescale. In this review, we present the nonadiabatic theory that describes such species kinetics and dynamics that initiate, drive, and sustain the decomposition of energetic materials. The quantum mechanical theory that models this behavior is discussed in detail with regard to the excitation of energetic and nonenergetic model systems. The multireference approach taken can distinguish similar energetic and nonenergetic molecules with respect to their decomposition behavior and generation of an initial product fragment. The theory described involves complete active space self-consistent field calculations of adiabatic potential energy surfaces that interact nonadiabatically at conical intersections (CIs). The CIs are thereby responsible for the ultrafast kinetics and dynamics that generate the initial fragmentation behavior of both energetic and nonenergetic molecules. This energy release mechanism has two important consequences for the decomposition of the molecular species of interest: (1) the molecule is excited to higher electronic states and thereby has of the order of at least 5 eV of energy to break bonds; and (2) through

Advances in Quantum Chemistry, Volume 69
ISSN 0065-3276
http://dx.doi.org/10.1016/B978-0-12-800345-9.00002-7

various CIs the energetic molecule can return to the ground electronic state (in <100 fs) at a very different part of its ground state potential energy surface so that unexpected new reactive radicals can be generated. We demonstrate such behavior for energetic and nonenergetic nitramines, furazans, tetrazines, imidazoles, and other energetic systems. Our approach readily distinguishes energetic from nonenergetic molecular behavior both theoretically and experimentally.

1. INTRODUCTION

Energetic materials are employed for the storage of energy in molecular systems to be utilized as fuels, explosives, and pyrotechnics at different times and places. These are unique systems characterized by high heats of formation (stored energy) and particular electronic structure. One can usefully compare energetic molecules with apparently similar species that are not classified generally as energetic; that is, energetic molecules, when in the condensed phase, will explode under appropriate conditions while "model" energetic species will not. Perhaps, the best example for such direct comparison is RDX [$(CH_2NNO_2)_3$, energetic] and *N*-nitropiperidine and (*o*, *m*, *p*) *N*,*N*-dinitropiperazine (nonenergetic model systems). These molecules are displayed in Fig. 2.1 along with many other such molecular species couples. The comparison between energetic and model or nonenergetic species is an important one because the difference is obviously essential in terms of mechanisms for energy release and for any theoretical model developed to explain energetic behavior: not only must a proposed quantum mechanical mechanism demonstrate energetic behavior as found experimentally, but it must also demonstrate that model molecules, with similar chemical structures that are not energetic, will possess a different decomposition mechanism that will not follow an energetic pathway. These are stringent requirements and constraints to place on a theoretical molecular mechanism, especially for large organic molecules with many and complex electronic potential energy surfaces (PESs) and potential reaction coordinates accessible in the decomposition/energy release process.

Release of the stored molecular energy is certainly phase dependent; that is, an energetic molecule isolated in the gas phase, in a thin film of pure material, or dissolved in a solvent, will not explore. True energetic behavior is typically displayed only in crystalline or concentrated solids. Nonetheless, the fundamental kinetics and dynamics of the decomposition, energy release

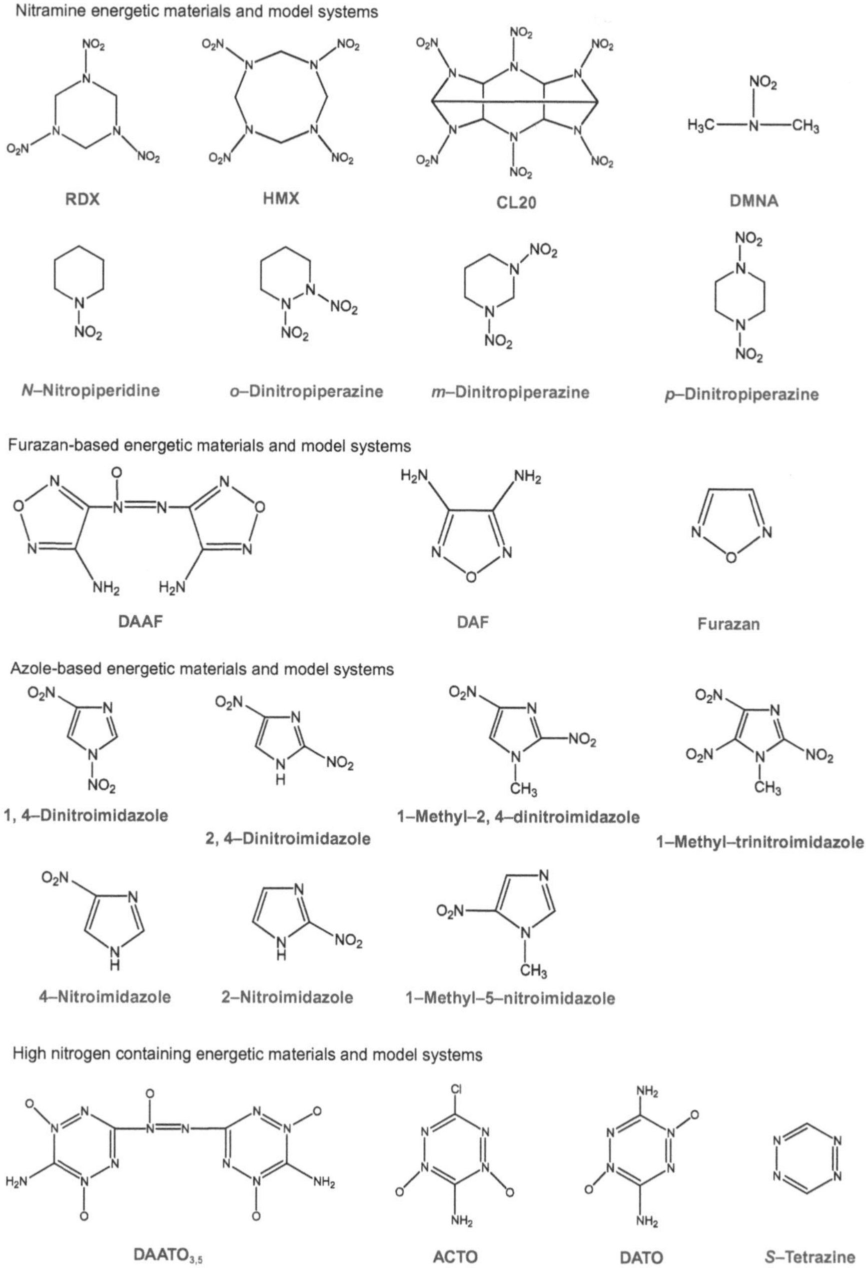

Figure 2.1 —Cont'd

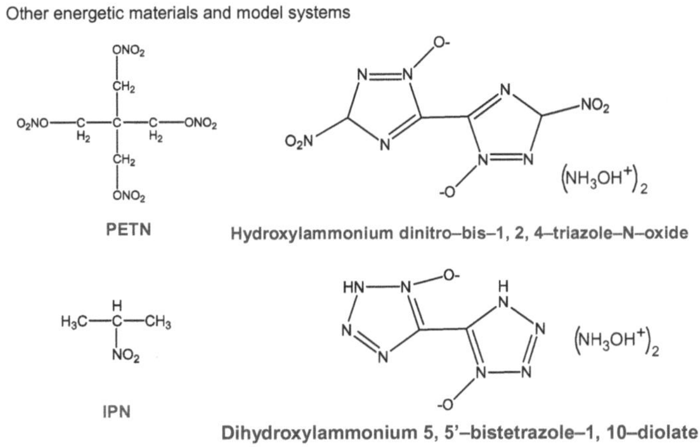

Figure 2.1 Structures of energetic materials (red) and model systems (blue). (For interpretation of the references to color in this figure legend, the reader is referred to the online version of this chapter.)

process must be molecular in nature prior to any intermolecular chain reaction that can occur in the crystalline or condensed phase. Thus, studying the molecular behavior of the identified, isolated energetic molecule (in the gas phase) is an important pursuit both experimentally and theoretically. In fact, only under such conditions do the experiment and quantum mechanical theory directly corresponds: the study of energetic materials at a molecular level is thereby an essential undertaking, if we are to understand and design them at a fundamental level.

Moreover, the goal of such studies is to present to synthetic chemists, who actually make the molecules of interest, a fundamental molecular mechanistic understanding of what is required to generate better (more energy storage, less sensitive, more environmentally benign, . . .) and new fuels, explosives, and pyrotechnics. Thus, we study energetic molecules and nonenergetic model molecules to learn about their detailed mechanistic, decomposition channels, and the relation of such channels to the physical and chemical properties of the individual molecules.

Since nearly all practical, secondary explosives are organic molecules, we must concern ourselves with the behavior associated with rather large systems with many heavy first and second row atoms (e.g., C, N, O, F, S, P, Cl, . . .). So what do we know about such systems, in general? Under typical initiation events, such as shocks, arcs, sparks, heat, pressure waves, laser pulses, . . ., energetic materials decompose. In fact, using a mortar and a pestle, many such compounds will emit light flashes. These flashes come from many molecular and atomic sources: $O^{\pm,0}$, $N^{\pm,0}$, $N_2^{\mp,0}$, $O_2^{\mp,0}$, $C_2^{\mp,0}$, the subject molecules themselves, small radicals, and reaction intermediates. Even for systems from which triboluminescence[1–10] is not obvious, excited electronic states can still be anticipated. Many mechanisms for triboluminescence have been suggested[11–19] but, in general, as crystal planes fracture, large electric fields can ionize and fragment even such tightly bound systems as N_2 (ionization energy ~15.6 eV and bond energy ~10 eV). Thus, even shock waves from gentle hand grinding can generate excited electronic valence, Rydberg, ionic, and atomic fragment states of organic molecules and small molecule inclusions. These excitations are a fact and cannot be ignored in consideration of the decomposition of energetic molecules and material. Of course, not all molecules that look similar to energetic ones are indeed energetic (e.g., RDX vs. a dinitropiperazine, Fig. 2.1) and we need to explain (theoretically) the energetic behavior for RDX and the nonenergetic behavior for its model systems. We have thus set ourselves the task of understanding the importance of and mechanisms for the decomposition of energetic and nonenergetic model molecules following excitation to higher electronic and eventually ionic states. Even if these excited state species are not the majority systems in the condensed phase, they can still play an essential role in the energy release process, as their products (e.g., NO, N_2, NO_2, . . .) can be highly vibrationally and translationally excited due to the eventual conversion of electronic excitation energy (5–10 eV) into vibrational and translational energy of small fragment molecules. Such "hot" fragments become very reactive in the condensed phase and can generate a chain of subsequent reactions to initiate explosive decomposition. Understanding the kinetics and dynamics of these mechanisms for energetic and model molecules will enable a fundamental explication of energetic behavior and its difference with respect to that of model systems. Thus, our program encompasses the experimental study of photoexcited, gas phase, cold, energetic, and nonenergetic model molecules to determine the initial steps in their dissociation: we identify the product molecule or radical and its rotational, vibrational, translational, electronic degrees of excitation, and its time evolution. From this information we determine, theoretically, a mechanistic pathway through the various PESs of the parent molecule that will generate such kinetics and dynamics. The essential theoretical effort is based on *ab initio* quantum chemistry calculations at a very high level (complete active space self-consistent field, CASSCF), and often with added second order perturbation theory (CAS PT2).

Our program is thereby both experimental and theoretical: experiments show us what is necessary to calculate and theory determines the kinetic mechanisms, dynamics, and reaction pathways and open channels based on the various PESs that must be accessed for the chemistry to occur. The initial reactant molecule is an isolated, gas phase, laser photoexcited ($S_n \leftarrow S_o$) parent molecule (energetic or nonenergetic) and the product species is a small molecule or radical (NO_X, NH_X, N_2, OH, CN, CH_X, HONO, ...) of known appearance time and energy distributions.

While the task may sound formidable, we are not alone in this effort; the much studied discipline of organic photochemistry has been well reported, reviewed, and is now a text book subject.[20–28] Organic molecule photochemistry is a broad and complex area bridging many complicated molecular phenomena and interactions between different electronic PESs: intermolecular vibrational energy redistribution (IVR), Fermi's Golden Rule, RRKM theory, PES calculations, PES crossings, conical intersections (CIs), nonadiabatic, non-Born–Oppenheimer (non-BO) interactions and transitions, steepest descent paths, reaction coordinates, density of states, etc. In this review, we will discuss and refer to many of these ideas and topics to explain the experimental results for the systems at hand.

So what must theory and *ab initio* calculations generate to explain the behavior of both energetic and nonenergetic molecules with regard to their decomposition following (laser) electronic excitation? The excited singlet electronic states mostly occur in the UV range of the spectrum below ~270 nm. Thus, laser excitation at 260–190 nm typically excites S_n, $n=1$–5, states of $\pi\pi^*$ and $n\pi^*$ character at the aromatic or NO_x moiety. (The tetrazine moiety as a high N-content system is an exception to this generalization: it absorbs initially in the visible range.) Following this event (in systems containing NO or NO_2 moieties), the initial product is NO, appearing in less than 100 fs, showing ultrafast kinetics for either energetic or nonenergetic molecules. The dynamics are such that the NO product has a cold rotational distribution ($T_{rot} \sim 20$–30 K), a hot vibrational distribution ($T_{vib} \sim 1300$–2000 K), and a high translational energy[29] for an energetic parent molecule. For nonenergetic molecules, the same set of NO results are typically $T_{rot} > 100$ K, $T_{vib} < 10^3$ K, and low translational energy. From these data for an energetic molecule, one can extract a qualitative model for the initial appearance of the product NO (or other) radical:

1. The ultrafast kinetics (<100 fs) cannot be explained simply by an intramolecular vibrational redistribution (IVR), internal conversion, Fermi's Golden Rule, density of states based process to convert the electronic excitation energy to vibrational, bond breaking energy by an RRKM-related process;

2. The most likely form of kinetics deals with CIs and non-BO, nonadiabatic coupling between PESs from $S_n \rightarrow S_{n-1} \rightarrow \cdots$;
3. The high vibrational excitation of the NO product from energetic parent molecules probably implies that the chain of events through CIs is $S_n \rightarrow S_{n-1} \rightarrow \cdots \rightarrow S_0$;
4. Bond breaking in S_o with all ~5 eV of electronic excitation converted to vibrational excitation of the parent energetic molecule;
5. The weak rotational excitation for the NO product is likely indicative of a near linear geometry at the S_0 fragmentation transition state or a final reaction coordinate on S_0 that does not place a torque on the departing initial NO product; and
6. The remaining energy goes to excitation of the parent fragment in S_o and/or translational energy in the light product di- or few-atomic.

Considering a nonenergetic parent molecule, the behavior is quite similar up to a point:

1. The process is still ultrafast, implying similar $S_n \rightarrow \cdots \rightarrow S_{n-x}$, but since T_{vib} is low for these nonenergetic systems the initial release of product might occur from an excited $S_{1,2,\ldots}$ state;
2. The high T_{rot} for NO in this case implies a torque generating final transition state and/or reaction (steepest descent) coordinate; and
3. The parent molecule would retain significant electronic excitation that would lead to a translationally and vibrationally cold small molecule/radical fragment.

The focus of theory is to generate a detailed, atomic/molecular level, quantum mechanical channel or pathway that can completely describe these events in chemistry, time, and energy distribution (mechanisms, kinetics, dynamics).

Apparently the different behavior of two chemically similar molecules, one energetic and the other nonenergetic, is buried in the details of the nonadiabatic interactions between the adiabatic (BO) PESs at the CIs. As these surfaces interact, transition states, barriers, steepest descent paths, reaction coordinates are created that control the ultrafast (<100 fs) chemistry and the energy distribution in the fragmentation kinetics. A molecule is thereby energetic or nonenergetic according to this process. From our experiments and theory to date, the distinction between energetic and model (nonenergetic) species can be drawn based on these ideas: differently put, energetic and nonenergetic materials are distinguished by the energy distribution of their initial product radicals following electronic excitation. Theory must give this result and communicate it chemically for synthesis of new and better energy storage systems.

A general point concerning modern synthesis of better and new energetic molecules can be made.[30,31] Three strategies have been taken for these synthetic efforts:

1. Combination of fuel and oxidizer in the same molecule—furazans, nitramines, and nitro aromatics are reasonable examples;
2. Ring and cage-strained systems—nitrocubanes and CL20 are good examples; and
3. High nitrogen content and high heat of formation compounds with the hope of generating an N_2 gas product—tetrazine and tetrazine-based energetics are possible examples of such systems, but see Ref. 32.

CIs, as presented above, play a central role on the electronic to vibrational energy conversion in organic molecular photochemistry. First, they are responsible for ultrafast IVR/internal conversion ($S_n \rightarrow \cdots \rightarrow S_0$) within the time domain 10^{-14} s. Second, they ensure the cascade from S_n to S_0 through nonadiabatic coupling of PESs. Third, they bring the molecule to explore different parts of the PESs along pathways of steepest descent and various reaction coordinates removed from the S_0 equilibrium (Franck–Condon, FC) position. These pathways often enable nonobvious chemistry, such as isomerizations rather than weakest bond rupture on S_0, or bond rupture on S_n. The steepest descent coordinates follow pathways with the smallest barriers rather than those with the weakest bonds.

Before discussing the main point of this review, the theory we have employed to address the above results, issues, and phenomena, we should give a short history of the idea of excited electronic state contributions to the decomposition of energetic materials. To the best of our knowledge, the first reference to this topic is by Ferd Williams in the early 1970s.[1–4] Williams points out that excited states of decomposing energetic materials can be generated in crystals by general initiation processes. He further discusses the adiabatic quantum mechanics of different potentials. This work was amplified by early studies of Sharma and coworkers.[5,6] In the last decade or so, a number of authors have considered generation of excited electronic states in molecular organic crystals. These include studies by the groups of Kuklja,[14,15,33–38] Fried,[11,39] Manaa,[13,40,41] Rice,[42,43] and Goddard,[44,45] which deal with single PESs (HF and DFT level calculations) and normal mechanisms for molecular organic crystal behavior; for example, excitation and exciton transport, shear plane reactive dynamics, electron generation, and HOMO–LUMO gap modulation. Triboluminescence behavior of organic molecular crystals does not appear to have been specifically considered for these systems.

In this review, we will discuss the isolated molecule, excited electronic behavior of energetic and model systems as appropriate for a modern organic photochemical event. Such considerations mandate multireference generation of adiabatic PESs that interact at CIs and the nonadiabatic couplings between such surfaces as they approach one another. We will first enumerate the general ideas, approaches, and issues for such situations and then discuss individual classes of organic species in more detail to show specific calculation from our program of study. These groups of systems will include nitramines, furazans, azoles, tetrazines, high N content species, and new systems. We emphasize that our group is a user of existing theoretical methods, not a developer; so this review will, taking examples from our work, present how to apply present state of the art, theoretical, mechanicist, organic photochemistry, based on high level quantum chemistry algorithms, to understand the excited electronic state chemistry of energetic materials.

2. GENERAL THEORETICAL APPROACH

As pointed out earlier, the subject of organic photochemistry is well established for molecular systems: this is our interest for the topic at hand. We wish to apply the basic ideas in the area of excited state organic reaction mechanisms and chemistry to our topic of energetic material decomposition. The basic behavior to be described follows these general lines: the molecule is left in an excited singlet state S_n in a nonequilibrium position governed by the Franck–Condon principle from the S_0 state; the molecule begins a steepest descent path (governed by $\langle\psi_f|\frac{\partial}{\partial R}\psi_i\rangle = \langle\psi_f|\frac{\partial H}{\partial R}|\psi_i\rangle[E_i(R) - E_f(R)]^{-1}$) in PES S_n toward the new S_n equilibrium position, converting electronic energy at the FC S_n position to vibrational energy; this molecule is isolated in the gas phase and must conserve energy, so the original ~5 eV of the photon excitation energy is stored in the excited molecule throughout the process; if the steepest descent path in S_n (the reaction coordinate) does not include a conical interaction, the molecule will stay on S_n and eventually emit or relax classically on the 10^{-9} to 10^{-11} s timescale; this path can involve barriers and CIs on the various PES S_{n-x} until the S_0 PES is reached at a considerable distance along the various reaction coordinates from the FC S_0 equilibrium point; excited state S_{n-x} chemistry can occur depending on the exact pathway and barriers, transition state, and CIs; on the S_0 PES with ~5 eV of now vibrational energy, the molecule can surmount many barriers and can execute various chemical rearrangements and fragmentations through different transition states; and finally the entire S_n to S_0 transition, through any number of CIs occurs in ~10^{-14} s and the initial radical fragments are released. Clearly, this process can only be modeled with a highly correlated (both static and dynamic), multireference algorithm that can account for degeneracy at special points for the PESs and will break the bonds required for agreement with experiment.

The approach we have employed is to generate PESs along various reaction coordinates with a CASSCF or CASPT2 algorithm, an IRC (intrinsic reaction coordinate) routine to connect reaction intermediates and transition states, and a scan routine (to generate a PES) also to follow reaction, steepest descent pathways to the final chemistry. These routines are labeled as in the Gaussian 09 program suite and the user's reference manual online at Gaussian.com. A CASSCF calculation (e,o) consists of a complete configuration interaction calculation using all orbitals and electrons in a chosen subspace (active space) of m electrons (e) and n orbitals (o). The distribution of occupied and virtual canonical orbitals are also chosen for this space: natural bond orbitals (NBO) with occupations from 2 to 0 or natural orbitals are also often the best choice for a CASSCF or CAASPT2 calculation as the involved orbitals are most easily determined from them. The difficulty of applying this technique lies in this orbital choice: the choice seems clear for a π electron description of a single aromatic like benzene (6 electrons, 6 orbitals) because the π,π^* orbital electrons are well separated from the σ system, but for a calculation of the neutral and ion states of a sugar molecule, for example, the choice can be very nonobvious. A practical maximum CASSCF calculation will be less than CASSCF (16,14), but for the calculation of CASSCF frequencies and CIs a practical limit is CASSCF (8,8) in Gaussian 09/03. The best approach to such calculations is to begin with an RHF/STO-3G calculation and choose canonical orbitals from this atomic localized level. An orbital visualization program is very useful in this effort. Following a CASSCF/STO-3G, one can proceed to a 4-31G and 6-31G(d) calculation, at each level checking the orbitals to determine if they are still appropriate. Canonical orbitals, NBOs, and NOs are very useful here to calculate excited states with CIs in this procedure; a state averaged calculation is required with equal weights for both states. The states will interact strongly for CIs at appropriate geometries or coordinates. To locate lowest energy CI, the "opt=(conical)" option is used. For CASSCF frequencies, the maximum size that Gaussian 09 will accept is CASSCF (8,8). One must be quite careful here to get the right space included for the calculation at all levels. We will discuss this further below with specific examples for a few different systems of interest. More general information concerning CASSCF calculation mechanics can be found at the Robb web site[46,47] and the Gaussian 09 web site.

2.1. Conical intersections

A CI is a concept and an interaction that has been identified in the early days of quantum mechanics.[48,49] The solution to the complete electron-nuclear Schrödinger equation, within the Born–Oppenheimer approximation (BOA), generates a molecular wave function, or the PES, that is adiabatic. The BOA makes the assumption that nuclear motions for normal modes of vibration (harmonic approximation) are so slow, and that the forces generated by these motions are so small, that they do not affect the electronic states of the molecule. These PESs are generated by solving the electronic Schrödinger equation at many different molecular geometries.

The meaning of the term adiabatic in this context is that an electronic state S_0 or S_1 can change character as the molecule follows a reaction coordinate, staying at a specific state (say S_1). A particular bonding pattern that does not change as the coordinate evolves corresponds to a diabetic state.[47] Since RHF and CASSCF wave functions are products of this "fixed nuclei" H^o, solutions of these equations are adiabatic states. The individual $S_0, \ldots, S_n$ are adiabatic multiconfiguration state functions that are generated through CASSCF calculations.[22–24,26,27] A CI is a point or seam at which two (or more) adiabatic surfaces are degenerate or can cross. This concept even applies to extreme nuclear motion, such as chemical reactions. Transitions between these adiabatic PESs can occur by nonadiabatic, non-BO terms that have been left out of the H^o: these can include spin–orbit intersections, an harmonic terms in the vibrational potential energy expansion, nuclear kinetic energy terms, or nuclear velocity terms. Any reaction that involves more than one PES is considered to be nonadiabatic. The essential points for us are that (1) CIs are the only mechanism that can lead to ca. 10^{-14} s chemical reactions following initial excitation from S_0 to S_n, and (2) CIs dominate the reaction mechanism and lead to ground and excited electronic state chemistry that is very different from Franck–Cordon, S_0 equilibrium state thermal chemistry. The topography of these multidimensional PESs is complex and difficult to visualize mostly because a 30 or so dimensional surface is difficult to visualize even without often intersecting and interacting PESs. Nonetheless, a reaction (steepest descent) pathway, barriers, transition states, and reaction intermediates can be computed and constitute a reaction mechanism through many excited PESs. The forces that operate on a molecule as it explores a multidimensional PES are given by the steepest descent slope on the surface or the PES gradient (velocity) at that position (e.g., $\langle\psi_\mathrm{f}|\frac{\partial H}{\partial R}|\psi_\mathrm{i}\rangle\left[E_i(R)-E_f(R)\right]^{-1}$).

Our discussions of CIs are quite qualitative at this point, but much more detail and mathematical rigor can be found in many of the review articles and text books referenced.[20–28] The original literature can be found referenced in many of these articles and goes back to Wigner and von Neumann in 1929.[48] See especially the work of Robb's group,[46,47] Yarkony's group,[50–53] Michl and coworkers,[22] and Nikitin.[20,21] A thesis from our group can also be quite helpful for detailed calculations of mechanisms.[54]

3. REACTION MECHANISMS FOR ENERGETIC MOLECULE DECOMPOSITION FOLLOWING ELECTRONIC EXCITATION

We have a very good understanding now from both theory and experiment of the molecular excitation process and of the ultrafast decomposition processes that follow the excitation for any organic molecule, whether energetic or a model energetic species. In this section, we present the work we have done on various energetic systems: nitramines and nitro-containing molecules, furazans, azoles, azines, high nitrogen content species, and new systems. These molecules are presented in Fig. 2.1. The role of CIs in the energy release chemistry will be a major theme for all these systems.

Before beginning with the individual species we have studied under isolated conditions, a few comments can be addressed to condensed phase quantum mechanical studies of organic crystals. Reference[55] discusses the bonding energy of small dimers that fundamentally represent the different varieties of intermolecular interactions: $(H_2O)_2$, $(NH_3)_2$, $(CH_4)_2$, $H_2O(NH_3)$, $CH_4(NH_3)$, and $(FHF)^-$. Calculations are presented for different algorithms from HF, DFT to MPX, CCSD(T), and others employing different usual basis sets (e.g., cc-pVXZ, X=D, T, Q, 5). While many nuances are found for the details of the calculations, the important conclusions reached are the following: (1) the cc-pVXZ Dunning basis sets can be used to establish basis set convergence for all the calculational methods; (2) B3LYP density functional and MP2 wave function approaches have similar basis set superposition error (BSSE, counterpoise corrected) characteristics; (3) energies of hydrogen-bonded systems are not sensitive to various basis sets if a BSSE correction is employed; (4) very large basis sets are needed to remove BSSE and converge binding energies for van der Waals interactions; and (5) DFT methods cannot be well used for dispersion interactions. Thus, quantum chemical calculations of condensed phase behavior of energetic systems present real challenges and can be very expensive.

With this caveat, we now review the "photophysics" and "photochemistry" (i.e., physical and chemical behavior following electron excitation) of organic energetic and model molecules.

3.1. Nitramines (DMNA, RDX, HMX, CL20)

3.1.1 DMNA

DMNA is the very convenient and accessible model system whose calculational approach we will outline in some detail in order to demonstrate what is necessary to compute reaction coordinates with PES interactions (CIs), transition states, reaction intermediates, and a steepest descent pathway. More complete discussions and details can be found in Ref. 56 and the PhD thesis of A. Bhattacharya (AB; CSU, Chemistry, 2009 ref. 54): our presentation here follows these original sources. Electronically excited DMNA is the reactant in this example and NO is the ground state product. A number of other "pathways" or "reaction coordinates" can be envisioned but experiments demonstrate that these two species must be our major concern theoretically. Elucidating the details of the trajectory constitutes a mechanism for the reaction. One must first establish the structure of S_n DMNA as it begins to undergo transitions to lower electronic states and eventually to the state (S_0, S_1, ...) in which it generates the NO product. Based on our prior discussion here and in Ref. 56 and Ref. 54 CIs between PESs are essential to locate along with various transition states and reaction intermediates. We begin with electronic excitation of DMNA, proceed to lower lying CIs, at which point nonadiabatic transitions between adiabatic PESs occur ($S_n \rightarrow S_{n-1}$, ..., S_1 or S_0). The quantum chemistry suite of programs Gaussian 09 (others such as GAMES, MOLPRO, ORCA, MOLCAS, ..., too) have these capabilities. To locate such special points on PESs we perform the following steps: predict starting geometry for CI; compute and examine canonical orbitals for an RHF/STO-3G level needed to select an active space for a CASSCF calculation; perform a CASSCF/STO-3G calculation for the ground state, using basis sets, in order, STO-3G,

4-31G, and 6-31G(d); perform a state-averaged CASSCF/6-31G(d) with 0.5 weight for each; check active space to determine it is still valid and run a CASSCF calculation to search for CIs.

DMNA undergoes a nitro-nitrite isomerization through a $(S_2/S_1)_{CI}$ following excitation to the FC point on the S_2 PES. Figure 2.2 shows this reaction pathway (arrows) and gives the critical points on the S_2, S_1, S_o PESs, and a slice along various reaction pathways: Fig. 2.2 represents a mechanism for the decomposition of DMNA following electronic excitation to generate NO. The decomposition occurs on the S_1 PES. Note that the horizontal axis of Fig. 2.2 is a "minimum energy steepest descent path, reaction coordinate", the details and composition of which evolve with the geometries as the reaction mechanism proceeds. The discussion below deals with the localization of the minimum energy CIs along this pathway for the nitro-nitrite isomerization reaction coordinate: once these critical points are generated, the reaction path is found through the IRC or scan algorithm (Gaussian 09).

The search for CIs begins on the S_0 PES along the nitro-nitrite isomerization channel, as the transition state geometry for this transition, predicted on the ground state surface, should be a good starting geometry

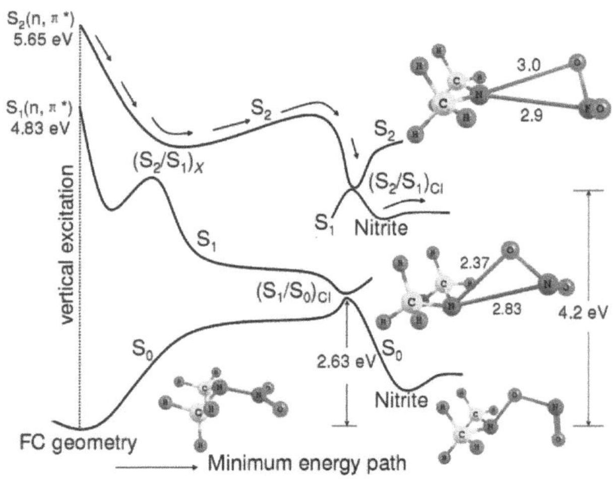

Figure 2.2 One-dimensional projection of the multidimensional electronic potential energy surfaces of DMNA computed at the CASSCF (10,7)/6-31G(d) level of theory. Note that the $(S_2/S_1)_X$ conical intersection is strongly coupled and that the adiabatic energy gap near this conical intersection is computed to be 485 cm^{-1}. $(S_2/S_1)_{CI}$ and $(S_1/S_0)_{CI}$ are relatively weakly coupled conical intersections. Path arrows represent the classical trajectory of the plausible excited electronic state decomposition mechanism of DMNA. (For color version of this figure, the reader is referred to the online version of this chapter.)

for the $(S_n/S_m)_{CI}$s. This is perhaps the most difficult part of the calculation of Fig. 2.2. The transition state for this isomerization on S_0 turns out to be a good starting geometry for $(S_2/S_1)_{CI}$ and $(S_1/S_0)_{CI}$. This nitro-nitrite isomerization on S_0 is a thermodynamically allowed reaction which results in a weak NO—NO bond that easily ruptures. The nature of this transition state has been a topic of some debate: loose vs. tight. A loose transition state has a long N—N bond distance before the NO_2 rotates and a tight transition state has the NO_2 rotation occurring at a short N—N distance. These transition states are located with an MP2/6-31G(d) calculation, but only the loose transition state can be minimized at a CASSCF (10,7)/6-31G(d) level. Both transition states should be located first at the MP2/6-31G(d) level with a starting geometry obtained with a visualization program such as Chem Craft or Gausview. These calculations will generate configuration state functions and (RHF) orbitals for the CASSCF calculations. (The "No sym" key word is essential here because Gaussian (03,09) reorients the molecule for the particular point group and will destroy the active space if this feature is not disabled.) The RHF orbitals are chosen as 5 occupied and 2 virtual for the (10,7) CASSCF. From this point with the S_0 PES transition state found, a starting point for the S_1/S_2 CIs is sensible. The S_1 and S_0 states can be optimized for a CASSCF (10,7) and the $(S_1/S_0)_{CI}$ is determined [opt=(conical)], as shown in Fig. 2.3. The energy gap between S_1 and S_0 at the $(S_1/S_0)_{CI}$ is calculated to be $\sim$10 cm^{-1}. This CI looks similar to the structure of the loose nitro-nitrite isomerization transition state.

Next, the $(S_2/S_1)_{CI}$ can be located, typically near the FC point on the S_2 PES, again, orbitals for the calculation come from an RHF/STO-3G level and a CASSCF (14,11) calculation is performed with state averaging equally over the S_2, S_1, S_0 states. Once the S_2 wave function is determined (third root), the $(S_2/S_1)_X$ (Fig. 2.2) CI is determined. Once all these CIs are found, a full reaction path can be mapped through the IRC or scan algorithms.

Referring to Fig. 2.2, the $(S_1/S_0)_{CI}$ can now be calculated. This CI is achieved for DMNA excited to FC S_1. In this instance, the intersection occurs for the S_1, S_0 PESs such that DMNA returns the S_0 FC, through the steepest descent pathway based on a CASSCF (14,11)/6-31G(d) calculation. DMNA-optimized S_1 structure is found and a state-averaged CASSCF [opt=(conical)] calculation uses this geometry as a starting point. With these critical points on S_0, S_1, S_2 PESs, the PESs can now be mapped along the reaction steepest descent pathway to find the appropriate mechanism, for the excited electronic state driven $(CH_3)_2N—NO_2$ to $(CH_3)NO_2$ plus NO reaction.

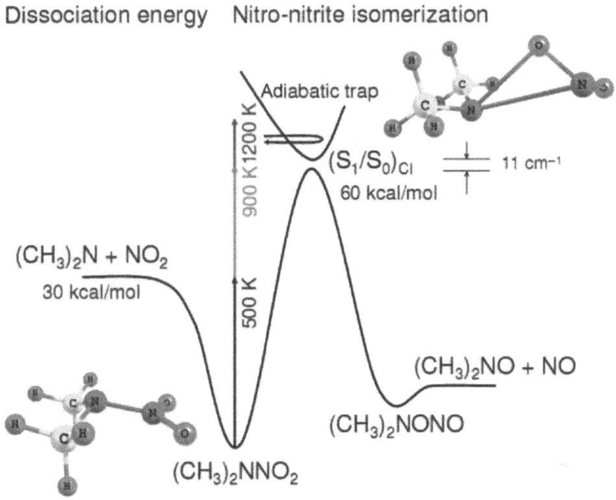

Figure 2.3 A schematic diagram of the N—NO_2 bond dissociation and nitro-nitrite isomerization pathways presented along two different directions with their respective energies of dissociation and activation computed at CASSCF (10,7)/6-31G(d) level of theory. Nitro-nitrite isomerization TS exists near the $(S_1/S_0)_{CI}$ conical intersection, and its energy is calculated to be about 3 kcal/mol below the energy barrier of the conical intersection. The adiabatic energy gap between the S_1 and S_0 surfaces near $(S_1/S_0)_{CI}$ is calculated to be 11 cm^{-1}. (For color version of this figure, the reader is referred to the online version of this chapter.)

With the above set of transition states, intermediates, reactants, products, and $(S_m/S_n)_{CI}$s, the stationary points on the PESs for a molecule, DMNA in this instance, are found. The PESs pathways or reaction channels or the reaction coordinates must be generated that will connect these positions (stationary points): that is, a particular set of intermediate/transition state/intermediate must be connected by an appropriate reaction coordinate. The quantum mechanical mechanism for a chemical reaction will be given as a cut through the PESs along various reaction coordinates that connect all these points. Between the reactants and products are transition states that are saddle points of higher energy on this pathway. The height of the energy barriers between the reactant and products controls the reaction kinetics and the energy difference between reactants and products gives the thermodynamics of the reaction. This entire description of the full PESs with the steepest descent pathways is referred to as transition state theory.[57–59] Stationary points on the PES imply $(\partial E/\partial Q_i) = 0$ (no forces) and $(\partial^2 E/\partial Q_i \partial Q_j) > 0$, and first-order saddle points have for one coordinate $(\partial^2 E/\partial Q_i \partial Q_j) < 0$, implying a maximum in this direction. A transition state for a reaction is a first-order saddle point in the reaction coordinate. Below we demonstrate this process for DMNA and present the reaction coordinate for $S_2 \rightarrow S_1$, nitro-nitrite isomerization, and for the $S_1 \rightarrow S_0$ transition through the $(S_1/S_0)_{CI}$ (see Fig. 2.2).

Consider first the nitro-nitrite isomerization pathway on the S_0 PES, characterized by Fig. 2.3. Many minima and transition states exist on S_0, each corresponding to a different reaction path or mechanism. We first search for the transition state structure to predict the activation energy barrier on S_0 for this reaction: needed are geometries for $(CH_3)_2NNO_2$ and $(CH_3)_2NONO$, corrected for zero point energy. As pointed out previously, this reaction on S_0 can have two transition states, (tight and loose) at the MP2/631G(d) level of theory, but only the loose state remains at the CASSCF (10,7)/631G(d) level. The activation energy for the loose transition state is 69 kcal/mol and for the tight transition state is 90 kcal/mol, both at the MP2/6-31G(d) level of theory. Employing the IRC algorithm in Gaussian 09, one can verify that the tight transition state connects the two above minima on the reaction path and explores the reaction path for the nitro-nitrite isomerization.

We can also explore the S_0 PES for the loose transition state; however, this is more difficult because the transition now occurs in two stages or coordinates. First, the N—NO_2 bond elongation occurs and second, the N—ONO bond is formed through rotation of the O—N—O moiety: so, both bonds, N—NO_2 and N—ONO, must be scanned. The N—NO_2 bond does not have a transition state, but the N—ONO bond does (the loose transition state). The N—NO_2 PES can be accomplished at the MP2 or CASSCF (14,11) levels of theory. Figure 2.4 shows the active space for the CASSCF calculation. The PES scan for the N—ONO bond of DMNA is computed at the CASSCF (12,9)/6-31G(d) theory level and its steepest descent path is given in Fig. 2.2.

The S_1 PES can now be explored with the IRC (or scan) algorithm: the starting geometry must be higher in energy than the rest of the reaction coordinate path. It could be the FC point, a transition state, or higher energy CI. To start at the FC point on S_1, the IRC calculations need initial force constants from a CASSCF calculation with not more than eight orbitals. From a CASSCF (14,11), a CASSCF (10,7) is employed to get force constants and the path from the FC to $(S_1/S_0)_{CI}$ is determined. All critical points on this path are computed at the CASSCF (10,7)/6-31G(d) level. Then the IRC algorithm can generate the pathway for S_0 from $(S_1/S_0)_{CI}$ to FC (the no reaction path), generating ultrafast internal conversion.

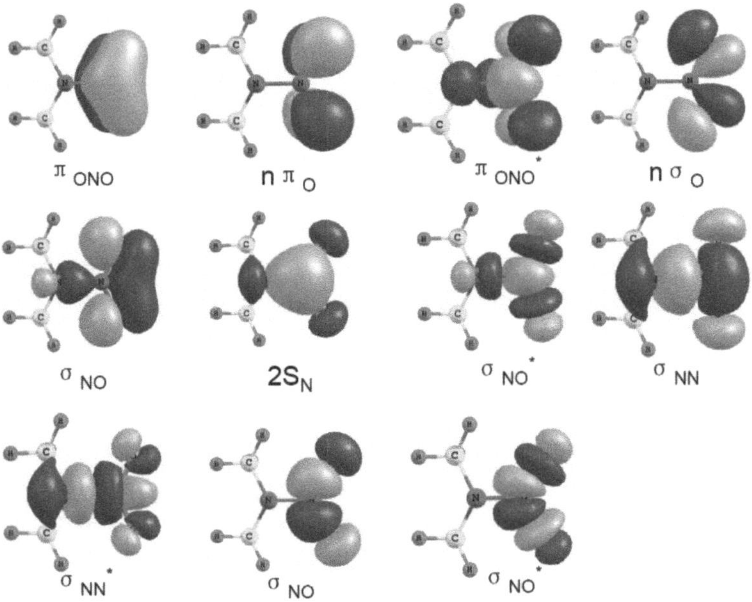

Figure 2.4 Orbitals used in the active space for CASSCF calculations for DMNA. These orbitals thus comprise the full active space (14,11) for DMNA (14 electrons in 11 active orbitals). (For color version of this figure, the reader is referred to the online version of this chapter.)

Both an IRC and a PES scan can be employed to find a reaction mechanism or steepest descent path for chemistry or internal conversion. Note, however, that analytical force constants are required, state averaging is also necessary, and the FC point in S_n is not a stable minimum or a transition state on the excited S_n. One needs to fix the IRC direction (reaction coordinate) for the particular desired path. The PES scan explores the surface for a particular normal reaction coordinate, such as a bond stretch, angle bond, or a dihedral angle change: this is only a suggested compound reaction coordinate that follows the steepest descent path in form. Activation energies for transition states and other critical points on a surface are calculated separately: the IRC or PES scan gives information about the reaction coordinate connecting these critical points on a PES.

In the following subsections and sections, we will give general descriptions of other calculations for energetic materials in much less detail because the explicit descriptions above are used throughout these general approaches to finding reaction mechanisms.

3.1.2 RDX

RDX is a real energetic material (Fig. 2.1) and one can consider it, and HMX and to a lesser extent CL20, as a cyclic polymer of DMNA ($[(CH_2)_2NNO_2]_3$). The RDX molecule is clearly much too complicated and large to approach head-on with a CASSCF calculation: if DMNA needs a CASSCF (14,11) calculation, then RDX would require a CASSCF (42,33) calculation and HMX would require a CASSCF (56,44) calculation to be treated at the same level of accuracy and detail as DMNA. Maybe someday, but not at present!

So we take advantage of symmetry and the fact that RDX and HMX are built from "uncoupled units" of $(CH_2)_2NNO_2$: we apply the ONIOM algorithm developed by Morokuma and coworkers.[60–65] ONIOM is the unlikely acronym for "our own N-layered integrated molecular orbital and molecular mechanics" method. Basically, this method divides the molecule into two parts: the first part is the "active site," in our case for RDX a $(CH_2)_2N_2O_2$ unit; and the second part is the rest of the molecule. The active site part of the molecule is treated at the full CASSCF (10,7) level of theory, as we know for DMNA in all its gory detail. The rest of the molecule is treated at a much lower level, for example, RHF, UFFMM (universal force field, molecular mechanics). In this way, the active site of RDX is localized on one N—NO_2 moiety and the remainder of the molecule is assumed inactive and only provides inductive or steric effects for the active localized excitation site. This is a perfect approach for RDX, HMX, CL20 type systems. These assumptions must be tested against the experimental results and CASSCF calculations must be designed (orbital choices and occupations) to apply to the known experimental results, as usual. The choice for the rest of the molecule is UFF or RHF. Both can be tried and yield roughly the same overall result for RDX. Thus, the total energy of a molecule PES for various electronic excitations is given by

$$E^{\mathrm{ONIOM}} = E_{\mathrm{AS}}^{\mathrm{HIGH}} + E_{\mathrm{TOT}}^{\mathrm{LOW}} - E_{\mathrm{AS}}^{\mathrm{LOW}}$$

in which $E_{\mathrm{AS}}^{\mathrm{HIGH}}$ is the active site CASSCF energy calculation, $E_{\mathrm{TOT}}^{\mathrm{LOW}}$ is the low-level energy calculation for the total molecule (UFF or RHF), and $E_{\mathrm{AS}}^{\mathrm{LOW}}$ is the low-level energy calculation for the active site. The Gaussian 09 program ONIOM algorithm does all this automatically following the choice of calculation types. For RDX the calculational difference for UFF and RHF is not significant and does not change the general conclusions. The energy gradient is the derivative of the E^{ONIOM} expression with respect to the reaction coordinate, and optimization is performed on the total individual PES gradients, for each PES. The details for the calculations and the comparison between RDX and DMNA under these calculational approaches are discussed in Ref. 66.

What we conclude from these energy release studies for DMNA and RDX is the following: DMNA undergoes a nitro-nitrite isomerization on its S_1 PES and generates an NO product that is vibrationally cold and rotationally hot; RDX undergoes a nitro-nitrite isomerization on its S_0 PES and generates an NO product that is vibrationally hot and rotationally cold. These decomposition mechanisms are applicable to the DMNA and RDX that have been molecular electronically excited to ca. 5 eV. The theoretical results agree with the experimental results completely. Both decomposition mechanisms begin at S_2 (FC), pass through various CIs (DMNA $(S_2/S_1)_{CI}$ and RDX $(S_2/S_1)_{CI}$, $(S_1/S_0)_{CI}$), and finally generate NO appropriately. The difference of course is that RDX converts all of its excitation ~5 eV to vibrational and translational excitation of the product NO, while DMNA only generates product NO with ~1 eV of excitation energy: both processes are nonadiabatic.

Comparison of the two different mechanisms for electronically excited DMNA and RDX reveals that the other nitramine groups of RDX influence the PESs such that a barrier on S_1 RDX can allow it to pass to the $(S_1/S_0)_{CI}$ and decompose on S_0. Note also that the interaction of the PESs (thus the CIs) is much different for the two systems, as shown in Fig. 2.5,[66] compared to Fig. 2.2. The topography of these surfaces in all their dimensions must be quite different due to the influence of the "rest" of the RDX molecule, even at the RHF and UFF levels. This approach not only shows how to understand large energetic molecules but also demonstrates that ONIOM and CASSCF methodologies and assumptions are compatible and importantly useful.

3.2. Furazans

The furazans, DAAF (energetic) and DAF, and furazan itself (nonenergetic) are presented in Fig. 2.1 and have a somewhat different energy release decomposition mechanism than that just presented for the nitramines. The furazan system is an interesting one because even though NO is still a final product for all molecules, both energetic and nonenergetic, the furazans display a very different set of PESs and a very different set of mechanisms. In this instance the NO generation involves two different CI

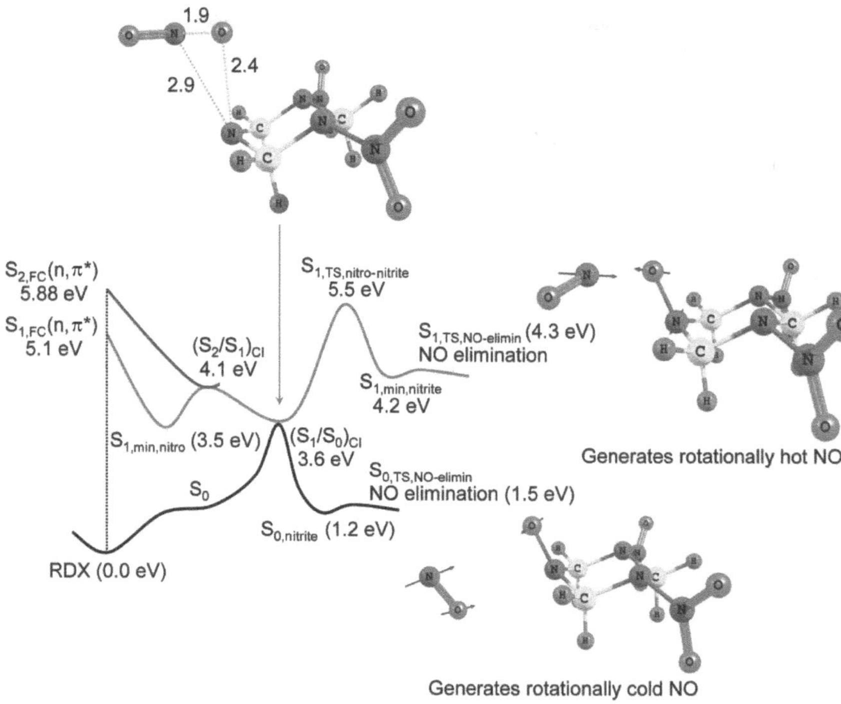

Figure 2.5 Schematic one-dimensional plot of the multidimensional electronic potential energy surfaces of RDX computed at the ONIOM [CASSCF-(10,7)/6-31G(d):UFF] level of theory. Blue arrows indicate the unstable normal mode of vibration associated with the respective transition state. Following electronic excitation to the S_2 state, the molecule follows the minimum-energy path: $S_2(FC) \rightarrow (S_2/S_1)_{CI} \rightarrow (S_1/S_0)_{CI} \rightarrow S_0(\text{nitrite}) \rightarrow$ NO elimination. (For interpretation of the references to color in this figure legend, the reader is referred to the online version of this chapter.)

pathways for furazan and DAF (depending on S_1 or S_2 excitation) but only one pathway for the energetic DAAF. The NO product from DAAF is rotationally cold and vibrationally hot independent of S_1 or S_2 excitation. For DAF and furazan, the rotational excitation of NO can be hot or cold depending on excitation wavelength, but the vibrational population is always low. All of this chemistry occurs on a timescale of less than 100 fs based on direct observation. We need to look for such results based on CASSCF calculations for the PESs.

The details of the theoretical approach can be found in Ref. 67 and we will present here only an outline of this interesting chemistry. CASSCF calculations for the energetic DAAF have not been undertaken as the molecule is too large: an ONIOM approach might be possible for DAAF but the azoxy-connecting group might render the isolated active site (a furazan moiety) assumption inoperative. We plan consideration of this possibility in the near future; additionally, the azoxy group could be another source of NO, but the excitation state might be a factor for this molecule.

A CASSCF (10,7) and a CASSCF (14,9) are employed for the excited state calculations for furazan and DAF, respectively. The active space includes all π and π^* bonding and antibonding orbitals, two in-plane N-nonbonding $n\sigma_N$ orbitals for furazan, and two additional nonbonding $n\pi_n$ orbitals for DAF.[67] These calculations locate the critical points on the PESs including $(S_1/S_0)_{CI,a}$, $(S_1/S_0)_{CI,b}$, and $(S_2/S_1)_{CI}$. Figure 2.6 summarizes these calculations for DAF and furazan.

Based on the computational results, two different nonadiabatic, radiationless dissociation pathways are possible for the decomposition mechanism of furazan and DAF from their $S_2(\pi,\pi^*)$ excited electronic states. One involves an almost barrierless nonadiabatic dissociation path from the FC point of the $S_2(\pi,\pi^*)$ state through the $(S_2/S_1)_{CI}$ CI followed by another nonadiabatic transition through $(S_1/S_0)_{CI,a}$ if the excitation energy is sufficient to surmount the $(S_1/S_0)_{CI,a}$ CI. The other involves an alternate nonadiabatic dissociation path from the FC point of the $S_2(\pi,\pi^*)$ state through the $(S_1/S_0)_{CI,b}$ CI after crossing the $(S_2/S_1)_{CI}$ CI if the excitation energy is lower than the $(S_1/S_0)_{CI,a}$ CI. In the first dissociation mechanism, following vertical electronic excitation to the $S_2(\pi,\pi^*)$ state, for both furazan and DAF, electronic energy can rapidly be thermalized through the $(S_1/S_0)_{CI,a}$ CI by populating out-of-plane vibronically active ring modes. This pathway must go via an adiabatic state switch from the spectroscopic (π,π^*) to the (n,π^*) state. This process can generate cold rotational distributions of the NO product as the unstable normal mode of vibration associated with respective transition states on their ground electronic surface does not possess any resultant torque acting on the NO moiety. In the second alternative channel, on the other hand, because electronic excitation of these molecules does not provide sufficient energy to surmount the $(S_1/S_0)_{CI,a}$ CI, electronic energy will eventually be thermalized by populating another vibronically active ring opening mode which can lead the deactivation along the $(S_1/S_0)_{CI,b}$ CI, yielding a rotationally hot distribution for the NO product as the unstable normal mode of vibration associated with the related TS on the ground electronic surface generates a torque acting on the NO moiety for the NO elimination pathway.

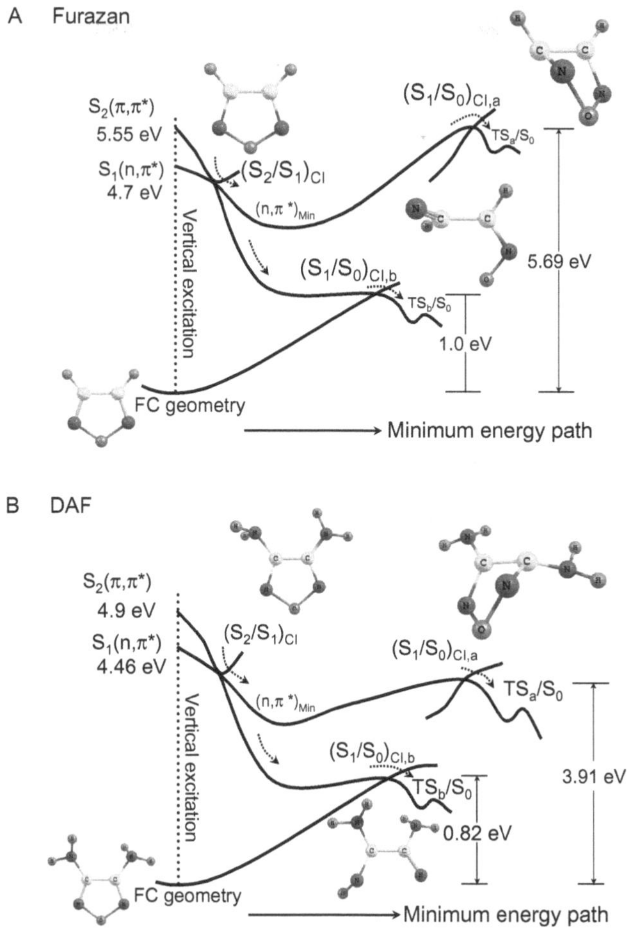

Figure 2.6 One-dimensional projections of multidimensional potential energy surfaces of furazan (A) and DAF (B) computed at CASSCF/6-31G* level of theory. Energies of the critical points and conical intersections are computed at CASMP2/CASSCF/6-31G* level of theory. The branching space formed by the gradient difference and derivative coupling vectors are not calculated here. Instead, the minimum energy paths leading to $(S_1/S_0)_{CI,a}$ and $(S_1/S_0)_{CI,b}$ are calculated by running an IRC calculation from respective CIs in a backward direction using the initial force constants calculated at the respective CIs. Two nonadiabatic dissociation channels along two different nuclear coordinates are indicated by dotted arrows. See text for further details. (For color version of this figure, the reader is referred to the online version of this chapter.)

The above result is included here to demonstrate how different a specific quantum mechanical mechanism can be and still generate a specific product like NO and how much we can learn about chemistry from such a set of open reaction channels.

From the behavior of the DAF and furazan nonenergetic model systems, we can conclude that the decomposition of the DAAF energetic molecule following electronic state excitation to S_1 and/or S_2 occurs through a single reaction channel or steepest descent coordinate, even though the decomposition for DAF and furazan do not. This implies that DAAF has PESs that have no significant barriers along the steepest descent pathways. Thus, a series of CIs brings the molecule to the S_0 ground state, away from the FC point, with all the initial electronic excitation energy ($\sim$5 eV) converted to vibrational excitation in S_0. This excitation energy can be employed in bond breaking and ring opening. The nonenergetic species split their decomposition chemistry between two pathways that may not generate an NO with sufficient vibrational and translational energy for further reactivity. The calculations for the furazan system (DAF, DAAF, furazan) represent the first attempt at finding a theoretically derived, excited state initiated reaction channel based on interacting PESs through a multireference (e.g., CASSCF) quantum mechanical, nonadiabatic algorithm for an energetic molecular system.

3.3. Tetrazine and tetrazine N-Oxide energetics (daato, data, acto)

As we proceed through these different energetic species, the general discussion will require fewer details and more reference to the original studies. We can understand now the general theme of these theoretically derived reactions mechanisms: interacting adiabatic PESs through nonadiabatic terms omitted from the simple, zero order, B.O. Hamiltonian. In large measure, pictures (slices) of the surfaces along the reaction coordinate and statements of the calculational levels employed should serve the general understanding for our purposes.

Two motivating factors have driven the synthetic community toward high nitrogen content energetic materials: (1) the importance of N_2 as a product with a very stable bond to generate a highly exothermic reaction for compounds with a positive ΔH_f°; and (2) N_2 would be an environment friendly final product for the decomposition chemistry. As we will soon see, neither of these hopes are supported by either the theory or experiments, in this instance.

These studies initially include the tetrazine N-oxide derivatives, DAATO, ACTO, DATO,[68] all of which are energetic molecules, presented in Fig. 2.1. The only model nonenergetic species we have studied in the general group is tetrazine itself and it will be discussed following description of the behavior of the three energetic systems. NO is observed to be the initial product of the decomposition reactions of DAATO, DATO, and ACTO following their electronic state excitation. The NO from ACTO and DAATO is rotationally cold and vibrationally hot, while the NO from DATO is rotationally hot and vibrationally cold. ACTO and DAATO are structurally different from DATO, with regard to the N-oxide ring positions, and this is suggested to be an explanation of the decomposition difference. Final transition state calculations are consistent with this dynamical difference for NO products. CASSCF/CASMP2 theory level calculations predict a ring contraction mechanism for generation of the initial NO product from these molecules, through a $(S_1/S_0)_{CI}$.

ACTO and DATO are calculated at the CASSCF (14,10)/6-31G(d) level and the chosen orbitals for the active space are given in Ref. 68. Recall that the active space orbitals must be chosen with the desired chemistry in mind; that is, the experiments guide the calculations and the calculations determine a reaction mechanism (i.e., a set of CIs, barriers, transition states along the reaction coordinate, steepest descent path). One need not do this if a full configuration interaction calculation can be done and all possible reaction coordinates can be explored. In such an instance, the lowest energy, most coupled, steepest descent reaction coordinate pathway would be found and identified as such. For a CASSCF calculation we must know the outcome required to choose the active space (i.e., the orbitals to be considered for the calculation). Figures 2.7 and 2.8 show, respectively, the calculated surface paths for $S_1(n\pi^*)$, $S_2(\pi\pi^*)$, $S_3(\pi\pi^*)$ excitation for ACTO and DATO. NO is generated on the S_0 surface, well away from the FC equilibrium point: in both cases, a ring opening/contraction mechanism is suggested that does not involve an N_2O intermediate. An N_2 product is not observed from either ACTO or DATO. CIs dominate the reaction kinetics and dynamics and the rotational and vibrational excitation of the product NO are governed by the final transition states and the reaction coordinates. The low vibrational excitation of NO from DATO may suggest that more energy is in its translational degrees of freedom or that more of its energy is used to open the ring. Detailed calculations for DAATO have not been attempted to date because of the size, complexity, low symmetry of the molecule (Fig. 2.1), and potential couplings of its moieties. We also searched for N_2 from DATO, ACTO, and DAATO. In each case the experiment was negative: N_2 is not a decomposition product from those energetic molecules following excitation to their valence electronic states.

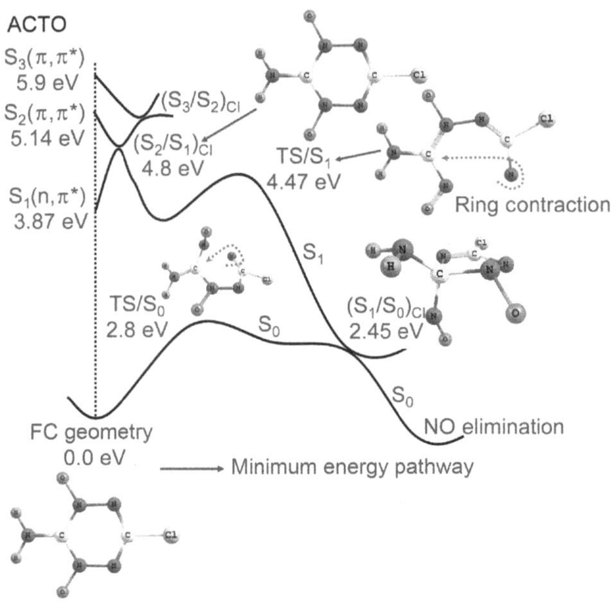

Figure 2.7 One-dimensional projection of the multidimensional singlet electronic potential energy surfaces of ACTO computed at CASSCF (10,8)/6-31G(d) level of theory. Geometry near the $(S_1/S_0)_{CI}$ supports a ring contraction mechanism for excited electronic state decomposition of ACTO, which finally results in NO elimination. Note that ring contracted form of ACTO near the $(S_1/S_0)_{CI}$ has a nearly linear C—NO moiety attached to the triazolering, which can finally generate a rotationally cold NO product. (For color version of this figure, the reader is referred to the online version of this chapter.)

The model or base system for those energetic molecules is tetrazine itself and we have studied the excited electronic state kinetics, dynamics, and chemistry of this system, as well.[32] Different pathways for tetrazine and its derivatives can be considered for decomposition (see Fig. 2.9).

Tetrazine has been known for close to 40 years to photodecompose into 2HCN and N_2, but the reaction mechanism and the dynamics for N_2 have not been specified, either experimentally or theoretically. Our experiments give the rotational distribution of N_2 from tetrazine to be cold (~20 K) but the vibrational temperature has not yet been determined due to wavelength constraints. The calculations given above for DATO and ACTO are applied to tetrazine (CASMP2/CASSCF (14,11)/6-31G(d)).

The mechanism that these calculations generate is depicted in Fig. 2.9. For this system, S_1 through S_4 are necessary to calculate at the CASSCF (14,11)/6-31G(d) and CASMP2 levels. Many CIs are generated for these PESs: $(S_4/S_3)_{CI}$, $(S_3/S_2)_{CI}$, $(S_2/S_1)_{CI}$, $(S_1/S_0)_{CI}$, and probably many others

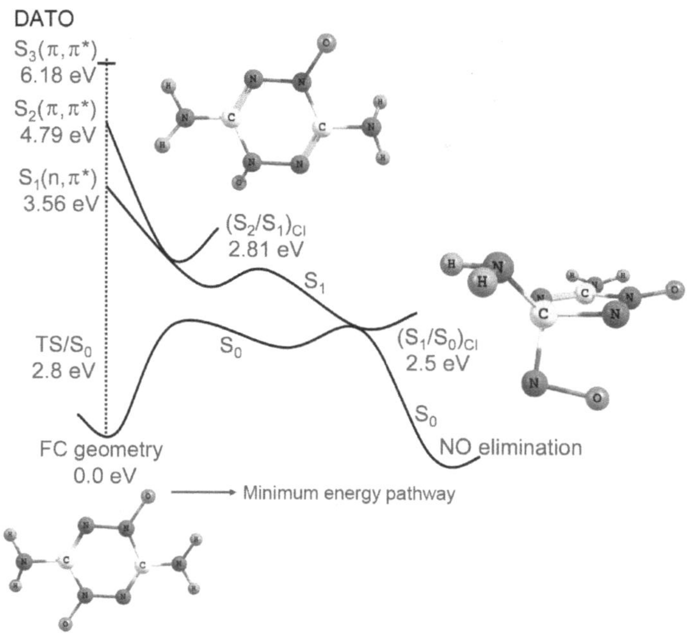

Figure 2.8 One-dimensional projection of the multidimensional singlet electronic potential energy surfaces of DATO computed at CASSCF (10,8)/6-31G(d) level of theory. Geometry near the $(S_1/S_0)_{CI}$ supports a ring contraction mechanism for excited electronic state decomposition of DATO, which finally results in NO elimination. Note that ring contracted form of DATO near the $(S_1/S_0)_{CI}$ has a bent C—NO moiety attached to the triazole ring, which can finally generate a rotationally hot NO product. (For color version of this figure, the reader is referred to the online version of this chapter.)

not uncovered are present. All of these are at points of near degeneracy for the PESs and are most likely strongly nonadiabatically coupled. Thus, tetrazine follows a minimum energy path through CIs from $S_4 \rightarrow \cdots \rightarrow S_0$ for the fragmentation reaction yielding 2HCN and N_2. This generates ~6 eV of vibrational energy in S_0 and a triple concerted dissociation is suggested. Little torque is generated on N_2 at the final dissociation S_0 transition state along the minimum energy reaction coordinate. Thus, N_2 is a product for the unsubstituted tetrazine, but the substituted energetic tetrazines (DATO, ACTO, DAATO) do not follow this pathway. This process is, as found in other cases, ultrafast ($\sim 10^{-14}$–10^{-13} s) due to the PES nonadiabatic couplings.

3.4. PETN [$C(CH_2ONO_2)_4$]

PETN is a very interesting molecule both because of its high symmetry and its ONO_2 bonding. PETN is an obvious candidate for the ONIOM method

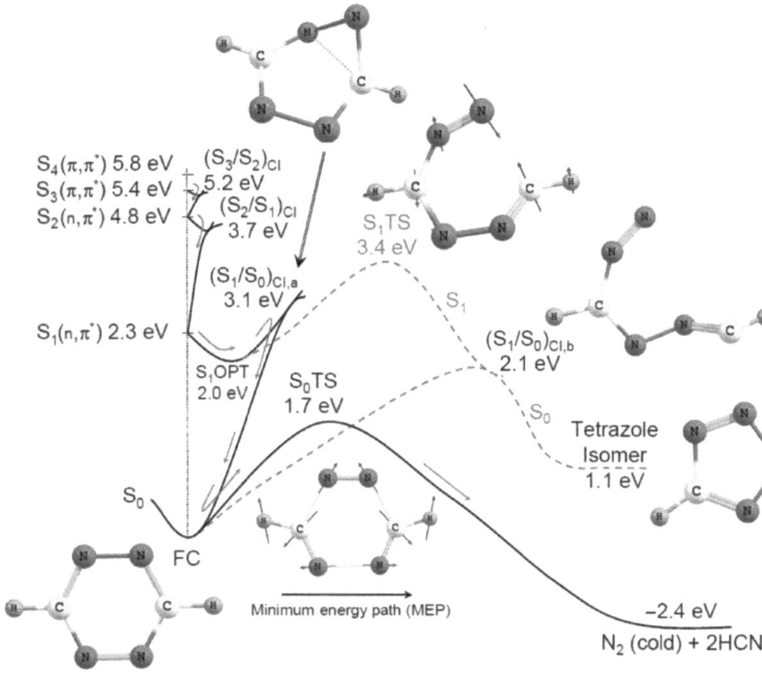

Figure 2.9 PESs of s-tetrazine obtained at a CASSCF (14, 11)/6-31G(d) theory level. Two pathways are possible following electronic excitation of this molecule at 203 nm: (1) internal conversion to the ground state through a half chair form of tetrazine ring followed by concerted triple dissociation producing rotationally cold N_2 and (2) C—N bond dissociation on the S_1 surface followed by nonadiabatic transition to S_0 on which an isomeric tetrazole derivative is formed. Arrows in the figure indicate the minimum energy pathway. Energies associated with the S_1 excited state are scaled by 0.7. Note that the N_2 molecule gets almost no torque along the S_0 dissociation minimum energy pathway. (For color version of this figure, the reader is referred to the online version of this chapter.)

and we employ an ONIOM [CASSCF (6,6)/6-31G(d):UFF] calculational level in an exact parallel to our treatment of RDX, etc., employing the Gaussian 09 program suite. No symmetry restrictions are employed throughout the calculational chain. Both IRC and scan algorithms are applied to map the surfaces following location of the CIs, transition states, and intermediates. The goal is to find the NO molecule product, generated by S_n electronic excitation at ca. 226, 236, and 248 nm with a cold rotational distribution and a hot vibrational distribution, which will then constitute a unique excitation energy independent, quantum mechanical, molecular mechanism for the PETN release of its stored chemical energy. The electronically excited PETN on the S_1 PES moves to the $(S_1/S_0)_{CI}$ to the ground electronic state, undergoes a nitro-nitrite isomerization on S_0, and generates the product NO with the observed dynamics. The calculated reaction path is given in Fig. 2.10. Structure for the various S_1/S_0 stationary points and CIs are given in Ref. 69. The pathway, reaction coordinate in this instance from the initial FC position on S_1 involves motion to the minimum of the S_1 state ($S_{1,min}$), surmounting the S_1 PES energy barrier, and finally reaching the $(S_1/S_0)_{CI}$, and nonadiabatically transitioning to the S_0 PES: this is the steepest descent reaction coordinate. After the molecule returns to S_0, a number of reaction channels are energetically available. The PETN molecule can then return to the equilibrium FC point or follow the steepest descent path to a nitro-nitrite isomerization and generation of NO. If the molecule returns to the FC point (over a 2.6 eV barrier), it can generate an NO_2 molecule with a ca. 1 eV barrier. On the other hand,

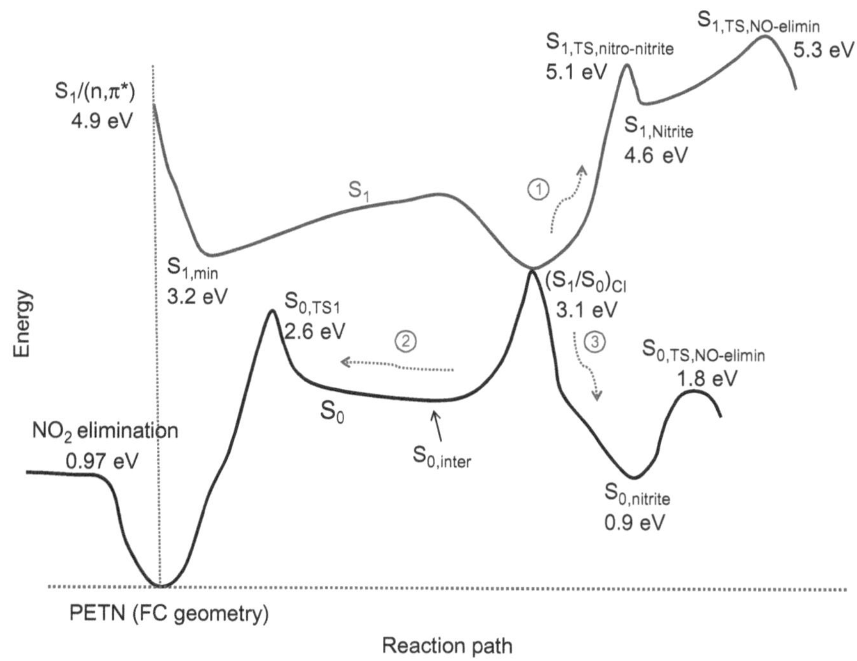

Figure 2.10 A schematic one-dimensional projection of the multidimensional potential energy surfaces of PETN computed at ONIOM (CASSCF (6,6)/6-31G(d):UFF) level of theory. The labeled relative CASSCF energies of the critical points (minimum, conical intersection, transition state) on the PESs with respect to the FC geometry (S_0) are not corrected for zero point energy. (For color version of this figure, the reader is referred to the online version of this chapter.)

following the $(S_1/S_0)_{CI}$, the reaction coordinate is barrierless to a nitro-nitrite isomerization and then NO is generated with a ~1.0 eV barrier at the transition state $S_{0,ST,NO\text{-}elimin}$. The unstable mode at this point puts little torque on the departing NO. All of this occurs in less than ca. 100 fs through the CIs. The similarity of behavior here with nitramines can be attributed to the fact that both N—NO_2 and O—NO_2 S_1 states are $n \rightarrow \pi^*$ excitations. The difference between energetic and nonenergetic species depends on S_0 or S_1 reaction coordinate decomposition, respectively.

3.5. Imidazoles: Mono- and dinitro

Mono-nitroimidazoles are all nonenergetic model systems for the energetic dinitroimidazoles.[70,71] The difference between those two sets of molecules is the same as we have noted above for the other energetic nitro-containing species: the energetic molecules, following electronic excitation to S_n PES (FC), all decompose on S_0 after a nitro-nitrite isomerization and generate NO with cold T_{rot} and hot T_{vib}; the nonenergetic molecules, under similar excitation conditions, render NO from their excited states (e.g., S_1) with T_{rot} warm and T_{vib} cold. In both instances, the kinetics and dynamics are governed by CIs and transition states on each PES. The specific mono-nitroimidazoles that are calculated for nonenergetic model systems are given in Fig. 2.1. The dinitroimidazoles are challenging to treat theoretically because the "active sites" are not separated or independent due to the aromatic π-ring system and because the excitation levels are two different types of excited states, for example, $\pi\pi^*$ and $n\pi^*$. Consider first the nonenergetic mono-nitroimidazoles. The calculations for the systems are accomplished in the now "usual way": a CASSCF (10,7)/6-31G(d) algorithm with active space orbitals two π-bonding ring orbitals, one NO nonbonding $n\sigma_{NO}$ orbital, one π nonbonding $n\pi_0$ orbital, one σ nonbonding $n\sigma_0$ orbital, one delocalized ONO π antibonding $\pi_{ONO}{}^*$ orbital, and one π antibonding ring π^* orbital. The first five orbitals are occupied and the last two are virtual. S_1 and S_2 states are $n\pi^*$ and S_3 is $\pi\pi^*$ in nature. For the ground state system, the CASSCF calculation shows that NO_2 and NO release require similar energies (~80 kcal/mol) and are thus competitive on S_0. These two pathways are located on very different parts of the S_0 PES, for different reaction coordinates, however.

The CASSCF scan and IRC calculations for the reaction coordinate on the S_0, S_1, and S_2 surfaces are similar for the three mono-nitroimidazole molecules and are presented in Figs. 2.11–2.13. Two possible

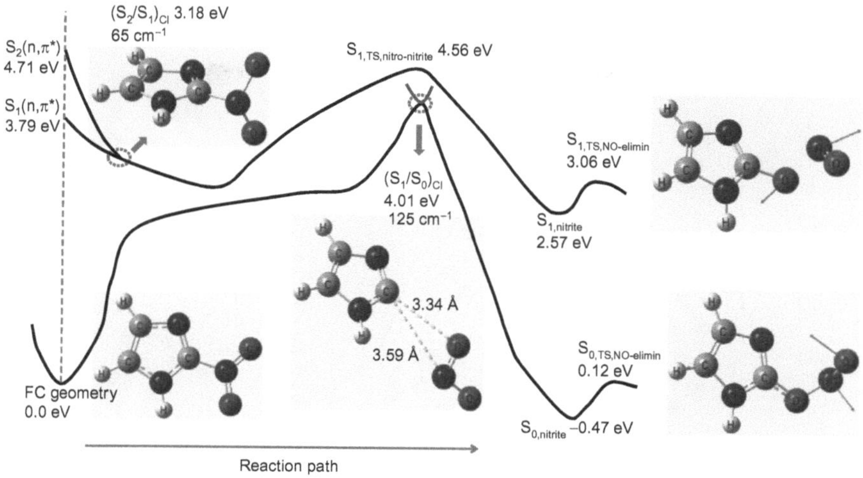

Figure 2.11 A schematic one-dimensional projection of the multidimensional potential energy surfaces of 2-nitroimidazole computed at the CASSCF (10,7)/6-31G(d) level of theory. The solid red line, which represents the S_1 electronic excited state, is generated by a different reaction coordinate. "FC geometry," which is a planar structure, is the optimized minimum energy structure of 2-nitroimidazole on S_0 state. "$(S_1/S_0)_{CI}$" represents the conical intersection between the ground and first excited electronic states. The adiabatic energy gap at this point is ca. 125 cm^{-1}. "$(S_2/S_1)_{CI}$" represents the conical intersection between the second and first excited electronic states. The adiabatic energy gap at this point is ca. 65 cm^{-1}. "$S_{1,TS,nitro\text{-}nitrite}$" is the transition state of the nitro-nitrite isomerization on the S_1 state. "$S_{0,nitrite}$" and "$S_{1,nitrite}$" are nitrite structures of 2-nitroimidazole after nitro-nitrite isomerization on S_0 and S_1, respectively. "$S_{0,TS,NO\text{-}elimin}$" and "$S_{1,TS,NO\text{-}elimin}$" are transition states for NO elimination from the nitrite structure of 2-nitroimidazole on S_0 and S_1, respectively. The arrows in the structures of the transition states show the reaction coordinate of the imaginary frequency. (For interpretation of the references to color in this figure legend, the reader is referred to the online version of this chapter.)

decomposition paths can be found for all three of these molecules: S_1 or S_2 generation of NO. S_1 dissociation is chosen as the most likely path for the following reasons: the $(S_1/S_0)_{CI}$s are weak due to the substantial (>100 cm^{-1}) adiabatic gaps between the two surfaces, and the NO product is not vibrationally very hot (<800 K) but is relatively cold (ca. 50 K) rotationally, as is consistent with the S_1 transition state ($S_{1,TS,NO\text{-}elimin}$). The imidazole aromatic ring orbitals seem to play a major role in the excited electronic state (S_1, S_2) PESs and generated CIs that are important for the overall mono-nitroimidazole dynamics. The general kinetics and dynamics for this series of model species seems to be independent of the C—NO_2 ring position.

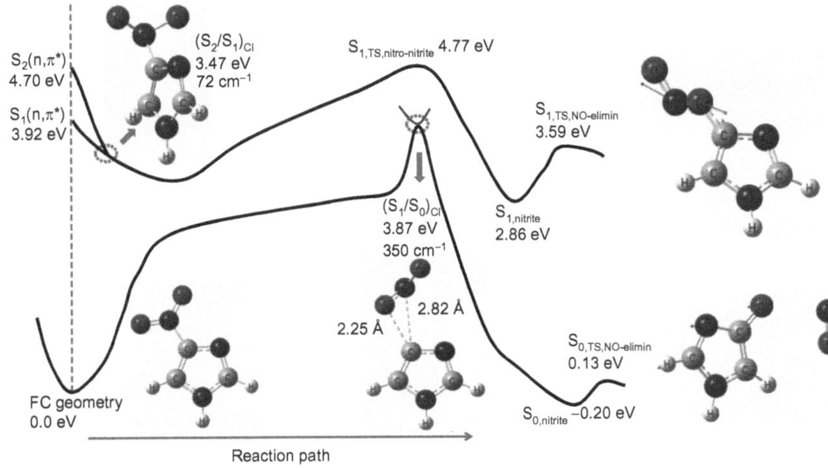

Figure 2.12 A schematic one-dimensional projection of the multidimensional potential energy surfaces of 4-nitroimidazole computed at the CASSCF (10,7)/6-31G(d) level of theory. The solid red line, which represents the S_1 electronic excited state, is generated by a different reaction coordinate. "FC geometry," which is a planar structure, is the optimized minimum energy structure of 2-nitroimidazole on the S_0 state. "$(S_1/S_0)_{CI}$" represents the conical intersection between the ground and first excited electronic states. The adiabatic energy gap at this point is ca. 350 cm^{-1}. "$(S_2/S_1)_{CI}$" represents the conical intersection between the second and first excited electronic states. The adiabatic energy gap at this point is ca. 72 cm^{-1}. "$S_{1,TS,nitro\text{-}nitrite}$" is the transition state of the nitro-nitrite isomerization on the S_1 state. "$S_{0,nitrite}$" and "$S_{1,nitrite}$" are nitrite structures of 4-nitroimidazole after nitro-nitrite isomerization on S_0 and S_1, respectively. "$S_{0,TS,NO\text{-}elimin}$" and "$S_{1,TS,NO\text{-}elimin}$" are transition states for NO elimination from the nitrite structure of 4-nitroimidazole on S_0 and S_1. The arrows in the structure of the transition states show the reaction coordinate of the imaginary frequency. (For interpretation of the references to color in this figure legend, the reader is referred to the online version of this chapter.)

The dinitroimidazoles are all energetic, and, in analogy to the DMNA/RDX situation, one can realize that the adjacent nitro moieties can have a significant effect on the S_0, S_1, S_2, S_3 PESs, especially with regard to CIs, barriers, and nonadiabatic couplings for the adiabatic states. Unfortunately, a large CASSCF would be required to express the true nature of the excited and even ground PESs: for example, a CASSCF (20,14) should be a good first guess. Another possibility would be an ONIOM calculation but the two "active NO_2 sites" are coupled by the ring π-system and thus are not isolated as required: the site and its remaining molecule strongly interact. A restricted active space calculation can also be considered, but the excited states are not all of the same character ($n\pi^*$, $\pi\pi^*$, etc.), and different (restricted) active spaces would need to be considered for the various PESs.

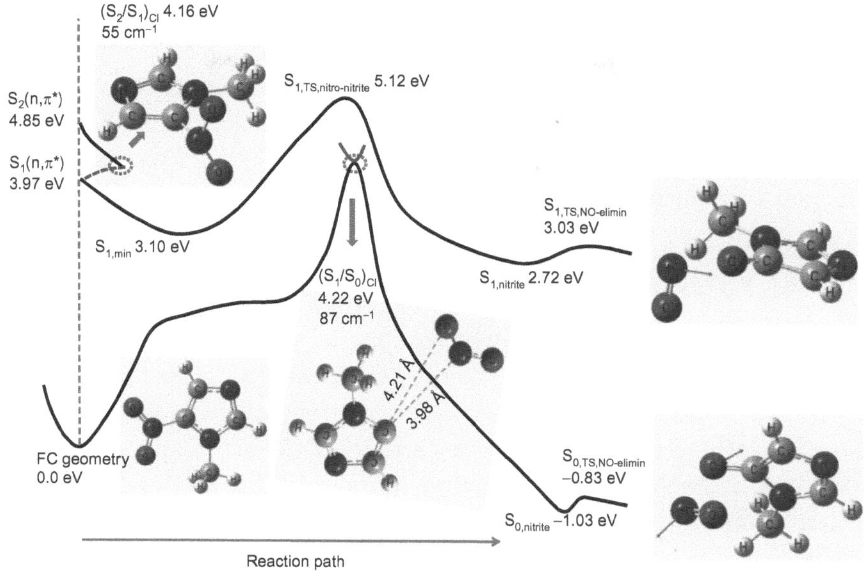

Figure 2.13 A schematic one-dimensional projection of the multidimensional potential energy surfaces of 1-methyl-5-nitroimidazole computed at the CASSCF (10,7)/6-31G(d) level of theory. The solid red line, which represents the S_1 electronic excited state, is generated by a different reaction coordinate. "FC geometry" is the optimized minimum energy structure of 2-nitroimidazole on S_0 state. "$(S_1/S_0)_{CI}$" represents the conical intersection between the ground and first excited electronic states. The adiabatic energy gap at this point is ca. 87 cm^{-1}. "$(S_2/S_1)_{CI}$" represents the conical intersection between the second and first excited electronic states. The adiabatic energy gap at this point is ca. 55 cm^{-1}. "$S_{1,TS,nitro\text{-}nitrite}$" is the transition state of the nitro-nitrite isomerization on the S_1 state. "$S_{0,nitrite}$" and "$S_{1,nitrite}$" are nitrite structures of 1-methyl-5-nitroimidazole after nitro-nitrite isomerization on S_0 and S_1, respectively. "$S_{0,TS,NO\text{-}elimin}$" and "$S_{1,TS,NO\text{-}elimin}$" are transition states for NO elimination from the nitrite structure of 1-methyl-5-nitroimidazole on S_0 and S_1, respectively. The arrows in the structure of the transition states show the reaction coordinate of the imaginary frequency. (For interpretation of the references to color in this figure legend, the reader is referred to the online version of this chapter.)

Experimentally, a unique excitation wavelength independent NO with cold rotations and hot vibrations is generated as dinitro imidazole energetics decompose following electronic excitation. The predicted reaction mechanism, now based on experience with the forgoing model and real energetic molecules, is that the molecule finds the steepest descent reaction coordinate through CIs and over barriers to S_0 and then undergoes a nitro-nitrite isomerization which leads to a transition state on S_0 that generates NO with the appropriate energy distribution. For C—NO_2 species, the NO_2 and NO reaction coordinates on S_0 at the FC point are energetically similar, but the N—NO_2 bond is much weaker and will generate the NO_2 molecule if only the S_0 PES(FC) is considered. The C—NO_2 species are more stable and the π-ring system enhances this stability and insensitivity to heat and impact/shock on S_0. The excited state reaction coordinates place the molecules far from the FC equilibrium point on S_0 and thus favor the nitro-nitrite NO reaction coordinate.

4. FUTURE DIRECTIONS, NEW SYSTEMS, CONCLUSIONS

Even though suggestions of the importance of excited electronic states for the release of stored chemical energy from energetic materials are now more than 40 years old, experiment and theory for such decomposition processes are relatively new. Multireference, state-of-the-art calculation beyond CASSCF, and CAS/PT2 will be a major focus for the foreseeable future: Bartlett's group is presently in the process of implementing a CCSD (T) multireference code that should be applicable to excited state energetic molecules, and their decomposition and their PES interaction behavior.

With regard to experiments, activity should increase for the synthesis of new energetic species, especially high N-content systems. In this regard, the new energetics recently published from the Klapötke laboratory are particularly interesting. They also seem to follow the general trend of generating NO following excitation with the now familiar energy distribution. Two such systems are presented as examples at the end of Fig. 2.1. Physical measurements on these new and many older systems all become more routine, and one should expect to see new and improved ultrafast (~10–100 fs) studies on the kinetic of energetic and model molecule decomposition. This in turn should stimulate more theory studies of the time evolution of both fragmentation reactions and energy dynamics following molecular excitation.

Interest is beginning to grow in the study of the next step, beyond the initial one, in the process of stored energy release from energetic species. For example, following the NO, N_2, NO_2, etc., initial step in the decomposition process, what happens to the large remaining radical, such as $[C_3H_6(N_2O_2)_2NO]^{\bullet}$ for RDX. Most of the molecular stored energy is still stored in this major parent fragment. The second, third, etc., step must be observed and calculated (maybe not in that order) to have a complete picture of the energy release processes involved. Moreover, many other excited states can be accessed in addition to the valence states discussed above: Rydberg and ion states can obviously be generated in the impact/excitation/fragmentation processes as one can learn from general triboluminescence experiments and observations.

Finally, following the study of molecular kinetic and dynamic behavior of energetic materials as isolated species, studies can be made of these molecules in pure and solvated clusters. Such clusters can be generated and found in the gas phase through the same experimental approach employed in our presently reported efforts. One can generate dimers, trimers, tetramers, etc. of these organic species and explore their excited state behavior, as well. Clusters can be made with energetic and nonenergetic species, such as $(RDX)_n$, $(RDX)_m(H_2O)_n$, $(RDX)_m(C_nH_{2n+2})_x$, $(RDX)_m(C_nH_{2n})_x$, etc., and thus many different types of mixed systems can be explored under experimental conditions for which substantial theory can be employed to analyze and predict results.

With regard to general conclusions that can be drawn from studies reviewed here, we can point to the following four major statements.

- First, nonadiabatic behavior, CIs, and full PESs dominate the fragmentation and decomposition dynamics and kinetics completely.
- Second, energetic and model system differences are subtle and depend on PESs, their interactions (CIs), their electronic composition (multiple configuration state functions), energy barriers, and transition states.
- Third, the fundamental theoretical model behavior applicable to all of these systems, based on electronic excitation, as the initial contributing factor to the decomposition of energetic materials, is organic photochemistry and photophysics.
- Fourth, most if not all model systems decompose on their excited state PESs, while all energetic species reach the S_0 state before decomposing and thereby convert all of their electronic excitation energy (typically >5 eV) into internal molecular vibrational energy employed to break bonds and generate vibrationally hot, fast, small, reactive radicals for further interactions and reactions.

ACKNOWLEDGMENTS

I would like to thank Drs. Z. Yu, B. Yuan, and A. Bhattacharya for many helpful comments and suggestions concerning this review and the reported studies. The research discussed in this article is supported by the Army Research Office for more than 10 years.

REFERENCES

1. Windawi, H. M.; Varma, S. P.; Cooper, C. B.; Williams, F. Analysis of Lead Azide Thin Films by Rutherford Backscattering. *J. Appl. Phys.* **1976**, *47*, 3418–3420.
2. Schanda, J.; Baron, B.; Williams, F. Low Temperature Photoluminescence of Lead Azide. *J. Lumin.* **1974**, *9*, 338–342.

3. Varma, S. P.; Williams, F. Infrared Absorption Spectra of Doped and Undoped Lead Azide. *J. Chem. Phys.* **1973**, *59*, 912–914.
4. Williams, F. Electronic States of Solid Explosives and Their Probable Role in Detonations. *Adv. Chem. Phys.* **1971**, *21*, 289.
5. Sharma, J.; Forbes, J. W.; Coffey, C. S.; Liddiard, T. P. The Physical and Chemical Nature of Sensitization Centers Left from Hot Spots Caused in Triaminotrinitrobenzene by Shock or Impact. *J. Phys. Chem.* **1987**, *91*, 5139–5144.
6. Sharma, J.; Beard, B. C.; Chaykovsky, M. Correlation of Impact Sensitivity with Electronic Levels and Structure of Molecules. *J. Phys. Chem.* **1991**, *95*, 1209–1213.
7. Dreger, Z. A.; Gruzdkov, Y. A.; Gupta, Y. M.; Dick, J. J. Shock Wave Induced Decomposition Chemistry of Pentaerythritol Tetranitrate Single Crystals: Time-Resolved Emission Spectroscopy. *J. Phys. Chem. B* **2001**, *106*, 247–256.
8. Tsuboi, Y.; Seto, T.; Kitamura, N. Laser-Driven Shock Wave-Induced Triboluminescence of an Organic Crystal: Toward a Semiquantitative Study. *J. Phys. Chem. B* **2003**, *107*, 7547–7550.
9. Aduev, B. P.; Aluker, E. D.; Kuklja, M. M.; Kunz, A. B.; Younk, E. H. Luminescence of Lead Azide Induced by the Electron Accelerator Pulse. *J. Lumin.* **2000**, *91*, 41–48.
10. Kuklja, M. M.; Rashkeev, S. N.; Zerilli, F. J. Ab Initio Calculations of the Electronic Structure of 1,1-Diamino-2,2.Dinitroethylene. In *Shock Compression of Condensed Matter—2003, Pts 1 and 2, Proceedings*; Furnish, M. D.; Gupta, Y. M.; Forbes, J. W. Eds.; American Institute of Physics: Melville, 2004; Vol. 706; pp 363–366.
11. Reed, E. J.; Joannopoulos, J. D.; Fried, L. E. Electronic Excitations in Shocked Nitromethane. *Phys. Rev. B* **2000**, *62*, 16500–16509.
12. Manaa, M.; Fried, L.; Reed, E. Explosive Chemistry: Simulating the Chemistry of Energetic Materials at Extreme Conditions. *J. Comput. Aided Mater. Des.* **2003**, *10*, 75–97.
13. Manaa, M. R.; Fried, L. E. Intersystem Crossings in Model Energetic Materials. *J. Phys. Chem. A* **1999**, *103*, 9349–9354.
14. Kuklja, M. M.; Stefanovich, E. V.; Kunz, A. B. An Excitonic Mechanism of Detonation Initiation in Explosives. *J. Chem. Phys.* **2000**, *112*, 3417–3423.
15. Kuklja, M. M.; Aduev, B. P.; Aluker, E. D.; Krasheninin, V. I.; Krechetov, A. G.; Mitrofanov, A. Y. Role of Electronic Excitations in Explosive Decomposition of Solids. *J. Appl. Phys.* **2001**, *89*, 4156–4166.
16. Kuklja, M. M.; Kunz, A. B. Ab Initio Simulation of Defects in Energetic Materials. Part I. Molecular Vacancy Structure in RDX Crystal. *J. Phys. Chem. Solids* **2000**, *61*, 35–44.
17. Tarver, C. M. Effect of Electric Fields on the Reaction Rates in Shock Initiating and Detonating Solid Explosives. In *Shock Compression of Condensed Matter—2011, Pts 1 and 2*; Elert, M. L.; Buttler, W. T.; Borg, J. P.; Jordan, J. L.; Vogler, T. J. Eds.; American Physical Society Publication, 2012; p 1426.
18. Kunz, A. B.; Beck, D. R. Possible Role of Charged Defects in Molecular-Solids. *Phys. Rev. B* **1987**, *36*, 7580–7585.
19. Tasker, D. G.; Dick, R. D.; Wilson, W. H. Mechanical Properties of Explosives Under High Deformation Loading Conditions. In *Shock Compression of Condensed Matter—1997*; Schmidt, S. C.; Dandekar, D. P.; Forbes, J. W. Eds.; American Institute of Physics: Melville, 1998; Vol. 429; pp 591–594.
20. Nikitin, E. E. *Theory of Thermally Induced Gas Phase Reaction*. Indiana University Press: Bloomington, IN, 1966.
21. Nikitin, E. E. *Theory of Elementary Atomic and Molecular Processes in Gases*. Clarendon Press: Oxford, 1974.
22. Klessinger, Martin; Michl, Josef *Excited States and Photochemistry of Organic Molecules*; Wiley-VCH: New York, 1995.
23. Turro, Nicholas J.; Scaiano, J. C.; Ramamurthy, V. *Modern Molecular Photochemistry of Organic Molecules*. University Science Books: Sausalito, CA, 2010.

24. Nakamura, H. *Nonadiabatic Transition: Concepts, Basic Theories and Applications*. World Scientific Publishing Company: New Jersey, 2012.
25. Platz, Matthew S.; Moss, Robert A.; Maitland Jones, J. *Reviews of Reactive Intermediate Chemistry*. John Wiley & Sons: Hoboken, NJ, 2007.
26. Olivucci, M. *Computational Photochemistry*. Elsevier: Netherlands, Amsterdam, 2005.
27. Baer, M. *Beyond Born-Oppenheimer: Electronic Nonadiabatic Coupling Terms and Conical Intersections*. John Wiley & Sons: Hoboken, NJ, 2006.
28. Holbrook, K. A.; Pilling, M. J.; Robertson, S. H. *Unimolecular Reactions*. John Wiley & Sons: Hoboken, NJ, 1996.
29. Im, H. S.; Bernstein, E. R. Photodissociation ofNO2 in the Region 217–237 nm: Nascent NO Energy Distribution and Mechanism. *J. Phys. Chem. A* **2002**, *106*, 7565–7572.
30. Fischer, N.; Gao, L.; Klapötke, T. M.; Stierstorfer, J. Energetic Salts of 5,5′-Bis(Tetrazole-2-Oxide) in a Comparison to 5,5′-Bis(Tetrazole-1-Oxide) Derivatives. *Polyhedron* **2013**, *51*, 201–210.
31. Fischer, D.; Klapötke, T. M.; Piercey, D. G.; Stierstorfer, J. Synthesis of 5-Aminotetrazole-1 N-oxide and Its Azo Derivative: A Key Step in the Development of New Energetic Materials. *Chem. Eur. J.* **2013**, *19*, 4602–4613.
32. Guo, Y.; Bhattacharya, A.; Bernstein, E. R. Decomposition of Excited Electronic State S-Tetrazine and Its Energetic Derivatives. *J. Chem. Phys.* **2011**, *134*, 024318.
33. Kimmel, A. V.; Sushko, P. V.; Shluger, A. L.; Kuklja, M. M. Effect of Charged and Excited States on the Decomposition of 1,1-Diamino-2,2-Dinitroethylene Molecules. *J. Chem. Phys.* **2007**, *126*, 234711.
34. Kunz, A. B.; Kuklja, M. M.; Botcher, T. R.; Russell, T. P. Initiation of Chemistry in Molecular Solid by Processes Involving Electronic Excited States. *Thermochim. Acta* **2002**, *384*, 279.
35. Aluker, E. D.; Krechetov, A. G.; Mitrofanov, A. Y.; Zverev, A. S.; Kuklja, M. M. Understanding Limits of the Thermal Mechanism of Laser Initiation of Energetic Materials. *J. Phys. Chem. C* **2012**, *116*, 24482–24486.
36. Aluker, E. D.; Krechetov, A. G.; Mitrofanov, A. Y.; Nurmukhametov, D. R.; Kuklja, M. M. Laser Initiation of Energetic Materials: Selective Photoinitiation Regime in Pentaerythritol Tetranitrate. *J. Phys. Chem. C* **2011**, *115*, 6893–6901.
37. Kuklja, M. M.; Rashkeev, S. N. Shear-Strain-Induced Structural and Electronic Modifications of the Molecular Crystal 1,1-Diamino-2,2-Dinitroethylene: Slip-Plane Flow and Band Gap Relaxation. *Phys. Rev. B* **2007**, *75*, 104111.
38. Kuklja, M. M. On the Initiation of Chemical Reactions by Electronic Excitations in Molecular Solid. *Appl. Phys. A* **2003**, *76*, 359–366.
39. Wu, C. J.; Fried, L. E. Ring Closure Mediated by Intramolecular Hydrogen Transfer in the Decomposition of a Push–Pull Nitroaromatic: TATB. *J. Phys. Chem. A* **2000**, *104*, 6447–6452.
40. Manaa, M. R.; Schmidt, R. D.; Overturf, G. E.; Watkins, B. E.; Fried, L. E.; Kolb, J. R. Towards Unraveling the Photochemistry of TATB. *Thermochim. Acta* **2002**, *384*, 85–90.
41. Manaa, M. R.; Yarkony, D. R. On the Role of Conical Intersections of Two Potential Energy Surfaces of the Same Symmetry in Photodissociation. 2. CH3SCH3 .fwdarw. CH3S + CH3. *J. Am. Chem. Soc.* **1994**, *116*, 11444–11448.
42. Rice, B. M.; Mattson, W.; Grosh, J.; Trevino, S. F. Molecular-Dynamics Study of Detonation. I. A Comparison with Hydrodynamic Predictions. *Phys. Rev. E* **1996**, *53*, 611–622.
43. Rice, B. M.; Mattson, W.; Grosh, J.; Trevino, S. F. Molecular-Dynamics Study of Detonation. II. The Reaction Mechanism. *Phys. Rev. E.* **1996**, *53*, 623–635.
44. Chakraborty, D.; Muller, R. P.; Dasgupta, S.; Goddard, W. A. Mechanism for Unimolecular Decomposition of HMX (1,3,5,7-Tetranitro-1,3,5,7-Tetrazocine): An Ab Initio Study. *J. Phys. Chem. A* **2001**, *105*, 1302–1314.

45. Chakraborty, D.; Muller, R. P.; Dasgupta, S.; Goddard, W. A. The Mechanism for Unimolecular Decomposition of RDX (1,3,5-Trinitro-1,3,5-Triazine). An Ab Initio Study. *J. Phys. Chem. A* **2000**, *104*, 2261–2272.
46. Robb, M. A. http://www.ch.ic.ac.uk/robb/casscf_benzene_handout.php.
47. Palmer, I. J.; Ragazos, I. N.; Bernardi, F.; Olivucci, M.; Robb, M. A. An MC-SCF Study of the S1 and S2 Photochemical Reactions of Benzene. *J. Am. Chem. Soc.* **1993**, *115*, 673–682.
48. Neumann, J. V.; Wigner, E. Zur Erklärung einiger Eigenschaften der Spektren aus der Quantenmechanik des Drehelektrons. *Zeitsch. Phys. A Hadrons Nuclei* **1928**, *49*, 73–94.
49. Kauzmann, W. *Quantum Chemistry*. Academic Press: New York, 1957; p 536.
50. Domcke, Wolfgang; Yarkony, David R.; Köppel, H. *Conical Intersections: Electronic Structure, Dynamics and Spectroscopy*. World Scientific Publishing: Singapore, 2004.
51. Yarkony, D. R. Diabolical Conical Intersections. *Rev. Mod. Phys.* **1996**, *68*, 985–1013.
52. Yarkony, D. R. Nonadiabatic Quantum Chemistry—Past, Present, and Future. *Chem. Rev.* **2011**, *112*, 481–498.
53. Köppel, H.; Barentzen, H.; Yarkony, D. R. *The Jahn-Teller Effect. Fundamentals and Implications for Physics and Chemistry*. Springer: Berlin, 2009.
54. Bhattacharya, A. Ph.D. Thesis, Chemistry, Colorado State University, 2009.
55. Rappé, A. K.; Bernstein, E. R. Ab Initio Calculation of Nonbonded Interactions: Are We There Yet? *J. Phys. Chem. A* **2000**, *104*, 6117–6128.
56. Bhattacharya, A.; Guo, Y. Q.; Bernstein, E. R. Experimental and Theoretical Exploration of the Initial Steps in the Decomposition of a Model Nitramine Energetic Material: Dimethylnitramine. *J. Phys. Chem. A* **2009**, *113*, 811–823.
57. Houston, P. L. *Chemical Kinetics and Reaction Dynamics*. McGraw-Hill: New York, 2001; p 102.
58. Steinfeld, Jeffrey I.; Francisco, Joseph S.; Hase, W. L. *Chemical Kinetics and Dynamics*. Prentice Hall: Englewood Cliffs, NJ, 1998; p 308ff.
59. Levine, Raphael D.; Bernstein, R. B. *Molecular Reaction Dynamics and Chemical Reactivity*. Oxford University Press: New York, 1987; p 187ff.
60. Svensson, M.; Humbel, S.; Froese, R. D. J.; Matsubara, T.; Sieber, S.; Morokuma, K. ONIOM: A Multilayered Integrated MO+MM Method for Geometry Optimizations and Single Point Energy Predictions. A Test for Diels-Alder Reactions and Pt(P(t-Bu)(3))(2)+H-2 Oxidative Addition. *J. Phys. Chem.* **1996**, *100*, 19357–19363.
61. Svensson, M.; Humbel, S.; Morokuma, K. Energetics Using the Single Point IMOMO (Integrated Molecular Orbital +Molecular Orbital) Calculations: Choices of Computational Levels and Model System. *J. Chem. Phys.* **1996**, *105*, 3654–3661.
62. Humbel, S.; Sieber, S.; Morokuma, K. The IMOMO Method: Integration of Different Levels of Molecular Orbital Approximations for Geometry Optimization of Large Systems: Test for n-Butane Conformation and S(N)2 Reaction: RCl + Cl. *J. Chem. Phys.* **1996**, *105*, 1959–1967.
63. Vreven, T.; Morokuma, K. On the Application of the IMOMO (Integrated Molecular Orbital Plus Molecular Orbital) Method. *J. Comput. Chem.* **2000**, *21*, 1419–1432.
64. Morokuma, K. New Challenges in Quantum Chemistry: Quests for Accurate Calculations for Large Molecular Systems. *Philos. Trans. R. Soc. Lond. A* **2002**, *360*, 1149.
65. Vreven, T.; Morokuma, K. Chapter 3 Hybrid Methods: ONIOM(QM:MM) and QM/MM. In *Annual Reports in Computational Chemistry*; Spellmeyer, D. Ed.; Elsevier: New York, 2010; Vol. 2; pp 35–52.
66. Bhattacharya, A.; Bernstein, E. R. Nonadiabatic Decomposition of Gas-Phase RDX Through Conical Intersections: An ONIOM-CASSCF Study. *J. Phys. Chem. A* **2011**, *115*, 4135–4147.

67. Guo, Y. Q.; Bhattacharya, A.; Bernstein, E. R. Excited Electronic State Decomposition of Furazan Based Energetic Materials: 3,3'-Diamino-4,4'-Azoxyfurazan and Its Model Systems, Diaminofurazan and Furazan. *J. Chem. Phys.* **2008**, *128*, 034303.
68. Bhattacharya, A.; Guo, Y. Q.; Bernstein, E. R. Unimolecular Decomposition of Tetrazine-N-Oxide Based High Nitrogen Content Energetic Materials from Excited Electronic States. *J. Chem. Phys.* **2009**, *131*, 194304.
69. Yu, Z. J.; Bernstein, E. R. Decomposition of Pentaerythritol Tetranitrate C(CH (2)ONO(2))(4) Following Electronic Excitation. *J. Chem. Phys.* **2011**, *135*, 154305.
70. Yu, Z.; Bernstein, E. R. Experimental and Theoretical Studies of the Decomposition of New Imidazole Based Energetic Materials: Model Systems. *J. Chem. Phys.* **2012**, *137*, 114303.
71. Yu, Z.; Bernstein, E. R. On the Decomposition Mechanisms of New Imidazole-Based Energetic Materials. *J. Phys. Chem. A* **2013**, *117*, 1756–1764.

CHAPTER THREE

Quantum-Chemical Modeling of Energetic Materials: Chemical Reactions Triggered by Defects, Deformations, and Electronic Excitations

Maija M. Kuklja[*,†,1]
[*]MSE Department, University of Maryland, College Park, Maryland, USA
[†]National Science Foundation, Arlington, Virginia, USA
[1]Corresponding author: e-mail address: mkukla@umd.edu; mkukla@nsf.gov

Contents

Advances in Quantum Chemistry, Volume 69
ISSN 0065-3276
http://dx.doi.org/10.1016/B978-0-12-800345-9.00003-9

Abstract

The chapter provides a brief overview of the current understanding of sensitivity of energetic materials to detonation, highlights the most recent advancements in quantum-chemical modeling and simulations of earliest chemical reactions, and emphasizes the role of structural and electronic defects and deformations in initiation of explosive decomposition of materials. An analysis of theoretical and computational results is presented in the context of available experimental data.

1. INTRODUCTION

Explosive decomposition consists of three main stages: initiation, combustion, and detonation. The two latter aspects are fairly well understood especially on the macro- and mesoscales. By contrast, a detailed understanding of the initiation process has yet to be established. It is clear that the initiation of chemistry in energetic compounds is a complex process, and hence progress in searching for new and improved energetic compounds has been rather slow. With recent advances in both computational capabilities[1–3] and ultrafast time-resolved spectroscopic diagnostics,[4] attention has begun to focus on the characterization of the physicochemical responses of specific energetic ingredients such as localized mechanical stresses,[5–9] crystal packing,[10–12] morphology,[13,14] and details of chemical decomposition reactions[15–17].

It is accepted that sensitivity to initiation of detonation (or a propensity of an energetic material to trigger a chemical reaction of explosive decomposition) depends upon many factors, including a chemical composition, bonding of functional groups, a distribution of the charge density, stoichiometry, molecular ordering, grain size, presence of imperfections, nanoparticles, interfaces, etc. However, correlations between the structure, composition, and sensitivity have yet to be revealed. Fundamental insights into these relationships are essential so that energetic materials' properties can be tailored to influence macroscopic responses. Understanding factors that govern sensitivity and at the same time provide high performance persists to represent an outstanding challenge.

In addressing a continuing need for improved energetic materials, one seeks optimal tradeoffs in energy content (a chemical composition), safety (a predictable response to the external stimuli), and cost (synthesis conditions). Since safety and performance of explosives are intimately tied to the decomposition process, detailed knowledge of the mechanisms is required; otherwise, the behavior of explosives remains disconcertingly empirical and hampers the most efficient use of existent materials and design of novel energetic systems.

Quality of energetic materials is often judged using two main parameters: sensitivity to detonation and its performance. Sensitivity is desirable to be low to make the material relatively stable to external stimulus, that is, controllable and able of triggering rapid dissociation only when needed and not accidentally. Performance, on the other hand, is to be high to provide a sufficiently large heat of the explosive reaction. These parameters (performance and sensitivity) do not necessarily correlate with each other and depend on many variables, such as the molecular and crystalline structures, history of samples, the particle size, the crystal hardness and orientation, external stimuli, aging, storage conditions, and others. Elucidation of bond dissociation mechanisms, energy localization, and energy transfer phenomena in the course of explosive decomposition of energetic materials are central for understanding, controlling, and enhancing the performance of these materials as fuels, propellants, and explosives. Mechanisms governing performance are fairly well understood whereas mechanisms of sensitivity are poorly known and need to be much more extensively studied. It is widely accepted though that the thermal decomposition reactions of the materials play a significant role in its sensitivity to mechanical stimuli and the explosive properties.

The decomposition of energetic materials can be initiated with a mechanical impact, thermal heating, laser initiation, shock wave, or a spark. Such events in solids generate molecules in highly excited vibronic and electronic states. Clearly, the decomposition of solid explosives under shock, spark, laser, or plasma ignition must include contributions from both ground and excited electronic states. Excitation in the UV can markedly reduce the power requirements for detonation of some secondary explosives. Therefore, establishing the initial steps of high explosive decomposition is an important goal to pursue. The decomposition triggered by excited electronic states is appealing because electronic excitation of the system to an unstable potential energy surface can result in a rapid dissociation of a molecule and a consequent chain reaction.

Research on detailed mechanisms that govern the sensitivity to detonation in energetic solids at multiple length and time scales is a focus of many individual scientists and research groups that are working to solve both practical and fundamental problems in theory and experiment. Design of novel materials with prescribed superior properties, the establishment of safe conditions for their synthesis and usage, applications of detonation techniques in other science and technology fields require knowledge of mechanisms of decomposition and qualitative data for the description of the processes over a wide range of temperatures, pressures, and different states of matter.

This brief review aims at highlighting the current understanding of the decomposition trends in high explosive materials gained from combined experimental and theoretical investigations. We focus on modeling of earliest stages (initiation) of chemical decomposition reactions by exploring effects of structural (e.g., vacancies, voids, surfaces) and electronic (e.g., electron and hole polarons) defects, deformations (e.g., shear-strain, dislocations), and charged states (e.g., ion radicals, charge transfer) on decomposition mechanisms and kinetics in energetic materials. We also recognize limitations in several newly synthesized energetic materials families and emphasize possible directions for further materials developments.

The presented review is addressed to a broad range of experimentalists and theorists from students to experts working in the field of energetic materials. To make the discussion especially useful for experimentalists, in particular, those dealing with the design and synthesis of new materials and with measurements of the decomposition kinetics, we emphasize the prospects provided by state-of-the-art *ab initio* calculations and indicate possible pitfalls of modeling approaches. Likewise, to reach theorists, in particular, those working on modeling of novel and existing high explosives, we underscore outcomes of *ab initio* calculations of defects and deformations that are relevant for a comparison with experiment and elucidation of the decomposition reaction mechanisms.

2. METHODS

2.1. Molecular calculations

Molecular calculations in most of our studies were performed using density functional theory (DFT),[18,19] as implemented in the GAUSSIAN 09 code.[20] We used a range of available functionals, including the PBE exchange correlation functional,[21] and hybrid B3LYP,[22,23] PBE0,[24] and M06[25] functionals, and empirical dispersion wB97XD[26] functional. Typically, we use 6–31+G(2df, p) or double-zeta cc-pVDZ Dunning's correlation consistent basis set[27] on atoms and settle on default values for convergence parameters for self-consistent field total energies and forces.

PBE functional is widely used for investigating solid-state reaction mechanisms[2] (it is implemented in most periodic computer codes because of its efficiency) and hence gas-phase calculations are required before any studies of decomposition pathways in practical materials are attempted (see, e.g., Refs. 5,7,14). Unlike PBE, hybrid B3LYP, PBE0, and M06 functionals consume significant time and computer resources but are usually considered relatively reliable for modeling reactions in gaseous phase (see, e.g., Ref. 28). Although a detailed analysis of DFT performance is beyond the scope of this review, we allow ourselves a few remarks regarding the use of various functionals. We note, for example, that while the PBE functional delivers activation barriers of the direct bond fission (such as $X–NO_2$, where $X=O$, N, or C) in acceptable agreement with the hybrid analog PBE0 and the empirically corrected wB97XD, it systematically underpredicts (by 6–30 kcal/mol) the barriers for the isomerization reactions involved a hydrogen or oxygen migration (see Section 3). This implies that the resulting conclusions regarding dominant decomposition pathways should be made with caution and whenever possible more than one computational technique should be used for modeling chemical reactions, especially those including concerted or step-wise molecular rearrangements. Nevertheless, a comparison of activation barriers to each other within the same functional typically allows for a consistent analysis and unambiguous conclusions regarding the nature of the dominant thermal decomposition mechanism in energetic molecules.

Vibrational frequencies have been calculated for relevant atomistic configurations to distinguish energy minima and transition states and to determine corresponding zero-point energy (ZPE) corrections. The stationary points corresponding to the energy minimum have been positively identified by having no imaginary frequencies and the transition states had exactly one imaginary frequency. An intrinsic-reaction-coordinate analysis was carried out by using the Hessian-based Predictor–Corrector integrator algorithm,[29,30] for each transition state to ensure that it was the true transition structure, connecting the desired reactants and products.

2.2. Chemical kinetics

Although a formalism of chemical kinetics has been long known,[31] actual calculations of reaction rates from first principles are still rare even for isolated molecules and practically not available for condensed matter processes. Often, reaction rates are taken from experiment. The problem with this is that experiments, for example, mass-spectroscopy measurements, register only an aggregate decomposition process and do not differentiate between particular pathways. In energetic materials, a distinction between possible channels is especially challenging because a coexistence of several pathways with close or nearly identical activation barriers is not unusual.

In our research, preexponential factors were calculated with conventional transition state theory (TST)[32] for the decomposition reactions that proceed through a formation of a transition state and with variational transition state theory (VTST)[33] for the homolytic cleavage pathways. A detailed methodology for obtaining preexponential factors and reaction rates was initially developed[34] for modeling the HMX gas-phase decomposition and then applied for HMX ideal bulk crystals and practical materials containing vacancies, voids, and surfaces.[14,35,36]

2.3. Periodic calculations

The ideal bulk and surface calculations were mostly performed in the GGA approximation with the PBE exchange correlation functional[21] and PAW pseudopotentials[37] as implemented in the plane wave code VASP.[38–40] In simulating an ideal crystal, we typically used $2 \times 2 \times 2$ Monkhorst–Pack k-point mesh and the kinetic energy cut-off was set to 800 eV. For larger supercells, which were used for modeling of defects and chemical reactions, the number of k-points needs to be prorated to keep the same density of k-point mesh and to make the total energies of studied systems comparable. Atomic coordinates and lattice constants were allowed to simultaneously relax without any symmetry constraints. The convergence criterion for electronic steps in calculations was set to 10^{-4} eV, and the maximum force acting on any atom was set not to exceed 0.02 eV/Å. We compare the calculated lattice constants to available experiments and earlier theoretical studies or results of other computational methods. For example, the calculated lattice parameters of PETN, $a=b=9.84$ Å, $c=6.95$ Å, were in reasonable agreement with experimental data ($a=b=9.38$ Å, $c=6.71$ Å)[41] and earlier theoretical estimates of Conroy et al. ($a=9.617$ Å, $b=9.612$ Å, $c=6.826$ Å)[42] add Sorescu et al. ($a=b=9.3348$ Å, $c=6.6500$ Å).[43] This illustrates quality of our modeling of the ideal crystalline system, which is a necessary step before any simulations of defects can be attempted.

Often, for modeling bulk reactions, a supercell needs to be modified to choose an appropriate crystal fragment, which is sufficiently large to accommodate reagents and products without inducing any artificial interactions in the system, and still practical to finish calculations at some reasonable computational time. For example, in PETN calculations, to model bulk reactions, the ideal crystal supercell was enlarged by choosing modified lattice vectors: $\vec{a} = \vec{a}_0 + \vec{b}_0$, $\vec{b} = -\vec{a}_0 + \vec{b}_0$, and $\vec{c} = 2\vec{c}_0$, with the resulting lattice constants of $a = b = c = 13.91$ Å and $\alpha = \beta = \gamma = 90°$.

Surface reactions were simulated in a slab model, in which the supercell typically consisted of the four-molecule-thick surface layer (appropriately cut out to create surfaces of interest) on top of a 10 to 30 Å of vacuum. The large size of the supercell ensured that spurious interactions between reactants and reaction products in neighbor cells are minimal. Surface supercell calculations we performed at the Gamma point only, with kinetic energy cut-off set to 600–800 eV, depending on the material and results of convergence tests. The convergence criteria for electronic and ionic steps were set to 10^{-4} eV and 0.02–0.03 Å/eV, respectively.

Minimal energy paths in the VASP periodic calculations were obtained with the nudged elastic band method[44] with six intermediate images. Atomic positions were relaxed using conjugate gradient and quasi-Newtonian methods within a force tolerance of 0.05 Å/eV.

We also used periodic calculations performed by the CRYSTAL06 code[45] to test or complement some of our predictions. The localized basis set based CRYSTAL (with a basis set of a linear combination of atomic orbitals-LCAO) and the GAUSSIAN code with periodic boundary conditions[46] proved to be useful in studies of molecular crystals[15] and complex oxides.[2,3] The advantage of the CRYSTAL code is an efficient realization of the hybrid functionals with calculations much faster than in the plane wave codes and a possibility to use a single isolated slab for surface calculations (which is an entirely 2D system), whereas plane wave codes require a 3D periodicity (i.e., slabs of a given thickness have to be periodically repeated also along the *z*-axis with a sufficiently large vacuum gap in-between).

2.4. Embedded cluster calculations

The hybrid "embedded cluster" and quantum mechanical/molecular mechanical (QM/MM) methods have long been used to study defects and reactions in solids involving charged and excited states (see, e.g., Refs. 47,48). They combine quantum-chemical methods for describing a part of the system with classical treatment of the rest of the system and are extensively used to describe crystalline as well as disordered solids,[49–51] and biomolecules[15,52,53] Surprisingly, applications of these techniques to molecular crystals are rare. This is in spite of the fact that the *inter*molecular interactions in many molecular materials are much weaker than *intra*molecular interactions and, therefore, these systems can be easily divided into regions, treated at different levels of theory and amenable to QM/MM. Recently, a QM/MM method has been applied to study the structure of a range of organic molecular crystals.[54] In the meantime, applications of this method to studying defects in molecular crystals, where it may have significant advantages with respect to molecular and periodic models, were not available until recently.[13,55–57]

We used an Embedded Cluster Method (ECM) implemented in the GUESS code[58] to model the electronic structure,[55] the hydrogen transfer,[57] the formation of polarons,[56] and isomer defects[13] in DADNE (Fig. 3.1). Within the ECM, the DADNE crystal is represented with a finite system, which is divided into a relatively small region of interest, described quantum mechanically (QM cluster), and the rest of the system treated classically using the shell model or rigid classical ions.[58] GUESS ECM self-consistently takes into account the interaction of the electrons and nuclei inside the quantum cluster with the rest of the host lattice, the perturbation of the lattice by the defect, and the reciprocal effect of the lattice polarization on the defect itself.[59,60] The challenging part of this approach is

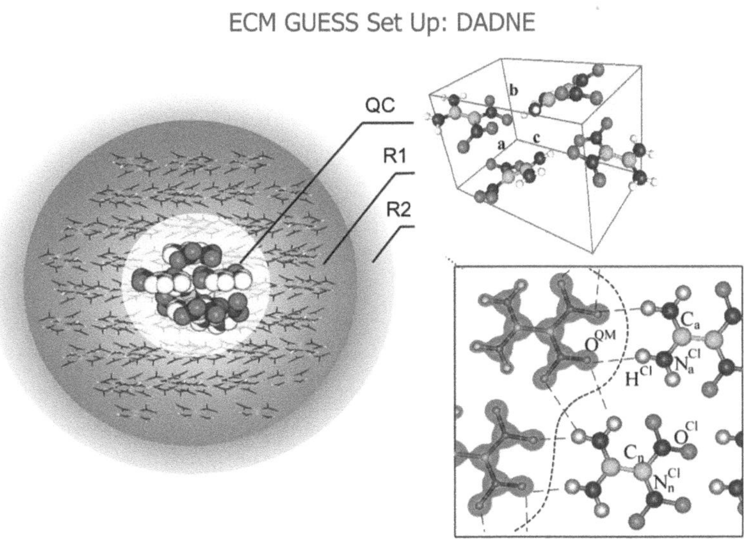

Figure 3.1 Embedded Cluster Method is illustrated for bulk and surface calculations. The simulated system contains a few hundred molecules in a Quantum Cluster and classical part (Region 1) and a few thousand atoms in Region 2. The relaxation of quantum cluster is fully taken into account. (For the color version of this figure, the reader is referred to the online version of this chapter.)

the description of the bonded interactions at the border between the QM and classical subsystems.[48,58] In molecular crystals, such as DADNE, the molecules are often linked together by the electrostatic and van der Waals interactions and by hydrogen bonds. The absence of bonded intermolecular interactions makes the description of the border between quantum and classical regions simpler than in the case of covalent or semicovalent systems where special semilocal embedding pseudopotentials are needed.[61]

Region I includes several hundred molecules with a QM cluster at the center. Positions of all atoms in the region I are optimized using the standard BFGS algorithm,[62,63] Region I is surrounded by a finite Region II, which is comprised of several thousand atoms, where atoms are kept fixed in their ideal positions in the crystalline lattice. The role of this region is to mimic the infinite crystalline environment and provide the correct representation of the electrostatic potential inside Region I.[64]

3. DECOMPOSITION OF GAS-PHASE MOLECULES

Obtaining mechanisms and kinetics from quantum-chemical modeling

The understanding of the thermal and kinetic aspects of gas-phase decomposition reactions is imperative in many fundamental and applied fields, including the environmental and atmospheric sciences, combustion and explosions, catalysis, matter under extreme condition, and planetary sciences. Obtaining reliable data to describe thermochemistry and kinetics is extremely complex for large molecules that are typical for energetic materials. The experimental determination of the activation energies is particularly difficult in high explosive materials because several possible reaction pathways coexist in the course of the thermal decomposition. *Ab initio* quantum-chemical calculations aimed to provide an interpretation of the experiment are often of little help if they produce only activation barriers and ignore the kinetics of the decomposition process. Although the theoretical methods to overcome this ambiguity do exist, their application is far from trivial, especially for large molecules and fast or complex decomposition processes.

Following the methodology developed for cyclotetramethylene-tetranitramine (HMX),[34] we briefly review here ground state chemical decomposition reactions of three most popular classes of nitro compounds, nitrate amines, nitrate esters, and nitrate arenes, creating a broader trend that likely to mirror a similar pattern by other energetic molecules.

3.1. Nitrate amines: A case study of β-HMX

The thermal and kinetic aspects of gas-phase decomposition reactions in all energetic materials are complex due to a large number of parameters, a variety of possible intermediates, and an overlap in thermal decomposition traces. A theoretical approach to study of the gas-phase thermodecomposition, including reaction energies, activation barriers, and reaction rates, is illustrated with the example of the β-octahydro-1,3,5,7-tetranitro-1,3,5,7-tetrazocine (HMX) molecule.[34] We initially studied three types of plausible candidate decomposition reactions characteristic of nitramines: (a) the HONO elimination, (b) the NONO rearrangement, and (c) the N–NO_2 homolysis.

We think that a β-HMX molecule serves as a good model system to study a gas-phase decomposition process in energetics because of several reasons. First, a β-HMX molecule is relatively big and contains functional chemical groups typical for a wide class of nitro-compound materials. Second, despite concerted efforts to extensively study its decomposition process, elucidation of the primary and secondary dissociation mechanisms remains a challenge for both experimental,[65–67] and theoretical exploration.[68–70] In experiments, it is difficult to distinguish between competing chemical reactions during the initial steps of the rapid decomposition of HMX as a global decomposition is measured. One has to speculate from the global kinetics and reaction products which reactions are taking place. As a result, the reported decomposition barriers vary by a factor of two to three and the preexponential factors in the reaction rates differ by orders of magnitude. More so, even the first step in the decomposition process is still a subject of debate. In addition, HMX is an important high energy density material for a broad range of technological applications spanning from high explosives and propellants to rocket engine fuels. Finally, the decomposition process of HMX is comprised of several seemingly coexisting pathways with a fairly small difference in activation barriers and significantly different nature of corresponding bond dissociation processes. The hydrogen transfer and nitro-to-nitrite isomerization reactions possess well-defined transition states while the direct N—NO_2 bond homolysis proceeds as a reaction with no activation barrier.

3.1.1 Reaction energies, activation barriers, and kinetics

We model a gas-phase HMX by placing a single β-HMX molecule in a $15 \times 15 \times 15$ Å^3 cell and optimizing the atomic positions with the VASP code. Calculations are performed only at the Gamma point. By using this model, we consider three chemical reactions as candidates for the initial step in the overall decomposition process: the hydrogen transfer with HONO elimination,[66,68,71] the nitro-nitrite isomerization (NONO rearrangement) with NO elimination,[72] and the direct N–NO_2 homolysis[66,68,71] as illustrated in Fig. 3.2. The HONO elimination is modeled by transferring one H atom to the axial NO_2 and relaxing the atomic positions. In order to simulate the NONO rearrangement, we extend the axial N—NO_2 bond and slightly rotate NO_2. After the relaxation, one of the O atoms from the NO_2 group binds to an N atom from the ring, forming the NONO isomer. The N–NO_2 homolysis is modeled by moving the NO_2 group away along the N—NO_2 bond direction by 5.7 Å. The N–NO_2 homolysis was studied in both spin-polarized and spin-nonpolarized computational schemes. The reaction energies of the HONO elimination, the NONO rearrangement, and the N–NO_2 homolysis are found by taking the differences of the total energies of the equilibrium HMX molecule and of the corresponding final state (HONO, NONO, or split-off NO_2). All three reactions are established to be endothermic with the energies of 2.42 kcal/mol, 19.2 kcal/mol, and 57.8, and 42.8 kcal/mol, respectively (see Table 3.1).

The activation barriers are defined as the differences between the total energies of the equilibrium HMX molecule and the corresponding transition states (unless it is a barrier-less reaction). The obtained activation energies of the HONO and NONO reactions are found to be 39.4 and 45.8 kcal/mol, respectively. The N–NO_2 homolysis pathway does not have

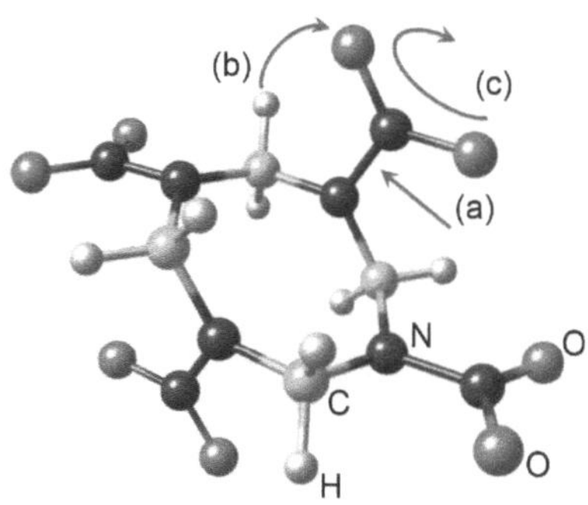

Figure 3.2 The structure of an HMX molecule is shown and three possible primary decomposition reactions are illustrated. The direct N-NO_2 homolysis (a) is simulated by the elongation of the N—N bond, the HONO elimination (b) is demonstrated as switching of the C—H···O bond to C···H—O bond, and the nitro-nitrite isomerization (c) is displayed by a combined rotation of the NO_2 group and the elongation of the N—N bond to form NONO. (For the color version of this figure, the reader is referred to the online version of this chapter.)

Table 3.1 Reaction energies, barriers and rates for HONO elimination, NONO rearrangement, and $N–NO_2$ homolysis decomposition of HMX

Reaction	Reaction energy (kcal/mol)	E_B (kcal/mol)	log A (s^{-1})
HONO elimination	2.4	39.4	14.3
NONO isomerization	19.2	45.8	14.9
$N–NO_2$ homolysis	42.8	38.1	17.9

Barrier heights include zero-point energy corrections.

a saddle point (Fig. 3.3); thus, the activation barrier is considered to be the same as the reaction energy, 38.1 kcal/mol (in the spin-polarized calculation).

The obtained activation barriers (including ZPE corrections) and Arrhenius factors are collected in Table 3.1. Figure 3.4 shows the logarithmic dependence ofk on 1000/T. As can be seen, at low temperatures, the hydrogen transfer mechanism has the activation barrier at 39.4 kcal/mol, very close to the $N–NO_2$ activation barrier, indicating that the HONO elimination could be among the earliest decomposition reactions in HMX. However, it is also a fairly slow reaction, which is reflected in its smallest preexponential factor at ~14 (s^{-1}), implying that it can be quickly overstepped by another faster reaction.

At all temperatures, the $N–NO_2$ homolysis prevails in the decomposition; its activation barrier at 38.1 kcal/mol is the lowest among the reactions probed here and its reaction constant at ~18 s^{-1} is large. The nitro-nitrite isomerization, NONO, with the activation barrier at 45.8 kcal/mol and the reaction constant at ~15 s^{-1}, is the slow process at all temperatures. This suggests that the NO elimination hardly contributes to the initiation of the decomposition of HMX, being always only a secondary reaction, which may proceed on the background of the primary decomposition mechanism. The obtained here results are consistent with previous theoretical studies, which also found that the $N–NO_2$ homolysis is the dominating reaction at low temperatures and the fastest at above 400 K.[73] Another study also found that the $N–NO_2$ is the dominating reaction under very high temperatures.[74]

Now, we link our results to the experiment and explain a fair amount of the dispersion in the experimental values that have reported the activation barriers for gas-phase HMX from 32 to 53 kcal/mol and (*logA)* of the reaction constants from 28 to 47 at 205–383 °C.[75] The discrepancies observed in the global kinetic measurements of the thermal decomposition of HMX

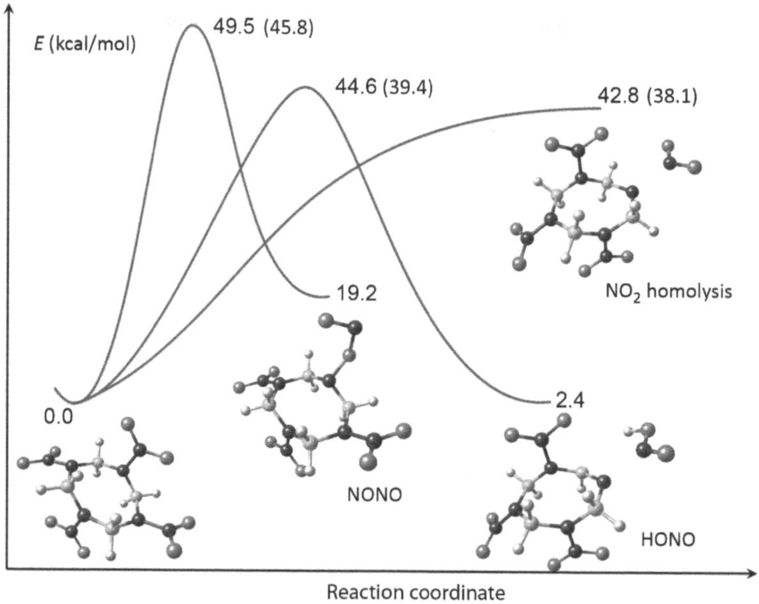

Figure 3.3 Minimal energy pathway for the dissociation reactions: HONO elimination, NONO rearrangement, and N–NO_2 dissociation. Barriers with zero-point energy corrections are shown in parentheses. (For the color version of this figure, the reader is referred to the online version of this chapter.)

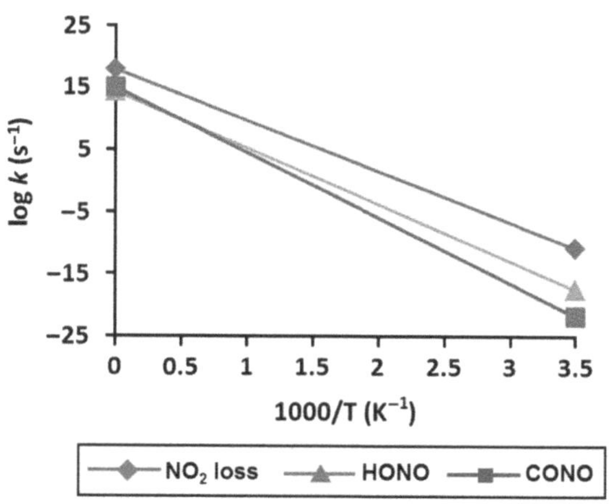

Figure 3.4 Reaction rates of HMX decomposition as a function of temperature. (For the color version of this figure, the reader is referred to the online version of this chapter.)

were attributed to the history and characteristics of the samples, variations in the experimental conditions,[75] different heating rates, and the potentially strong effects of autocatalytic reactions.[76] It is also known that clustering of HMX molecules occurs in the gas phase, even at low pressure,[77] so it was even proposed that isolated molecules are probably not decomposing.[75] Naturally, one has to keep in mind that the measurements reflect the rate of the overall decomposition process as opposed to a specific reaction as simulated in this study. Although experimental values of the HMX decomposition rates and activation barriers range widely, most data were noted to exhibit a linear relation between lnA and E_B, sometimes referred to as a "compensation effect."[75]

Following our results, the activation barriers for both HONO (39.4 kcal/mol) and N–NO_2 fission (38.1 kcal/mol) reactions fall in the experimentally observed range[33] for the activation barriers at 32–53 kcal/mol.[29] Similarly, the calculated reaction rate constants for HONO and N–NO_2 fission dissociation mechanisms are also found to be within the experimentally measured range.[29] However, the conclusions regarding the trends in relationship between the reaction constant and the activation barrier for the individual reactions fall short in supporting "the chemical compensation effect."

A combination of quantum chemistry and the theory of transition states in its standard and variational versions was applied here to simulate the entire process of gas-phase thermochemistry, including understanding the nature of potential decomposition mechanisms at the earliest stages and deriving the corresponding reaction energies, activation barriers, and the reaction rates. It was established that the decomposition process in β-HMX is determined by a competition of at least two chemical reactions at any given temperature, with a visible dominance of the direct N–NO_2 homolysis while the HONO-isomerization still continues in the background being the slower dissociation mechanism. The obtained conclusions are in agreement with the earlier theoretical studies and the range of experimental data for the overall HMX decomposition.

Our calculations illustrate the importance of including the Arrhenius prefactors in the quantum-chemical consideration of decomposition mechanisms of complex molecules. The HONO elimination and the N–NO_2 homolysis have almost the same activation barriers and they would appear to be strongly competing dissociation mechanisms in HMX. However, the reaction rate calculations show that the N–NO_2 in the triplet state is 3–5 orders of magnitude faster reaction than the HONO elimination. This implies that the HONO elimination will play a minor role in the decomposition yielding the governing role to the N–NO_2 homolysis.

The computational strategy developed in this study and the performed analysis can be extended to not only other materials, but also to much more intricate solid-state processes, which will facilitate the revelation of important details of the extremely complex decomposition of condensed molecular materials and the behavior of materials under extreme conditions.

3.2. Nitrate esters: A case study of PETN

PETN is an important high explosive and pharmaceutical, which belongs to the nitrate ester class of organic compounds. In its most stable form, PETN appears as white crystals assembled into a tetragonal structure with $P\text{-}42_1c$ symmetry[78] and two molecules per unit cell with S_4 molecular point group[41] (Fig. 3.5). As a pharmaceutical, PETN falls within the same therapeutic class as nitroglycerine and is used as a the long-term therapy of coronary artery disease and congestive heart failure.[79,80] PETN is often called a benchmark energetic material.[81–83] The most stable of the nitrate esters PETN is the most sensitive secondary high explosive in its class (and hence it is used extensively as an initiating agent, or booster, in detonators[84]), while all materials that exhibit higher sensitivity than PETN are classified as primary explosives.

Generally, PETN can be quite easily detonated by shock, however, it was shown that the shock initiation of PETN strongly depends on the orientation of the crystal axes relative to the shock propagation vector, with several directions of impact that will not cause detonation, even when the shock wave fully crosses the crystal.[85] For example, strong light emission, identified as chemiluminescence from the (NO_2^+) nitronium ion,[86] was observed when the

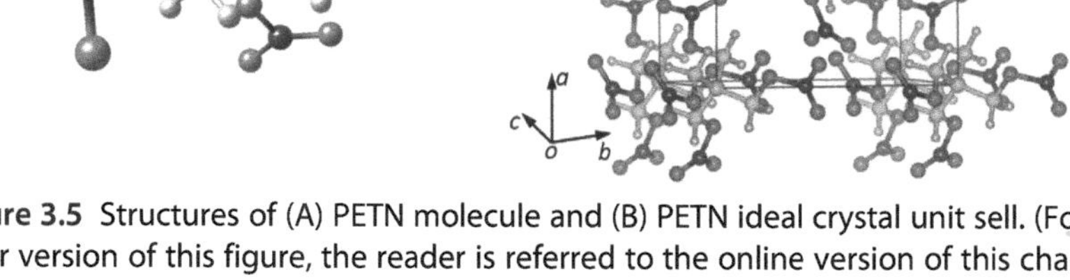

Figure 3.5 Structures of (A) PETN molecule and (B) PETN ideal crystal unit sell. (For the color version of this figure, the reader is referred to the online version of this chapter.)

PETN crystal was shocked along the (110) and (001) orientations.[87–89] In the meantime, shock-initiated detonation did not occur along the *c*-axis.[85,87–89] The first-principles molecular dynamics simulations[90] revealed that hypervelocity bimolecular PETN collisions in the (001), (011), and (110) directions are most sensitive with lowest threshold velocities, while collisions normal to the (010), (101), and (100) planes are most insensitive[90] and would require unphysical high velocities for chemistry to occur.

There is a large body of work performed on PETN (see, e.g., Refs. 91–94 and references therein). Thermal decomposition of solid PETN was suggested to proceed competitively with sublimation,[95,96] and activation energies E_a are largely scattered in a wide range of values falling between 30 and 70 kcal/mol,[95,97–105] with the most reliable data falling into the interval from 32.6 to 47.3 kcal/mol[95,101–105] and commonly reported products of CO, CO_2, NO, N_2O, CH_2O, HCN, HNCO.[106] With the large dispersion of data primarily attributed to differences in experimental settings, conditions under which experiments were performed, and preparation of PETN samples, most researchers agree that the initial decomposition step is the $O{-}NO_2$ bond homolysis.[95,97,106] Recently performed series of experiments on laser irradiation of gas-phase[107] and solid PETN samples mixed with light-sensible additives[108] suggested new ways of decomposition through resonance photoinitiation, while at the same time, posed new questions regarding initiation chemistry at large. The ground state decomposition chemistry[109] and electronic structure[110] of PETN were recently rigorously studied.

We explored[109] seven most plausible decomposition mechanisms that simulate breaking of possible chemical bonds in an isolated PETN molecule, as illustrated in Fig. 3.6. Those include (a) the homolytic cleavage of the $O{-}NO_2$ bond (Eq. 3.1), which is commonly assumed as the primary dissociation step,[106,111,112] (b) the elimination of nitrous acid (HONO) (Eq. 3.2), which is usually considered as a competing reaction to the NO_2 loss in nitro compounds,[112,113] (c) the nitro-nitrite rearrangement (OONO)[107,114,115] (Eq. 3.3), (d) the so-called γ-attack, that is, an encounter of γ oxygen atom and the central C atom (Eq. 3.4),[116,117] (e) the homolytic C—O bond cleavage (Eq. 3.5), and (f) two variations of the homolytic C—C bond cleavage (Eqs. 3.6 and 3.7):

$$C(CH_2ONO_2)_4 \rightarrow (CH_2ONO_2)_3C-CH_2-O\bullet + \bullet NO_2 \tag{3.1}$$

$$C(CH_2ONO_2)_4 \rightarrow (CH_2ONO_2)_3C-CHO + HONO \tag{3.2}$$

$$C(CH_2ONO_2)_4 \rightarrow (CH_2ONO_2)_3C-CH_2-O-ONO \tag{3.3}$$

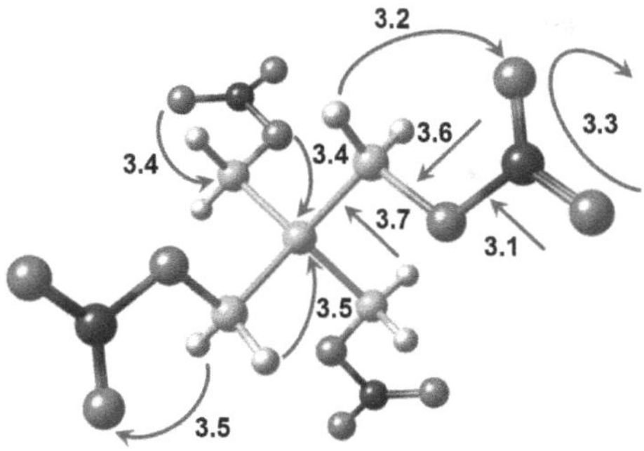

Figure 3.6 Schematic representation of PETN thermal decomposition pathways; the numbers correspond to the numbers of equations. (For the color version of this figure, the reader is referred to the online version of this chapter.)

$$C(CH_2ONO_2)_4 \rightarrow (CH_2ONO_2)_3C-O-CH_2-ONO \quad (3.4)$$
$$C(CH_2ONO_2)_4 \rightarrow (CH_2ONO_2)_3C-H_2C\bullet + \bullet ONO_2 \quad (3.5)$$
$$C(CH_2ONO_2)_4 \rightarrow (CH_2ONO_2)_3C\bullet + CH_2O + \bullet NO_2 \quad (3.6)$$
$$C(CH_2ONO_2)_4 \rightarrow (CH_2ONO_2)_3CH + HONO + CO \quad (3.7)$$

The obtained activation barriers and preexponential factors are collected in Table 3.2.

Table 3.2 shows that the NO_2 loss (Eq. 3.1) requires 34.9–36.6 kcal/mol, depending on the computational method used. The reaction proceeds with no barrier hence the value of the calculated reaction energy is also assigned to the activation energy. All three DFT approximations, the pure PBE and hybrid PBE0 and wB97XD functionals, give close energies of 34.9, 35.8, and 36.6 kcal/mol, respectively, which differ from each other less than 2 kcal/mol. The obtained O—NO_2 bond dissociation energies (BDE) are in good agreement with the most recent differential scanning calorimetry (DSC) measurement of 32.6 kcal/mol[105] and with a range of earlier experimental studies, 35.0,[104] 39.5[101,118] and 46.0[95] kcal/mol. The calculated O–N BDE are also consistent with the reported earlier computational values of 35.8,[117] 37.7,[90] 39.0,[117] 39.2,[119] and 39.9 kcal/mol, given the differences in the employed computational techniques. The estimated preexponential factors of ~17–18 are in reasonable agreement with relevant experimental values of 14.9,[105] 15.6 (PETN decomposition in melt and in a trotyl solution),[97] 16.1 (PETN in 5% phthalate solution)[101] and 19.8 s^{-1} (PETN in melt).[101] The observation that the endothermic O–NO_2 homolysis reaction exhibits the lowest calculated activation barrier, the highest preexponential factor, and the highest reaction

Table 3.2 The calculated activation barriers E_a (kcal/mol), zero-point energy corrected barriers E_{ZPE} (kcal/mol), and preexponential factors, log A (s^{-1}) of the PETN decomposition reactions in the gas phase

	PBE			PBE0			wB97XD		
Reaction	E_a	E_{ZPE}	log A	E_a	E_{ZPE}	log A	E_a	E_{ZPE}	log A
(1)NO_2 loss	41.8 [41.8][a]	36.6	17.5	40.2 [40.2]	34.9	7.7	41.2 [41.2]	35.8	17.7
(2)HONO	33.6 [−18.6]	28.8	13.4	45.3 [−14.0]	40.5	13.3	47.3 [−18.6]	42.7	12.9
(3)OONO	52.3 [34.3]	48.1	15.1	73.0 [37.0]	68.7	15.2	76.6 [34.3]	72.7	14.4
(4)γ-Attack	67.1 [−12.3]	63.1	14.5	84.1 [−8.7]	80.5	13.8	88.7 [−12.3]	85.3	13.4
(5)ONO_2 loss	73.3 [73.3]	67.7	–	81.6	74.6	–	86.0 [86.0]	79.0	–
(6)C–C: (CH_2O+NO_2)	– [51.1]	–	–	82.6 [54.4]	76.0	–	88.4 [56.0]	81.9	–
(7)C–C: (CO+HONO)	56.7 [−22.3]	50.0	16.5	80.6 [−14.9]	74.2	15.6	86.4 [−22.3]	80.7	14.6

[a]The corresponding reaction energies are shown in brackets.

rate (Fig. 3.7) of all probed mechanisms confirms that the homolytic O—NO_2 bond cleavage is the most favorable decomposition channel of PETN from both an energetic and kinetic points of view.

The HONO elimination (Eq. 3.2) requires 40.5 (PBE0)–42.7 (wB97XD) kcal/mol, a slightly higher (by ~6 kcal/mol) activation energy, than the O—NO_2 bond scission (Table 3.2). The obtained barrier is still within the experimental range and in general agreement with other theoretical results, 36.2[117] and 39.2[117] kcal/mol. As expected,[120,121] the PBE calculations underestimate the HONO barrier height (by ~12–14 kcal/mol), as compared to PBE0 and wB97XD. Neglect of this known trend of pure DFT methods to underpredict the activation barriers for proton migration reactions versus data yield by hybrid functionals may lead to a possible erroneous conclusion regarding dominance of the HONO-isomerization reaction as PBE gives the energy ~8 kcal/mol lower than O–NO_2 BDE (Table 3.2). On one hand, the activation barrier, 40.5–42.7 kcal/mol, preexponential factor, ~13 (Table 3.2), and the reaction rate (Fig. 3.7) indicate that the HONO is less favorable and much

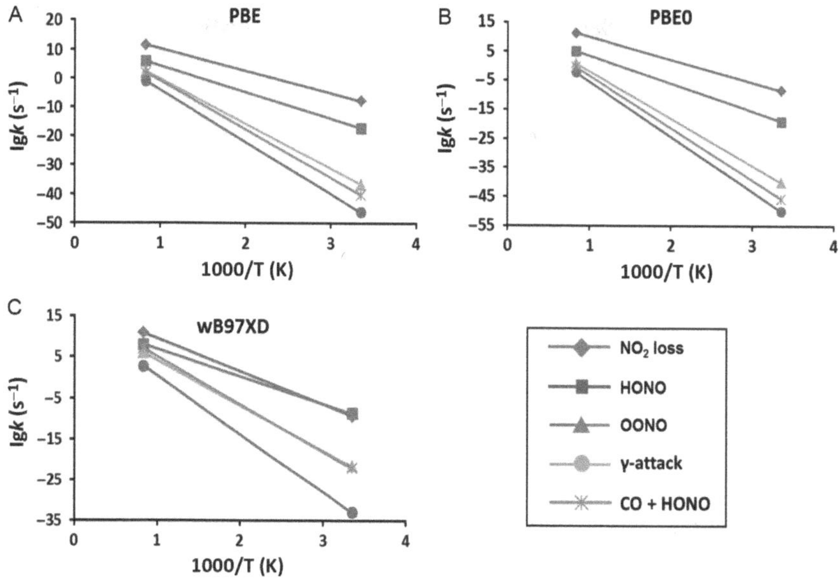

Figure 3.7 Reaction rates for the decomposition channels of gas-phase PETN calculated with (A) PBE, (B) PBE0, and (C) wB97XD functionals. (For the color version of this figure, the reader is referred to the online version of this chapter.)

slower reaction than the NO_2 loss, and hence it is triggered somewhat later in time (i.e., at higher temperature) than the O–NO_2 cleavage. On the other hand, the HONO elimination is an exothermic reaction, releasing 14.0–18.6 kcal/mol (Table 3.2), and as such should play a crucial role in facilitating and catalyzing the further decomposition of neighbor molecules.

The activation barrier, computed for the OONO rearrangement (Eq. 3.3), 68.7 kcal/mol (PBE0) and 72.7 kcal/mol (wB97XD), is 34–37 kcal/mol higher than the energy of the NO_2 loss (Table 3.2). The obtained barriers are in agreement with the value 71.5 kcal/mol,[107] calculated in another study based on the GAUSSIAN code. In contrast to the report by Yu and Bernstein,[107] which suggests that the OONO-isomerization is a two-stage process with the reaction energy of 20.7 kcal/mol, our modeling shows that the reaction proceeds in onestep with the reaction energy of34.3–37 kcal/mol. Note that the OONO-isomerization activation energies practically coincide with those of the nitro-nitrite rearrangement in nitromethane (67 kcal/mol[122] and 69 kcal/mol[114]) and DADNE (66.4 kcal/mol)[123] and appear noticeably higher than those in nitroethylene (57.9 kcal/mol),[124] nitrobenzene (61.1 kcal/mol,[125] 63.7 kcal/mol[126]) and trinitrotoluene (54.9 kcal/mol).[127] We conclude that the nitro-nitrite isomerization with the consequent NO release, which is proposed to play an important role in the decomposition of optically excited PETN,[107] should be ruled out as a major decomposition mechanism on the ground state potential surface.

The γ-attack reaction path (Eq. 3.4) yields the activation barriers of 80.5 (PBE0) and 85.3 (wB97XD) kcal/mol and the reaction heat of 8.7 (PBE0) and 12.8 (wB97XD) kcal/mol (Table 3.2). In spite of its exothermicity, this pathway requires 35–50 kcal/mol higher energy than the NO_2 loss and therefore is an unfavorable decomposition channel. Both the obtained activation barrier and reaction energy are found to be in good agreement with the previously estimated values of 80.1 kcal/mol for the barrier and 13 kcal/mol for the reaction energy.[117]

Consistent with other computational reports,[90,117] the energy, 67.7 (PBE), 74.6 (PBE0), and 79.0 (wB97XD) kcal/mol, required for the C—ONO_2 bond break leading to the NO_3 loss (Eq. 3.5, Table 3.1) is significantly higher than the activation of the O–NO_2 homolysis leading to the NO_2 loss (Eq. 3.1, Table 3.2). Similarly to the O–N fission reaction, the C–N BDE obtained with all three functionals agree reasonably well, with the smaller difference of less than 5 kcal/mol between PBE0 and wB97XD and the larger discrepancy of ~11 kcal/mol between PBE and wB97XD. The calculated barriers are in good correspondence with 73.3 kcal/mol found with B3LYP,[117] 82.2 kcal/mol delivered by M06,[117] and 79.3 kcal/mol resulted from PBE.[90] The high activation energy of the C—ONO_2 bond homolysis defines this mechanism as a low probability channel of PETN decomposition.

Finally, we consider peculiarities of the C–C homolysis (Eqs. 3.6 and 3.7). Contrary to expectations, the intrinsic-reaction-coordinate-based downhill procedure, implemented in GAUSSIAN code, fell short finding a definite pathway corresponding to the C—C bond breaking. This implies that the homolytic C–C cleavage does not proceed in accordance with an obvious equation $C(CH_2ONO_2)_4 \rightarrow (CH_2ONO_2)_3C^{\cdot} + {}^{\cdot}CH_2ONO_2$. Two other processes were found instead. While we were able to estimate an activation barrier for the reaction generating $(CH_2ONO_2)_3C^{\cdot}$ and ${}^{\cdot}CH_2ONO_2$ moieties, which turned out to be 76.0 (PBE0) and 81.9 (wB97XD) kcal/mol, the configuration of the system was metastable. The geometry relaxation led to $(CH_2ONO_2)_3C^{\cdot}$, CH_2O, and NO_2, as described in Eq. (3.6), with the reaction energy of 51.1 kcal/mol (PBE), 54.4 kcal/mol (PBE0), and 56.0 (wB97XD) kcal/mol. These energies are consistent with previous estimates.[90,117] According to Liu et al.[117], the calculated 2D scan of the C—C and O—NO_2 bond lengths produced the CH_2O and NO_2 as the main products formed during the C—C bond scission, which required 41.7 (B3LYP)[117] and 49.1 (M06)[117] kcal/mol. Landerville et al.[90] suggested that the C—C bond was the second weakest bond in PETN molecule (after O–NO_2)[90] fission of which required 60 kcal/mol. Although our calculations were not able to localize the transition state structure (similarly to Liu et al.[117]) and obtain definite activation barriers for this reaction, we can safely assume that the activation barrier of the C—C bond scission, yielding CH_2O and NO_2, should fit in the range of 51.1–81.9 kcal/mol, and would require a similar or higher energy than the calculated reaction energy (Table 3.2).

An alternative concerted mechanism of the HONO elimination accompanied the C—C bond scission (Eq. 3.7), which generates $(CH_2ONO_2)_3CH$, HONO, and CO as the final products, also has a high activation barrier of 74.2 (PBE0) and 80.7 kcal/mol (wB97XD). The resultant high energies do not favor these complicated decomposition mechanisms (Eqs. 3.6 and 3.7) and point to the C—C bond fission only as a possible secondary reaction.

Overall, the calculated activation barriers, reaction energies (Table 3.2), and reaction rates (Fig. 3.7) indicate that the homolytic cleavage of O—N bond releasing NO_2 as a primary product (Eq. 3.1) is the predominant pathway in the thermal gas-phase decomposition of PETN molecules. This reaction exhibits the lowest barrier (~35–36 kcal/mol), the highest preexponential factor (~log18 s^{-1}), and proceeds with the highest rate among all probed reactions. The only other process that can be considered as a competing, or rather complementary, mechanism is the HONO elimination (Eq. 3.2). Although it is activated by a slightly higher barrier (by ~5–6 kcal/mol) and progresses with a much lower reaction rate than the O–NO_2 homolysis, the exothermic character of this reaction determines a notable role that the HONO plays in facilitating the further bond dissociations and largely being responsible for triggering an autocatalytic stage of the global decomposition. This role, which may appear minor in the gas phase, is especially meaningful for the condensed phase processes (*vide infra*, see Section 5). Due to the high activation barriers and low reaction rates, other possible bond dissociations (Eqs. 3.3–3.7) could contribute to the decomposition process only at much later stages when the temperature becomes sufficiently high.

3.3. Novel nitro-arenes: Case study of BNFF derivatives

3,4-Bis(4-nitro-1,2,5-oxadiazol-3-yl)-1,2,5-oxadiazole-*N*-oxide (BNFF, also referred to as DNTF) has received considerable attention due to its good thermal stability, low melting point (110 °C), high density (1.937 g/cm^3), and moderate sensitivity.[128–133] Newly synthesized BNFF analogs, 3,4-bis

(4-nitro-1,2,5-oxadiazol-3-yl)-1,2,5-oxadiazole (BNFF-1, also known as LLM-172) and 3-(4-amino-1,2,5-oxadiazol-3-yl)-4-(4-nitro-1,2,5-oxadiazol-3-yl)-1,2,5-oxadiazole (ANFF-1, also known as LLM-176), were recently suggested to be good candidates for insensitive high explosives (Fig. 3.8). Low melting temperatures of BNFF-1 (85 °C) and ANFF-1 (100 °C) coupled with their predicted good stability in the melt make these materials attractive as melt-castable explosives. Forecasted distinctive physical and detonation properties of BNFF inspired a series of computational studies of thermodynamics and reactivity of various furazan derivatives,[134–137] and optically excited furazan-based molecules,[138,139] A detailed study of these novel energetic molecules was recently published.[140] The geometric structures of the BNFF-1 and ANFF-1 molecules in equilibrium obtained with the M06 method are depicted in Fig. 3.9.

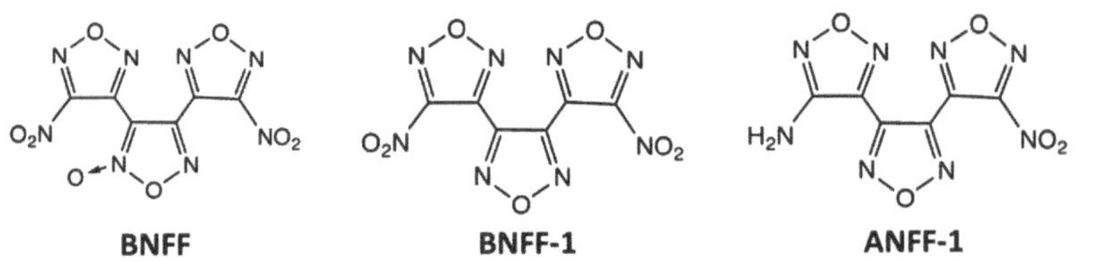

Figure 3.8 Structures of 3,4-bis(4-nitro-1,2,5-oxadiazol-3-yl)-1,2,5-oxadiazole-*N*-oxide (BNFF), 3,4-bis(4-nitro-1,2,5-oxadiazol-3-yl)-1,2,5-oxadiazole (BNFF-1), and 3-(4-amino-1,2,5-oxadiazol-3-yl)-4-(4-nitro-1,2,5-oxadiazol-3-yl)-1,2,5-oxadiazole (ANFF-1).

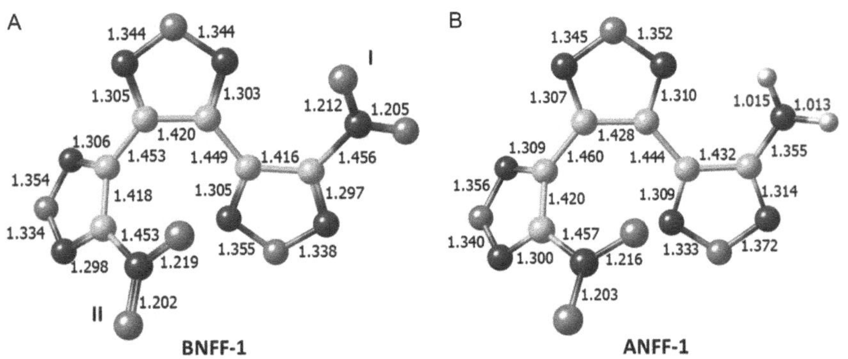

Figure 3.9 The geometry structures of BNFF-1 (A) and ANFF-1 (B). The nonequivalent nitro groups in BNFF-1 are denoted with (I) and (II). The nitro group (I) is substituted with an amino group to form ANFF-1. The shown bond distances (Å) correspond to the M06 level of theory. (For the color version of this figure, the reader is referred to the online version of this chapter.)

Both BNFF-1 and ANFF-1 molecules have fairly similar nonplanar structures. The calculated equilibrium bond distances of the BNFF-1 molecule are very close to those of ANFF-1. Figure 3.9 demonstrates that a substitution of the nitro group in BNFF-1 by the amino group in ANFF-1 affects only the fragment connected to the nitro group (the C5–C6–N5–O4–N4 ring and C4—C5 bond distances) and introduces negligible changes in the rest of the molecule. The C5—C6, C6—N5, N5—O4 bond distances in ANFF-1 are slightly elongated and O4—N4 and C4—C5 bond lengths are slightly decreased as compared to BNFF-1. The torsion angles φ(C4C3C2C1) and φ(C3C4C5C6) corresponding to a rotation of the ring fragments about C2—C3 and C4—C5 bonds in BNFF-1 are 51.3° and −162.9°, respectively. A nearly planar fragment (φ(C3C4C5C6) = − 177.8°) of the ANFF-1 molecule consists of the central and amino-furazan rings; the nitro-furazan fragment is rotated about C2—C3 bond by 53.9°. All geometry parameters obtained within different DFT functionals employed agree well with each other with the average discrepancy between bond distances in ANFF-1 calculated at M06 and B3LYP not to exceed 0.6%.

Thermal stability of the BNFF-1 and ANFF-1 molecules was explored through modeling the five most plausible decomposition mechanisms,[140] schematically illustrated in Fig. 3.10. The homolytic cleavage of the C—N bond (reaction 1, Table 3.3), leading to NO_2 loss, is usually assumed to be the most favorable decomposition channel of nitro compounds. The

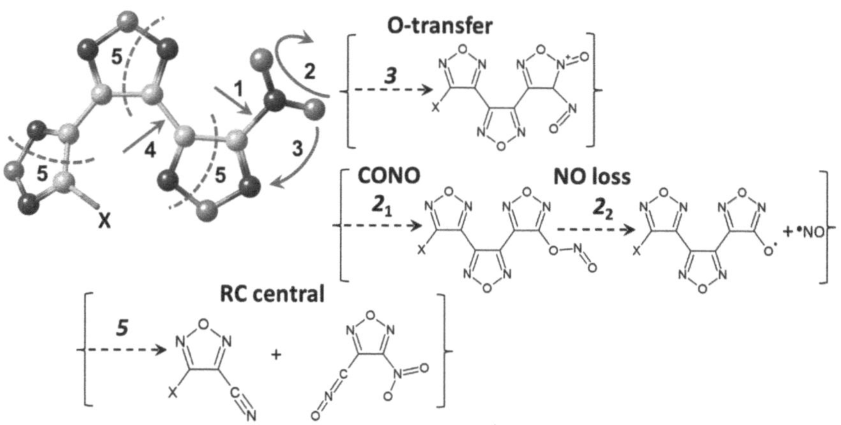

Figure 3.10 The sketched decomposition mechanisms represent five selected initiation reactions in BNFF-1 and ANFF-1. The X stands for a nitro group in BNFF-1 and an amino group in ANFF-1. (For the color version of this figure, the reader is referred to the online version of this chapter.)

two-step reaction of the nitro-nitrite rearrangement (CONO-isomer formation), proceeding via a pseudo rotation of the nitro group (reaction 2, Table 3.3), results in a release of NO and is often considered as an alternative primary dissociation reaction of nitro-arenes. The oxygen transfer involves the bond switching of the oxygen atom from the nitro group to the ring nitrogen (reaction 3, Table 3.3). We also probed the homolytic cleavage of the C—C bonds that connect two rings (reactions 4, Table 3.3) and a fragmentation of the molecules through the furazan ring cleavage (RC, reactions 5, Table 3.3).

The homolytic C–NO_2 break was simulated following the standard Eq. (3.8):

$$X-(C_2N_2O)_3-NO_2 \rightarrow X-(C_2N_2O)_3\bullet + NO_2\bullet, \qquad (3.8)$$

where X stands for a nitro group in BNFF-1 and an amino group in ANFF-1. According to our calculations, the reaction barrier in BNFF-1 requires 57.6–63.1 kcal/mol depending on the computational method used, indicating a slightly higher activation barrier for the nitro group (II) than for the nitro group (I), shown in Table 3.3. The difference in obtained energies is, however, very small and does not allow for a definite discrimination of one nitro group versus another. All three DFT functionals, PBE, M06, and B3LYP give consistent energies, which are very close to each other, with the M06 value being the highest (63.1 kcal/mol for nitro group (I) and 65.6 for the nitro group (II)), the B3LYP energy being the lowest (57.6 and 60.0 kcal/mol, respectively), and PBE energy falling right in the middle (61.1 and 62.2 kcal/mol, respectively). The similar energies obtained for the ANFF-1, 59.5–61.5 kcal/mol are even closer to each other, showing practically no dispersion. The estimated preexponential factors point towards the NO_2 loss as the fastest overall initiation reaction among all investigated mechanisms (Table 3.3).

Experimental and theoretical data on BNFF-1 and ANFF-1 are limited at the time being, therefore we will link our results to existing relevant literature on other nitro-compounds and general trends reported in earlier studies.[141–143] The calculated activation energies of the C–NO_2 homolysis of ANFF-1, 59.5–61.1 kcal/mol, and BNFF-1, 57.6–63.1 kcal/mol, rest at the lower end of the range of the typical dissociation energies of C—NO_2 bonds, 61–70 kcal/mol, determined for nitrofurazan (56.8 kcal/mol),[144] C-nitro derivatives of triazole (~67 kcal/mol),[145] a wide variety of nitroaromatics,[141] including TATB (59–64 kcal/mol[142,146] ~70 kcal/mol,[147,148]), DADNE

Table 3.3 Calculated activation barriers E_a (kcal/mol), zero-point energy corrected barriers E_{ZPE} (kcal/mol), and preexponential factors log A (s^{-1}) of BNFF-1 and ANFF-1 decomposition reactions

	PBE			M06			B3LYP		
Reaction	E_a	E_{ZPE}	log A	E_a	E_{ZPE}	log A	E_a	E_{ZPE}	log A
BNFF-1 (LLM-172)									
1 NO_2 loss (I)[a]	64.9	61.1	17.6	67.0	63.1	17.2	61.6	57.6	17.6
NO_2 loss (II)[a]	66.3	62.2[b]	18.1	69.7	65.6	17.9	64.3	60.0	18.2
2 CONO (I)	48.3	46.1 (−13.3)	13.2	55.4	53.2 (−12.8)	13.2	55.4	53.2 (−14.6)	13.2
CONO (II)	50.2	47.9 (−11.6)	13.8	58.2	55.9 (−10.0)	14.0	–	–	–
NO loss (I)[c]	34.1	30.9	–	27.9	25.0	–	22.8	19.8	–
NO loss (II)	33.9	30.8	–	27.9	24.9	–	–	–	–
3 O-transfer (I)	71.6	69.4 (17.9)	12.9	82.3	80.3 (24.3)	12.7	80.1	77.9 (22.7)	12.6
O-transfer (II)	74.8	72.4 (13.0)	13.0	86.8	84.6 (25.1)	12.5	84.3	81.9 (23.1)	12.7
4 C–C (I)	127.1	122.6	–	129.5	125.4	–	125.3	120.9	–
C–C (II)	129.3	124.8	–	131.2	127.1	–	126.6	122.2	–
5 RC central	45.7	42.9 (27.6)	15.3	50.5	47.6 (27.3)	14.0	50.4	47.5 (23.3)	15.1
RC (I)	48.3	45.4 (36.6)	14.9	53.7	50.7 (37.5)	14.0	54.0	50.8 (34.1)	14.8
RC (II)	48.3	45.3 (36.4)	15.2	55.2	52.1 (38.0)	14.3	54.3	51.1	15.0

Continued

Table 3.3 Calculated activation barriers E_a (kcal/mol), zero-point energy corrected barriers E_{ZPE} (kcal/mol), and preexponential factors log A (s^{-1}) of BNFF-1 and ANFF-1 decomposition reactions—cont'd

	PBE			**M06**			**B3LYP**		
Reaction	E_a	E_{ZPE}	**log** *A*	E_a	E_{ZPE}	**log** *A*	E_a	E_{ZPE}	**log** *A*
ANFF-1 (LLM-175)									
1 NO_2 loss	65.4	61.3	18.1	65.2	61.5	18.0	63.8	59.5	18.3
2 CONO	54.3	51.9 (−9.8)	13.8	62.6	60.3 (−8.2)	13.7	61.6	59.1 (−10.7)	13.9
NO loss[c]	32.9	29.8	–	26.8	23.8	–	21.6	18.6	–
3 O-transfer	77.0	74.6 (18.4)	13.4	88.6	88.5 (25.0)	12.9	85.8	83.5 (23.1)	13.1
4 C–C (NO_2)	130.3	125.8	–	132.1	128.1	–	127.5	123.2	–
C–C (NH_2)	134.3	129.8	–	136.4	132.4	–	132.2	127.7	–
5 RC central	45.5	42.7 (27.5)	15.2	50.6	47.6 (24.6)	14.9	50.0	47.1 (23.8)	15.0
RC (NO_2)	48.5	45.5 (32.4)	15.2	54.9	51.8 (30.3)	14.2	54.4	51.3 (31.1)	15.1
RC (NH_2)	50.6	47.3 (40.6)	15.4	53.7	50.8 (39.7)	14.6	55.6	52.2 (36.3)	15.8

[a]The nonequivalent nitro groups in BNFF-1 are denoted with (I) and (II) and correspond to Fig. 3.9.
[b]Reaction energies are shown in parentheses. For the barrier-less reactions of the homolytic bond cleavage, the reaction energy coincides with the bond dissociation energy.
[c]The NO loss is a secondary reaction step following the CONO isomerization.

(58–67.0 kcal/mol,[123] ~70 kcal/mol[149]), and a series of related aminotrinitrobenzene compounds.[142,143]

The CONO isomerization, the precursor of the NO loss, proceeds via a pseudo rotation of a nitro group accompanied by breaking the C—NO_2 bond and creating the C—ONO bond (Eq. 3.9):

$$X-(C_2N_2O)_3-NO_2 \rightarrow X-(C_2N_2O)_3-ONO. \qquad (3.9)$$

The CONO isomerization in BNFF-1 appears to require a somewhat lower activation barrier (by 4–15 kcal/mol depending on the method) than the NO_2 loss reaction pathway (Table 3.3), and the NO loss needs an even lower energy of 19.8–30.9 kcal/mol. Besides, the CONO isomerization is an exothermic reaction, releasing heat of 12.8–14.6 kcal/mol (Table 3.3).

Although the nitro group (I) shows a slightly smaller activation barrier and a slightly higher reaction heat than the nitro group (II), the observed difference is too small to claim a distinction. In the meantime, the dispersion of energies resulting from different functionals deserves some attention. PBE gives the lowest energy, 46.1 kcal/mol, for the CONO isomerization versus 53.2 kcal/mol obtained with both M06 and B3LYP. PBE also yields the highest energy for the NO loss, 30.9 kcal/mol, versus 25.0 kcal/mol obtained with M06 and 19.8 kcal/mol obtained with B3LYP (see the reaction 2 in Table 3.3). Given that M06 and B3LYP produce the same value of 53.2 kcal/mol and taking into account that PBE tends to underestimate molecular rearrangements,[28] the difference of 4–10 kcal/mol between the CONO-isomer formation and the NO_2 loss reactions appears more realistic than 4–15 kcal/mol.

Similar results are obtained for the ANFF-1 molecule with close activation barriers of the CONO-isomerization stage ranging from 51.9 to 60.3 kcal/mol, an appreciable difference in energies of the NO loss stage (18.6–29.8 kcal/mol), and the reaction heat of 8.2–10.7 kcal/mol.

The calculated activation energies of the nitro-nitrite isomerization of BNFF-1 and ANFF-1 are consistent with the barrier of the CONO formation in DADNE (66.4 kcal/mol,[123] 59 kcal/mol[150]) and nitromethane, (55.5 kcal/mol, estimated from infrared multiphoton dissociation experiments,[151] and 51.7 kcal/mol, obtained from *ab initio* calculations[152]). The similar rearrangement in nitroethylene was found to require about 57.9 kcal/mol.[124] Close barriers were also obtained in C-nitro isomers of triazole (60.1–65.2, 63.5 kcal/mol)[145] and nitroaromatics, nitrobenzene (61.1 kcal/mol,[125] 63.7 kcal/mol[126]) and trinitrotoluene (54.9 kcal/mol).[127] The BNFF-1 and ANFF-1's behavior (Table 3.3) resembles the nitroethylene's decomposition trend (with CONO isomerization requiring ~15 kcal/mol less than the energy needed to cleave the C—NO_2 bond)[124] and somewhat differs from that ofDADNE, which exhibits nearly isoenergetic reactions of the nitro-nitrite isomerization and the C–NO_2 cleavage.[123] Another difference is that the CONO isomerization in nitrobenzene is an endothermic reaction (1.1[125] and 4.0 kcal/mol[126]) whereas in DADNE it is only weakly exothermic with the energy gain of 4 kcal/mol[123,150] versus relatively higher heat produced in BNFF-1 (12.8–14.6 kcal/mol) and in ANFF-1 (8.2–10.7 kcal/mol, Table 3.3), similar to trinitrotoluene (8.1 kcal/mol[127]).

An analysis of the kinetic parameters, generated by both PBE and M06, predicts a competition between the CONO formation and the C–NO_2

fission in BNFF-1 over a wide temperature range while kinetics obtained from B3LYP suggests that the C–NO_2 homolysis is favored over all other probed mechanisms (Table 3.3 and Fig. 3.11). Unlike BNFF-1, only PBE supports the CONO isomerization while both M06 and B3LYP prefer the C–NO_2 break in the initiation of decomposition of ANFF-1. Hence, the CONO preference achieved from the energetic considerations is practically negated once the kinetics is taken into account.

We emphasize here that energetic considerations should be coupled with kinetics to understand the thermal decomposition, especially, when several chemical pathways coexist. Thus, the exothermic nitro-nitrite isomerization and the NO loss require a relatively low energy but the preexponential factor of this dissociation reaction is estimated to be much smaller than that of the NO_2 loss (Table 3.3). Hence, our analysis suggests that the C–NO_2 fission is

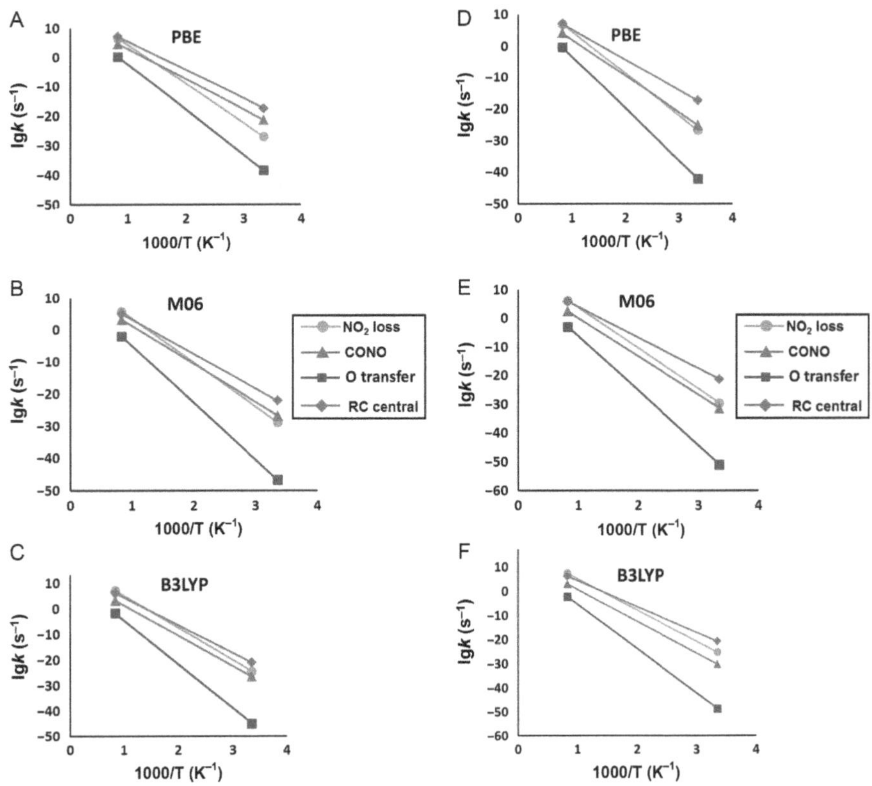

Figure 3.11 Reaction rates of the simulated decomposition channels of BNFF-1 (A–C) and ANFF-1 (D–F) molecules. (For the color version of this figure, the reader is referred to the online version of this chapter.)

the fastest dissociation process while the NO loss contributes to decomposition as a slow background reaction. Interestingly, neither of these two reactions is the dominant initiation mechanism in the decomposition of BNFF-1 and ANFF-1 (*vide infra*).

Oxygen Transfer, another type of isomerization, results in a formation of nitroso-furoxan isomers of BNFF-1 and ANFF-1 displayed in Fig. 3.4D–F (Eq. 3.10).

$$X-(C_2N_2O)_3-NO_2 \rightarrow X-(C_2N_2O)_2-C(NO)OCN(O) \quad (3.10)$$

The O-transfer mechanisms in both molecules have significantly higher activation barriers (~70–80 kcal/mol in BNFF-1 and ~75–89 kcal/mol in ANFF-1) than the corresponding homolytic cleavage of the C—NO_2 bonds and the CONO-isomerization reactions and exhibit fairly low pre-exponential factors (reaction 3 in Table 3.3 and Fig. 3.10). The difference obtained for the different nitro groups is again small. Hence, this reaction should be ruled out as a possible candidate to initiate the decomposition of either BNFF-1 or ANFF-1.

The homolytic C–C fission in the molecules (Eqs. 3.11a and 3.11b) are also energetically unfavorable reaction pathways with the calculated dissociation energies of ~121–125 kcal/mol in BNFF-1 and ~123–128 kcal/mol in ANFF-1, which are much higher than the corresponding energies required for breaking the C—N bonds (reactions 4, Table 3.3). These estimates agree with the C–C dissociation energy of a DADNE molecule (120.2 kcal/mol).[123]

$$X-(C_2N_2O)_3-NO_2 \rightarrow X-(C_2N_2O)_2\bullet + \bullet(C_2N_2O)-NO_2 \quad (3.11a)$$
$$X-(C_2N_2O)_3-NO_2 \rightarrow X-(C_2N_2O)\bullet + \bullet(C_2N_2O)_2-NO_2 \quad (3.11b)$$

The difference in energy between splitting off of the peripheral nitro-furzan rings in BNFF-1 does not exceed 2 kcal/mol. The distinction between the nitro-furazan ring's loss and amino-furazan ring loss is comparable, ~4 kcal/mol, signifying a measure of the substitution effect (reaction 4, Table 3.3). Hence, the splitting off the rings would not contribute to the initial stages of the overall decomposition of either molecule.

A heterocyclic RC: Opening of the heterocyclic ring may be in principle achieved through breaking one of the ring's bonds (see, e.g., Ref. 144). We explored concerted fragmentation mechanisms of BNFF-1 and ANFF-1 molecules (path 5 in Fig. 3.10 and reaction 5 in Table 3.3) in which a cleavage of the central furazan ring (Eq. 3.12a) or one of the peripheral rings

(Eqs. 3.12b and 3.12c) proceeds through a simultaneous fission of the N—O and C—C bonds:

$$X-(C_2N_2O)_3-NO_2 \rightarrow X-(C_2N_2O)CN+CNO-(C_2N_2O)-NO_2 \quad (3.12a)$$

$$X-(C_2N_2O)_3-NO_2 \rightarrow XCNO+CN-(C_2N_2O)_2-NO_2 \quad (3.12b)$$

$$X-(C_2N_2O)_3-NO_2 \rightarrow X-(C_2N_2O)_2-CN+CNO-NO_2 \quad (3.12c)$$

The transition state structures, shown in Fig. 3.12, indicate an elongation of N—O (~1 Å) and C—C (~0.6 Å) bonds in the furazan rings. The calculations reveal that the RC reactions in BNFF-1 require 42.9–51.1 kcal/mol of energy. The endothermic cleavage of the central ring (Fig. 3.3) in BNFF-1 requires 42.9–47.6 kcal/mol, with the lowest barrier (42.9 kcal/mol) obtained with PBE and very close barriers of 47.6 and 47.5 kcal/mol obtained with M06 and B3LYP, respectively. This energy is slightly (~3 kcal/mol) lower than the barriers for the opening of peripheral rings (reaction 5 in Table 3.3). An energy dispersion obtained with different functionals is ~5 kcal/mol, with a systematic difference observed between PBE and hybrid methods and almost identical energies received from M06 and B3LYP. The energy difference between outer nitro-furazan rings does not exceed 1 kcal/mol.

Consistent activation barriers (42.7–55.6 kcal/mol) and reaction energies are obtained for ANFF-1. The same trends are seen: the central ring opening needs a lower energy than the peripheral rings, and all three methods produce corresponding energies with the lowest barrier (42.7 kcal/mol) delivered by PBE and nearly identical barriers (47.6 and 47.1 kcal/mol) obtained from M06 and B3LYP. The difference in energy between cleavage of the nitro-furazan ring and the amino-furazan ring is negligible.

Most importantly, the concerted ring opening reactions exhibit the lowest activation barriers among all mechanisms explored in both molecules, BNFF-1 and in ANFF-1. These values agree with the activation energy of 42.3 kcal/mol obtained from DSC measurements of BNFF.[133] Pre-exponential factors (log A (s^{-1}) ~15, Table 3.3) of the RC mechanisms are still lower than the NO_2 loss but visibly higher than the CONO isomerization. The estimated reaction constant is consistent with the log A (s^{-1}) = 13.68, reported for BNFF.[133] Kinetic behavior is illustrated in Fig. 3.6, which shows that the fragmentation through the central ring opening dominates the decomposition initiation in both molecules at low and moderate temperatures (regardless of the computational method used).

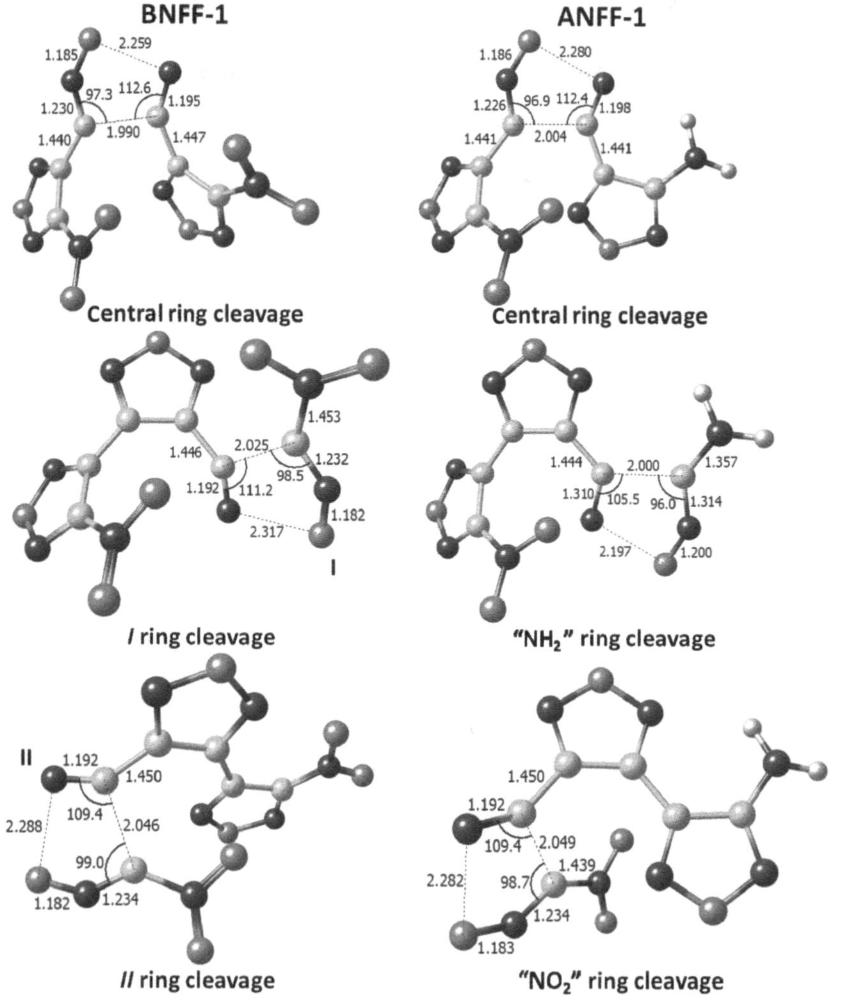

Figure 3.12 The structures of transition states for ring cleavage reactions in BNFF-1 and ANFF-1. The bond distances (Å) are obtained with M06 functional. (For the color version of this figure, the reader is referred to the online version of this chapter.)

The NO_2 loss, while being the fastest reaction, prevails only at high temperatures. The exothermic CONO-isomerization pathway makes a modest contribution to decomposition at the earliest stages.

3.3.1 Comparing BNFF-1 and ANFF-1

Next, we analyze the difference between the BNFF-1 and ANFF-1 decomposition processes, or discuss the substitution effect. In both molecules, the cleavage of the central ring seems to be the most favorable dissociation process from an energetic point of view (42.9–47.6 kcal/mol in BNFF-1 vs. 42.7–47.6 kcal/mol in ANFF-1), with a negligible difference between ANFF-1 and BNFF-1. The energy of the C—NO_2 bond dissociation in ANFF-1 (61.1 delivered by PBE, 63.1 by M06, and 57.6 kcal/mol by B3LYP) also shows only a small change as compared to BNFF-1 (61.3, 61.5, and 59.5 kcal/mol, respectively). The C—NO_2 bond preexponential factors evidence that the detachment of NO_2 from BNFF-1 and ANFF-1 will proceed at comparable rates and faster than other reactions. At the same time, dissociation energies coupled with kinetic parameters point towards the ring opening as a dominating dissociation process at low and moderate temperatures while the C–NO_2 homolysis becomes important at high temperatures.

The most notable effect of the BNFF-1's nitro group replacement with the ANFF-1's amino group appears in the barrier heights and reaction energies of the CONO decomposition channel. The CONO isomerization of ANFF-1 (51.9 kcal/mol found by PBE, 59.1 kcal/mol by B3LYP, and 60.3 kcal/mol by M06) requires ~6 kcal/mol higher energy in comparison to BNFF-1 (46.1 kcal/mol obtained by PBE and 53.2 kcal/mol by M06 and B3LYP) while the exothermicity of this pathway is reduced by ~4 kcal/mol (Table 3.3). Due to a somewhat higher activation barrier and somewhat lower reaction heat of ANFF-1 than those of BNFF-1, the contribution of the slow exothermic CONO-isomerization reaction in BNFF-1 is expected to be more pronounced as compared to ANFF-1. However, we suggest that this process has only a minor effect on the overall thermal dissociation and hence the presence of nitro or amino groups in these heterocyclic molecules does not make a significant difference.

Summing up, modeling of candidate decomposition channels of new energetic compounds BNFF-1 and ANFF-1 revealed that the cleavage of the central furazan ring dominates the overall decomposition initiation of both BNFF-1 and ANFF-1 molecules, and breaking the peripheral rings requires only a slightly higher energy. The obtained decomposition activation barriers of both BNFF-1 and ANFF-1 agree with DSC measurements of BNFF. The next important decomposition mechanism is the homolytic C—NO_2 bond cleavage. Despite being the fastest reaction, the C–NO_2 homolysis prevails only at high temperatures due to a notably higher activation barrier than that of the ring opening process. The two-step NO loss in both molecules, while it exhibits a somewhat higher activation barrier than the ring opening reactions and may coexists with the primary dissociation channel, it proceeds at a much lower rate.

Thus, our calculations established that the decomposition scenario in heterocyclic nitro-furazan-based molecules differs from known relatively insensitive nitro-arenes, in particular, TATB. Unlike nitro-arenes, the thermal decomposition in BNFF analogs is strongly affected by the presence of heterocyclic rings and the ring's opening reactions trigger the overall bond dissociation. A further development of the dissociation processes is defined by the interplay of the slow exothermic CONO-isomerization and the fast C–NO_2 homolysis reactions. While the activation barrier of the NO loss requires a lower energy than the NO_2 detachment, the reaction kinetics happened to be orders of magnitude slower for the CONO isomerization than for C–NO_2 homolysis.

The substitution of one nitro group in BNFF-1 with the amino group in ANFF-1 does not affect the ring opening reactions, negligibly decreases the C—NO_2 bond strength, tends to increase the CONO-isomerization activation barrier and to reduce the reaction heat.

3.4. Overall trends in the decomposition of nitro molecules

Ab initio modeling of three classes of nitro molecules with different critical bonds (O–NO_2 in nitro-esters, N–NO_2 in nitramines, and C–NO_2 in nitro-arenes) allows us to distil a few important observations.

First, the most common major decomposition pathway in all three groups of molecules is the homolysis of the critical bond X–NO_2 (X=C, N, or O) and release of NO_2. Typically, in existing energetic molecules, this reaction has the lowest activation barrier, the highest preexponential factor, and the fastest rate. Novel heterocyclic nitro-arenes exhibit somewhat unusual behavior, with the dominating decomposition reaction of the furazan ring opening (rather than the NO_2 loss, as it was expected from a comparison with TATB). While the preexponential factor of C–NO_2 homolysis is still appreciably larger than that of other reactions in heterocyclic nitro-arenes, this pathway starts dominating over the ring opening reaction only at relatively high temperature.

Second, the decomposition process at the very early stages is defined by the interplay of at least two reactions. These are the NO_2 loss and the HONO elimination in nitro-esters and nitramines and the furazan ring opening and the NO_2 loss in heterocyclic furazan nitro-arenes. Note that in other nitro-arenes (e.g., TATB or DADNE), those two reactions are the NO_2 loss and nitro-nitrite-isomerization.[5–9,57,123]

Third, while *ab initio* calculations provide a powerful tool for interpretation of experiments, modeling of chemical decomposition mechanisms should be performed as comprehensively as possible as novel methods become available because incomplete data may lead to erroneous conclusions. For example, in the absence of preexponential coefficients and reaction rates, activation barriers are insufficient to make determination regarding the dominating or triggering decomposition mechanism in these complex molecules.

Finally, we would like to note that the conclusions obtained from modeling of gas-phase decomposition process, while provide valuable insight on overall chemistry in energetic molecules, cannot be simply extrapolated to materials because interactions between molecules are crucially important for decomposing crystals and practical explosive compounds. Hence, the next step in this research is solid-state periodic calculations involving modeling of similar chemical reactions in the crystalline environment.

4. CHARGED AND EXCITED STATES: NEW PHYSICS AND CHALLENGES

4.1. Dissociation mechanisms of excited and charged DADNE molecules

A chemical reaction of decomposition can start from the ground state of the system or from one of its excited vibrational or/and electronic states. Various mechanisms of initiation of the decomposition chemistry and further processes leading to products have been discussed over the years. For example, electronic transport,[153] a multiprocess photochemical model,[154] and an excitonic mechanism[17,155] were proposed as possible means of detonation initiation. Although experimental evidence that electronic excitations (electrons, holes, and excitons) play an important role in the initiation of chemistry in energetic materials is abundant,[107,108,156–165] details of processes involving electronic excitations are far from being understood. Since the first electronic structure calculations of ideal energetic materials,[166,167] and simple defects in them,[168–170] were performed, a great deal of studies became available. Despite this, even a general concept of electronic excitations being able to trigger explosive decomposition chemistry is not broadly accepted in the field of energetic materials (see, e.g., discussions in Refs. 108,165,171). High quality experiments combined with *ab initio* calculations are required to advance our understanding of ultrafast reactions and to uncover the particular role that electronically excited and charged states play in the initiation process.

The first *ab initio* study that methodically analyzed and compared energies of the decomposition reactions on ground and excited potential energy surfaces was performed for DADNE molecules.[123] Synthesized DADNE (also called FOX-7) was reported in 1998[172] and thought as a potential insensitive energetic material with superior properties.[173–175] Possible bond dissociation processes, including a rupture of C—NH_2, C—NO_2, and C=C bonds as well as CONO- and HONO-isomerizations in DADNE, illustrated in Fig. 3.13, have been studied by GAUSSIAN program to reveal the primary and secondary mechanisms of the material's decomposition.[123] The reaction energies and activation barriers have been calculated and compared for bond dissociation processes starting from the ground [R], ionized $[R]^+$, negatively charged $[R]^-$, and lowest excited $[R]^*$ states of a DADNE molecule (R stands for the molecule).

It was found that the excited states can not only change the energetics of chemical reactions by reducing the reaction barriers, but also change the type of the dominating chemistry from endothermic to exothermic. In particular, the positive electronic affinity of the DADNE molecule leads to the electron trapping, thus forming a negatively charged radical, which can decompose more rapidly than the neutral molecule. It was also established that the positively charged (ionized) state of the molecule has a double-well potential energy profile, which may open alternative pathways for the molecular decomposition.

A B

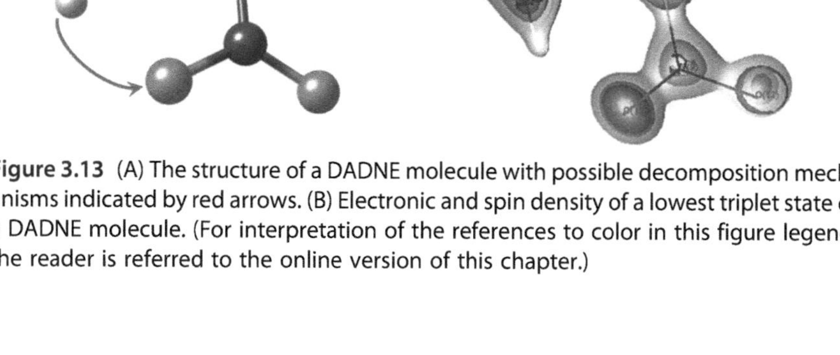

Figure 3.13 (A) The structure of a DADNE molecule with possible decomposition mechanisms indicated by red arrows. (B) Electronic and spin density of a lowest triplet state of a DADNE molecule. (For interpretation of the references to color in this figure legend, the reader is referred to the online version of this chapter.)

Two nearly isoenergetic processes that lead to the decomposition of DADNE are found. The C—NO_2 bond fission in the ground state of the neutral DADNE molecule requires the activation energy of 67 kcal/mol in agreement with the previous theoretical investigation.[149] Nearly the same energy, 66 kcal/mol, is needed for the nitro-nitrite isomerization, that is, a formation of the CONO group; this is consistent with the results of the earlier work.[150] Both of these energies fall into the range of the experimentally determined dissociation energies, 61–70 kcal/mol, for gas-phase nitro compounds.[143] The drawn conclusion suggested that the C–NO_2 scission and the C–NO_2 to CONO isomerization were two competing mechanisms of DADNE dissociation; this essentially reconciled the seemingly contradictory earlier theoretical predictions[149,150] and also helped to explain the experimental observations.[174,175] Other bond-fission reactions considered for C=C and C—NH_2 bonds in the neutral molecule can hardly take place at the early stages of the decomposition because of their high barrier energies.

More importantly, these calculations demonstrate that the excitation and charging of the molecule can have a dramatic effect on the decomposition process by facilitating some mechanisms of dissociation and precluding the others. For example, the formation of CONO isomers can take place in the neutral molecule, but the negatively charged DADNE molecule does not support nitro-nitrite rearrangement and therefore breaks this channel of decomposition. However, trapping an electron by the CONO-rearranged molecule leads to the spontaneous detachment of NO species with the energy gain.[123]

In general, a localization of an extra electron on nitro groups in DADNE leads to an imbalance of the interatomic interactions and results in bond weakening. Therefore, decomposition reactions proceed with smaller barriers than in the neutral state. This has been indeed demonstrated in our calculations on the examples of fission of C—NH_2, C—NO_2, and C=C bonds. Table 3.4 shows that the NO_2 loss reaction requires 67.0 kcal/mol if the molecule decomposes from its ground state while the activation barrier drops to 54.71 kcal/mol for the ionized molecule, to 32.97 kcal/mol for the molecule in the excited triplet state, and to 20.62 kcal/mol for the negatively charged ion. In addition, a hole trapped by the ionized DADNE makes this reaction exothermic, with the heat release (27.67 kcal/mol) that is sufficient to initiate a neighbor negatively charged ion with the activation barrier of 20.62 kcal/mol. Moreover, negative charging of DADNE makes the intramolecular hydrogen transfer and formation of HONO fragments possible and even exothermic while in the neutral case such a transfer does not take place at all.[123]

Table 3.4 Calculated dissociation energies $E_{dissociation}$ and activation barriers $E_{barrier}$ of the C—NO_2 bond fission in the ground state of DADNE molecule, negatively charged state $DADNE^-$, ionized state $DADNE^+$, and the lowest triplet state $DADNE^*$ are shown in kcal/mol

Reaction	$E_{dissociation}$	$E_{barrier}$
(1) DADNE → NO_2 + [$C_2NO_2N_2H_4$]	66.98	66.98
(2) DADNE → NO_2^+ + $[C_2NO_2N_2H_4]^-$	246.96	246.96
(3) DADNE → NO_2^- + $[C_2NO_2N_2H_4]^+$	128.88	128.88
(4) $DADNE^+$ → NO_2 + $[C_2NO_2N_2H_4]^+$	−27.67	54.71
(5) $DADNE^-$ → NO_2^- + [$C_2NO_2N_2H_4$]	14.48	20.62
(6) $DADNE^*$ → NO_2 + [$C_2NO_2N_2H_4$]	12.44	32.97

The decomposition of the ionized state of DADNE molecule via fission of C—NH_2 and C=C bonds is an unlikely process due to both the high reaction barriers (of the order of 100 kcal/mol) and the high ionization energy. Although C–NO_2 to C–ONO isomerization does not occur from an ionized state of the DADNE molecule, the ionization of the CONO configuration leads to a spontaneous NO detachment. The C—NO_2 bond rupture in the ionized state becomes an exothermic reaction. We note, however, that due to the high ionization potential of DADNE (8.31 eV),[123] the energy needed for decomposition is effectively higher than the calculated activation barriers. At the same time, due to the positive electron affinity of DADNE (1.26 eV),[123] indicating that it readily forms a stable negatively charged radical $(DADNE)^-$, the activation barriers to trigger decomposition in electron-rich environment are expected to be reduced.

Recent studies of PETN,[107–109,176] HMX,[177] and furazan-based molecules,[138] performed with high-level methods, are also consistent with these earlier results. Modern refined techniques for studying excited states, such as time-dependent DFT (TD DFT),[178–183] and configuration interaction formulated in terms of the complete active space self-consistent field (CASSCF),[184–187] along with its lately developed variations,[188–191] are able of providing more thorough information regarding decomposition pathways on the excited state potential surfaces.

The overall conclusion obtained (thus far) is that charged and excited states dramatically affect the decomposition initiation processes in energetic molecules by enabling some dissociation pathways and prohibiting or slowing dawn others and therefore all reaction parameters (activation barriers, reaction energies, reaction rates, and reaction products) strongly depend on the initial state of the decomposing molecule. This finding is extremely important because it implies that once we understand atomistic details of decomposition processes in energetic molecules and learn how to prepare desired conditions (molecular environment, e.g., excess or deficit of electrons or stimulate charge transfer), we will have tools to initiate chemical reactions with the degree of control which is unprecedented today. This, in turn, will open the door to tuning sensitivity of materials to initiation of detonation.

4.2. Modeling electron and hole polarons

A phenomena of electron and hole trapping in insulators represent fundamental challenge in both basic and applied sciences. Will charge carriers undergo free band propagation or will they instead become self-trapped or trapped by impurities is a basic issue. Polarons in molecular crystals have been a topic of numerous model studies before and after classical work by Holstein[192] and Emin.[193] However, many-electron *ab initio* calculations of polarons in these crystals are still limited (see, e.g., Refs. 155,194,195). An answer to a crucial question as to whether or not the carrier is localized depends strongly on the Hamiltonian used in such calculations. It is well established that calculations using the Hartree–Fock approximation tend to localize electrons and holes and these using DFT in LDA or GGA approximations tend to predict delocalized solutions,[196,197] Therefore making reliable predictions is difficult even when newly developed DFT functionals are used.

An advantage of using ECM here is that a single extra charge can be modeled in a relatively small QM cluster whereas the polarization of the rest of the solid is accounted for via polarization of the region I. Comparing with experiment requires calculating optical and magnetic characteristics of localized carriers, which is again easier within molecular methods implemented in ECM. Disadvantages are that in ECM an extra electron or hole is confined to QM cluster and also that perturbations induced by the boundary effects can trigger localization. Therefore this method requires calibration with respect to periodic calculations to reliably establish the character of career localization. It is best applied to calculating the spectroscopic properties and diffusion barriers of localized polarons.[49,56]

4.2.1 Periodic polaron calculations

The periodic CRYSTAL06 calculations employed $2 \times 2 \times 1$ and $2 \times 2 \times 2$ supercells containing 16 and 32 DADNE molecules, correspondingly. Using these supercells ensures good balance between computational effort and spacing of the periodic images. In addition, Durand–Barthelat effective core potentials with their associated valence basis sets (equivalent to the basis set used in the embedded cluster calculations) were used for the 1s shells of C and O to speed up the convergence process.[198] Eight *k*-points in a Monkhorst–Pack scheme were selected for the *k*-point mesh.

The electronic structure of a DADNE crystal inherits the main features of the molecular electronic structure. Indeed, the top of the valence band of the crystal is formed by carbon states and oxygen orbitals of nitro groups, when deeper valence band levels are formed by the amino groups and carbon atomic orbitals. The bottom of the conduction band is mainly formed by the orbitals of the nitro groups. These states form a narrow band, which is split from the next subband of the conduction band and causes the electron localization described below. In the perfect lattice calculations, an extra electron populates this LUMO state and is delocalized over the nitro groups of the molecules constituting the periodic cell. The excess charge was compensated by a charged background. However, if the lattice relaxation is allowed, the excess electron induces a pronounced lattice distortion accompanied by the electron localization. The extra electron charge is unequally distributed between the two neighboring molecules in adjacent layers. One of them has a larger amount of spin density than the other, and it is localized on the nitro groups of each of the two molecules 1 and 2 (see Fig. 3.14A).

The polaron-induced lattice deformation along the *z*-axis (perpendicular to the layers) is relatively small, however, the two electron-bearing molecules move apart. This induces an increase of the distance between the carbon atoms bonded to the nitro groups by about 0.5 Å. The intralayer relaxation exhibits smaller shifts of molecules (0.25 Å) around the two molecules. The intralayer hydrogen bonds between uncharged and electron-bearing molecules are shorten from the distances characteristic for the neutral crystal 2.04/2.01 to 1.79/1.82 Å. This is due to the attraction between the protons and the oxygen atoms bearing the extra negative charge. Due to the asymmetric character of the extra electron localization, the local geometries of the molecules 1 and 2 also differ. The O—N bond in the molecule 1 is elongated from 1.31 to 1.33/1.36 Å. For the molecule 2 with only about 30% of the extra electron, there is almost no change of

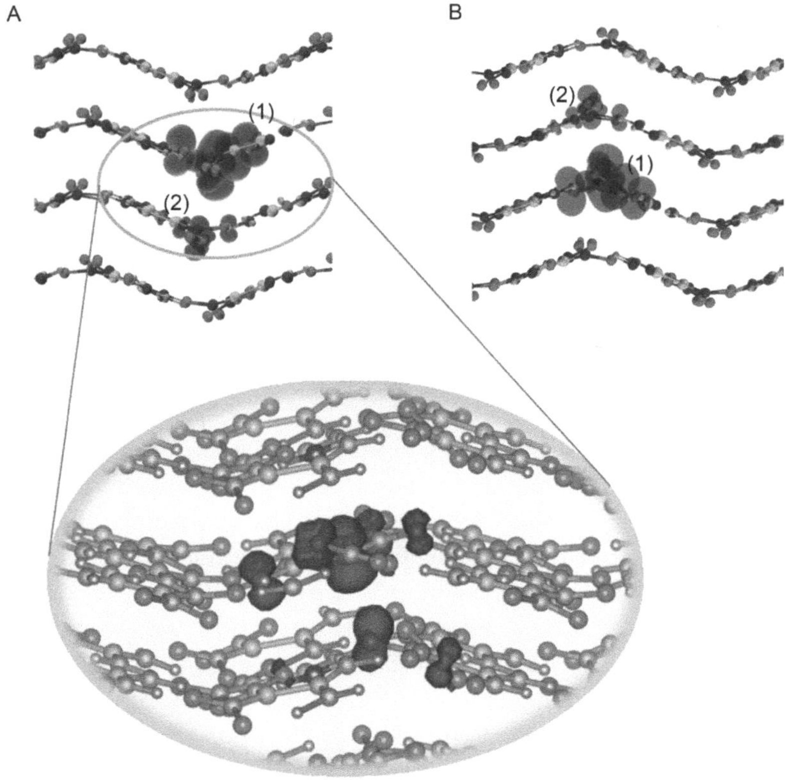

Figure 3.14 Spin-density plot of localized electron and hole polarons in a DADNE crystal. (A) An extra electron localizes over two molecules (1) and (2) in neighboring layers, with most of the electron density localized on molecular orbitals of nitro groups; (B) a hole, similarly to an electron, localizes over two molecules in neighboring layers to minimize the dipole moment of the system. (For the color version of this figure, the reader is referred to the online version of this chapter.)

N—O bond length. At the same time, the N—C bonds in both molecules become shortened from 1.4 to 1.35/1.39 Å, whereas the C=C bond length is decreased by 0.2 Å. The rotation angles of nitro groups with respect to the molecular plane increase on 12°/25° and 10°/13°. At the same time, ∠O–N–O and ∠N–C–N angles in both molecules are changed only within 2° with respect to the perfect geometry.

Ionization of perfect DADNE crystal leads to hole localization on the two neighboring molecules in adjacent layers. The spin density is again unequally distributed between the two molecules, with a larger density on the molecule 1 (see Fig. 3.14B). The hole polaron occupies nitro-carbon and nitro groups of molecule 1 and polarizes one nitro group of the molecule 2 (see Fig. 3.14B). This is accompanied by a split of a one-electron level, which is located inside the valence band in the perfect crystal, up from the top of the valence band by about 0.4 eV after the hole localization. The interlayer relaxation of the geometry along the z direction is negligible, except that the polaron-bearing molecules 1 and 2 shift towards each other: the distance between the carbon atoms in these molecules decreases from 4.41 to 4.12 Å, and the interlayer distance between the nitrogen atoms decreases by 0.42 Å. The intralayer relaxation is similar to that for the electron polaron: there are small interlayer displacements of molecules with respect to each other, which are facilitated by the flexibility of hydrogen bonds. Due to the lack of charge, the intermolecular attraction between the O atoms of molecules 1 and 2 and a respective proton of the nearest neutral molecule is weakened, which leads to a slight increase of the H–O distance to 2.06 Å. The intramolecular parameters are also affected by the presence of the trapped hole. The O—N bond lengths in the molecules 1 and 2 decrease by 0.1–0.2 Å, whereas the C—N bonds elongate by 0.3–0.4 Å, and the C=C distance increases by 0.2 Å. The intramolecular ∠O–N–O and ∠N–C–N angles change by only of 2–3°.

4.2.2 Embedded cluster model polaron calculations

In the ECM calculations, we used a QM cluster (see Fig. 3.1) that contained eight molecules surrounded by region I with the radius of 20 Å. Similarly to periodic calculations, the ECM results show that the extra electron is delocalized in the perfect lattice and becomes localized on the nitro groups of two molecules after the relaxation with self-trapping energy of about 0.5 eV. In ECM, the radius of the region where molecules are allowed to relax is much bigger than in the periodic calculations as it involves the whole region I. As has been shown,[199] a defect-induced relaxation in solids formed by flexible networks can propagate far from a defect site. An analysis of the lattice distortion caused by the electron trapping shows that atomic displacements as large as 0.05 Å are still induced at the distance of about 15 Å from the defect center, that is, much beyond the border of the QM cluster (11.5 Å). Not surprisingly, the lattice relaxation is mainly characterized by the deformation and rotation of flexible nitro and amino groups. In the case of the electron polaron, the largest atomic displacements of the order of 0.17 Å correspond to nitro groups. They are caused by the fact that the extra electron localized on nitro groups strongly repels the surrounding nitro groups because their oxygen atoms bear a negative charge. At the same time, the amino groups are less affected by polaron localization: their largest displacements do not exceed 0.11 Å. Carbon atoms are perturbed the least and displaced only by ~0.04 Å.

The character of the intralayer relaxation is similar to that obtained via the periodic model. Indeed, hydrogen bonds between uncharged and electron-bearing molecules are shortening to 1.77/1.79 Å. Due to the presence of the localized negative charge, the local geometry of polaron-bearing molecules characterizes by shortening of C—N bonds on 0.1 Å, and the elongation of O—N bonds on 0.4 Å. However, the rotation angles of nitro groups with respect to the molecular plane increase on 5°/20° and 6°/10°, that is, less pronounced then in the periodic model.

ECM modeling of a hole polaron has also shown its delocalization in the perfect lattice. Relaxation of the lattice leads to localization of the polaron on two molecules in adjacent layers. However, the self-trapping energy of a hole, 1.2 eV, is larger than that found for electron polaron. The hole trapping induces a long-range relaxation of the lattice up to 11 Å with displacements of the molecules with respect to their initial positions. The distortion of polaron-bearing molecules is expressed by a rotation of nitro groups and slight elongation of C—C bonds. These results demonstrate that the boundary conditions in ECM do not affect the character of electron or hole localization. They do not, however, provide a proof of electron localization as the extent and character of electron localization is known to depend on the amount of Hartree–Fock exchange included in the hybrid B3LYP functional. We note that for smaller gap insulators, such as HfO_2,[49] B3LYP tends to give more reliable predictions of the character of electron or hole localization than for wider gap oxides, such as MgO or SiO_2.[49]

To test whether the charge localization obtained in DADNE is an artifact of the B3LYP functional, we carried out similar calculations for tetracene. This crystal has a very similar band gap (about 3 eV) and tetracene molecules also have double C=C bonds, however, no lattice polaron (i.e., electron or hole localization) has been observed in this material.[200] Therefore, if using B3LYP would lead to artificial polaron localization, this would indicate that our predictions for DADNE could be also flawed. We used a 1×2×1 B3LYP optimized supercell, with the parameters a=8.60 Å, b=13.11 Å, c=12.26 Å. A distortion (elongation) in one of the C=C bonds was introduced as a precursor for the polaron state. The two C atoms have different environments, one of being bonded to C and H atoms, when the other being attached only to C atoms. An additional electron was added to the supercell to produce the polaron state, and the excess charge was compensated by a charged background. Then, the system was allowed to relax to the minimum total energy. The resulting structure shows the electron delocalized among all the C atoms in the structure and no lattice polaron was formed, in agreement with experimental data.[201] This result suggests that the electron and hole localization obtained in DADNE using the B3LYP functional could be indeed a real effect.

4.2.3 Significance

The results of this work demonstrate that the ECM can serve as a useful tool for modeling chemical reactions, defect structures, and defect-induced processes in molecular crystals. It allows us to take into account both the short-range interaction between molecules and the long-range effects of the electrostatic potential and the defect-induced lattice distortion using a combination of quantum-chemical methods and classical force-fields.

ECM was applied here to study the proton transfer and electron and hole localization in DADNE. The problem of the hydrogen (or proton) transfer is as important as it is complicated and is far from solution and completion, equally in energetic solids and in other materials.[57,202] Our research sheds some light on the mechanisms of the proton transfer and the HONO-isomer formation in the chemical decomposition of DADNE. In particular, we demonstrated that the HONO formation should not be considered as a precursor for the thermal decomposition of crystalline DADNE. This suggests that the observed NO products in the course of thermal decomposition of solid DADNE[172,174] are likely to arise from the nitro-to-nitrite isomerization or from excited state dissociation pathways,[203] and H_2O stems from secondary reactions. The obtained conclusions confirm that intermolecular interactions in the lattice are very important for describing these processes in a solid state.[57,202]

Our periodic and ECM calculations suggest that both electrons and holes can self-trap in DADNE. We predict that in the static approximation using B3LYP functional, an extra electron or a hole can break the static lattice translational symmetry, becoming localized at its self-generated local distortion. Although we did not study the polaron hopping between equivalent neighboring lattice sites, the relatively large self-trapping energies of about 0.5 eV suggest that it is a likely mechanism of conductivity at room temperature.

Application of ECM to a broad class of molecular materials can help to establish uncovered yet details of the initiation of chemistry under a variety of conditions, including the presence of lattice imperfections, elevated pressure and temperature, laser initiation, and high radiation fields.

5. CHEMICAL REACTIONS IN CONDENSED ENERGETIC MATERIALS: UNCERTAINTIES AND INSIGHTS

Although many scientists, C. Tarver, A. Dremin, and T. Brill among others, pointed to the urgency and deficiency of the knowledge about the interplay of the X—NO_2 bond scission and the –NO_2 to –ONO isomerization, especially in the condensed phase energetic materials some 20 years ago, not much progress has been achieved since then. New first principle understanding of specific reactions in solid state has been gained slowly and incrementally rather than systematically. Most research on decomposition of materials has been performed with classical methods, such as molecular dynamics and its variations, which, while delivered some useful information, face natural difficulties and lack the electronic resolution. Most of quantum-chemical studies that have sufficient accuracy and precision in reproducing chemical reactions are accomplished on molecules, which are not applicable to practical materials as they do not include intermolecular interactions, a morphology of real crystals, containing defects and deformations, as well as collective phenomena in their many appearances. This study contributes to a better understanding of the interplay of chemical reactions triggered in a decomposing molecule in a crystalline environment.

5.1. Shear-strain deformations in DADNE and TATB crystals

We briefly review the comparative first-principles analysis of the initiation of chemistry of a fairly prospective[174] energetic material DADNE and the stability benchmark TATB[5–9,204,205] Simulations of chemical reactions of decomposition in ideal, defect-containing, and deformed crystalline DADNE and TATB suggest that the shear-strain deformation plays a crucial role in defining the sensitivity of explosive crystals to initiation and strongly depends on the interactions between molecules and the crystalline lattice.[5–9] Using quantum-chemical calculations, we were able to catch the important difference, which shear-strain inflicts on the chemical properties of these two materials. By focusing on the molecular nature of the shear-strain-induced chemistry, we found that energetic barriers for DADNE decomposition decrease due to shear while those for TATB are not affected by this deformation. This implies that the shear strain may have an essential role in defining sensitivity to initiation in the materials.

The perfect structure of a DADNE molecular crystal consists of nearly planar molecules, which are hydrogen bonded into corrugated sheets, and is described by the $P21/n$ space group with four molecules (56 atoms) per unit cell[172,173] (Fig. 3.15A). The unit cell of TATB is triclinic with $P\bar{1}$ symmetry and contains two molecules (48 atoms)[206] (Fig. 3.15B). The weak bonding between layers is largely defined by van der Waals interactions.

Features of solid state behaviors of the energetic materials were simulated from the physical chemistry point of view. We demonstrated that the decomposition of condensed high explosives is fundamentally cooperative phenomena comprehensive understanding of which requires an integrated approach by combining state-of-the-art quantum-chemical solid-state calculations and high quality experiments. In particular, we focus here on exploring the interplay of early decomposition mechanisms in the perfect and imperfect crystals and establish that shear-strain-induced deformations are likely to control the initiation of detonation in DADNE and TATB. We also touch upon aspects of autocatalytic reactions and show that an analysis of the crystalline structure, defects and deformations, and modifications of materials properties induced by them is imperative for understanding the nature of hot spots and sensitivity to detonation issues in general.

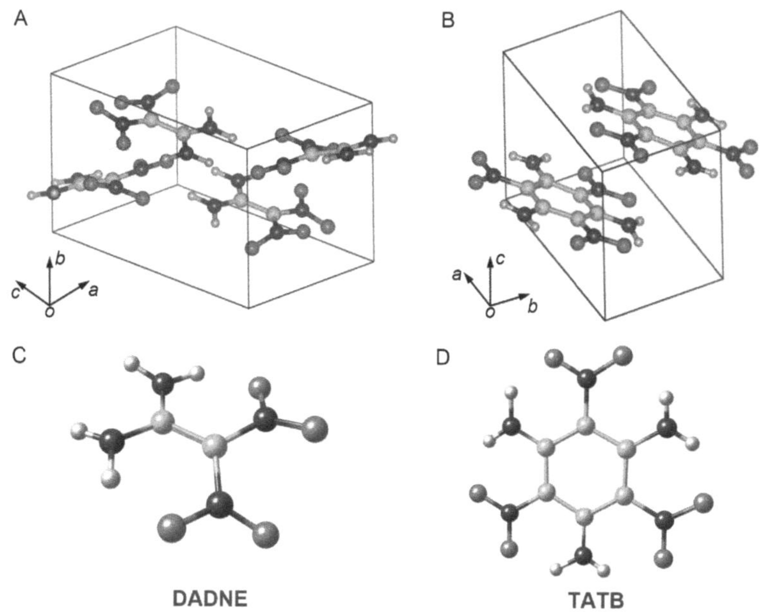

Figure 3.15 The ideal crystalline (A) and (B) and molecular (C) and (D) structures of DADNE and TATB. The C atoms are shown in gray, O in red, N in blue, and H in while. (For interpretation of the references to color in this figure legend, the reader is referred to the online version of this chapter.)

5.1.1 Modeling shear-strain deformation in DADNE and TATB

This work was largely motivated by observations of the anisotropic response of pentaerythritol tetranitrate (PETN) to shock loading.[87–89] Single PETN crystals shocked normal to the (100) and (101) planes were found to be insensitive while those shocked normal to the (110) and (001) planes were found to be relatively sensitive. A numerical analysis using molecular mechanics indicated that steric hindrance, or the interaction between molecules, was more severe for the (110) and (001) directions leading to an enhanced energy transfer and greater reactivity, as a result. This was verified by additional experiment[207] and semiempirical calculations implicating shear stress as contributing to the sensitivity when shocked normal to the (110) planes and thermal initiation when shocked normal to (100).

In order to simulate shear-strain deformations in the DADNE and TATB crystals, we constructed supercells formed by six molecular-layer slabs separated by a 10-Å layer of vacuum. The shear strain was introduced by shifting the three upper layers of the slab relative to the three lower layers in the directions parallel to the crystalline slip plane. An arbitrary shift vector in the *ac*- plane in DADNE is: $\Delta = \gamma_a a + \gamma_c c$ (Fig. 3.16A); for TATB, the similar formula is written for the *ab*-plane: $\Delta = \gamma_a a + \gamma_b b$ (Fig. 3.16B). The activation barriers for the decomposition and structural rearrangements of molecules placed at the interfacial layers (where the shear strain reaches its maximum) were calculated using the nudged elastic band approach for each Δ (Fig. 3.16C).

In simulating the three main decomposition mechanisms, which are well supported by earlier work[57,143,208,209]: the C–NO_2 homolysis, nitro-nitrite isomerization, and proton transfer (Fig. 3.16C), it was found that all three mechanisms are likely to take place in TATB with CONO isomerization being only slightly more favorable (the activation barrier of 55 kcal/mol) than HONO isomerization (62 kcal/mol). The C–NO_2 homolysis is activated by the visibly higher energy barrier of 100 kcal/mol. The experimental activation energy of the global decomposition of solid TATB ranges from 59 to 63 kcal/mol[142,143] and, generally, the C–NO_2 dissociation ranges from 61 to 70 kcal/mol for many nitro-componds.[142,143] Interestingly, all three mechanisms are almost insensitive to the applied shear-strain deformation. Unlike TATB, only two mechanisms likely occur in DADNE, the C–NO_2 homolysis and nitro-nitrite isomerization, while the proton transfer is found to be unfavorable.

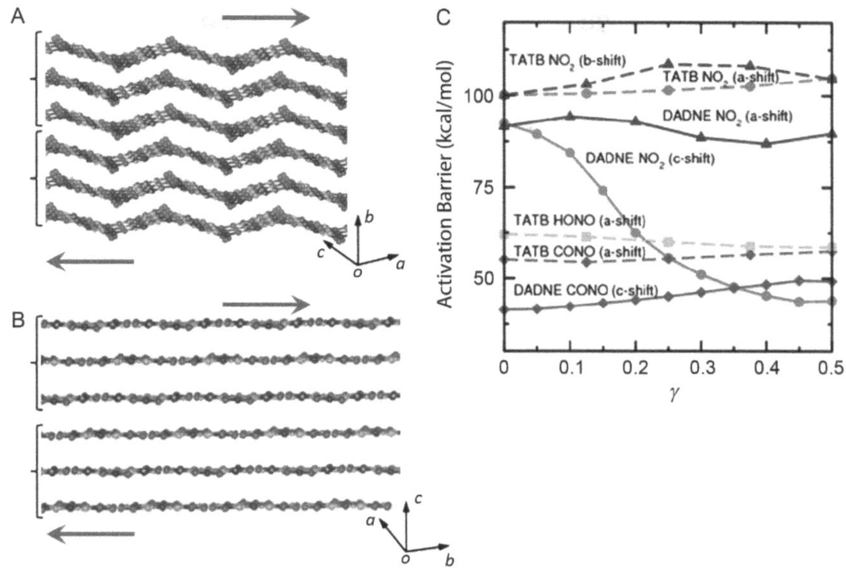

Figure 3.16 Relaxed 6-layer (3 + 3) slab of the DADNE (A) and TATB (B) molecular crystals. Red arrows schematically show the directions of the applied shear-strain deformation in which the three upper layers are shifted relatively to the three lower layers. (C) Activation barriers of C–NO_2 homolysis in DADNE (solid lines) and TATB (dashed lines) as a function of shift vectors γ for various shift directions (*c* and *a* directed shifts are shown for DADNE and *a* and *b* for TATB). (For interpretation of the references to color in this figure legend, the reader is referred to the online version of this chapter.)

This is in agreement with the earlier study,[57] where the authors explained that the proton transfer took place in TATB but not in DADNE due to the differences in the electrostatic potential profile and molecule–lattice interactions in the materials. Figure 3.16C demonstrates that the CONO pathway is activated in DADNE by 42 kcal/mol of energy while the C–NO_2 homolysis requires about 94 kcal/mol of energy. It also shows that the C—NO_2 bond break is strongly dependent upon a shear strain in the *c* direction. The C–NO_2 dissociation barrier drops from 94 for molecules in an ideal material ($\gamma_c = 0$ and 1) to 44 kcal/mol for molecules exposed to the maximal shear stress (at $\gamma_c = 0.45$, Fig. 3.16C). The barrier is much less sensitive to the *a* shift (Fig. 3.16C). At the same time, the CONO formation, which is essentially an intramolecular process in DADNE, only slightly increases with the interface deformation. The experimental overall decomposition of DADNE is characterized by the estimated activation energy of 58 kcal/mol.[174]

An increase of the shear-strain level in DADNE, leading to a reduction of the C–NO_2 homolysis activation barrier, positions the C–NO_2 pathway above CONO pathway. This, in fact, means that certain spots in DADNE crystals, which contain a high concentration of dislocations, grain boundaries, stacking faults, or other deformed regions induced by the shear strain, cause the dominating decomposition chemistry to switch from the slow CONO isomerization to the fast C—NO_2 bond rupture. This also suggests that both reactions contribute to the global decomposition of the material. Based on energetic considerations, molecules placed in the ideal part of the crystal favor the CONO isomerization dissociation pathway while molecules placed on or near defects/deformations prefer the C–NO_2 pathway. This example illustrates how the mechanical stress in DADNE controls the decomposition chemistry and changes the initiation process by facilitating specific reaction channels and terminating others. More importantly, TATB does not exhibit this kind of behavior, which is probably among reasons of its remarkable stability. Hence, a comparative analysis of the structure and mechanochemistry can provide useful insights on the sensitivity of materials to the initiation of detonation.

5.2. Autocatalytic aspects of early decomposition in ideal and deformed DADNE and TATB

Autocatalytic mechanisms, that is, the processes in which intermediate products of the chemical decomposition become catalysts themselves and accelerate further decomposition of the material, are among the least understood and the least studied aspects of the initiation.[142] Although it looks obvious that autocatalysis may significantly contribute to the initiation of detonation, investigations of these phenomena represent extraordinary challenges both experimentally and theoretically, and studies of autocatalytic aspects of initiation are largely lacking.[142]

To detect autocatalytic trends (or the lack of those), we probed several consequent dissociation acts, which result in the cumulative appearance of NO_2 groups in the interstitial space between molecular layers in the crystalline structures of TATB and DADNE (Table 3.5). We modeled how interactions among the products of the initial bond dissociations and the molecular crystalline framework advance the decomposition process. Only a few secondary reactions originating from the NO_2 and NO fragments confined in the crystalline lattice were considered.

We found that the presence of interstitial NO_2 and NO groups affects DADNE and TATB in different ways. We observed that the NO_2 group,

Table 3.5 Reaction energies of interstitial NO_2 and NO molecules with each other and with the crystalline network (the index "s" means "solid," that is, belonging to the network) at shear-strain interfaces in DADNE and TATB crystals

Reaction	Configuration of product(s)	Reaction energy (kcal/mol)	
		DADNE	TATB
(1) $2NO_2 \rightarrow N_2O_4(I)$	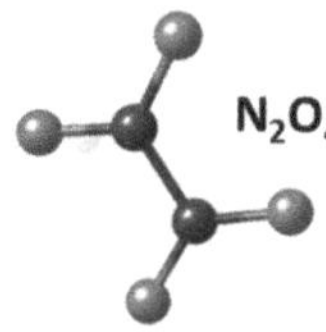 N_2O_4 (I)	−13.8	−6.9
(2) $2NO_2 \rightarrow N_2O_4(II)$	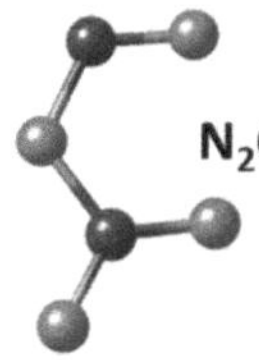 N_2O_4 (II)	−18.4	−9.3
(3) $N_2O_4(II) \rightarrow NO + NO_3$	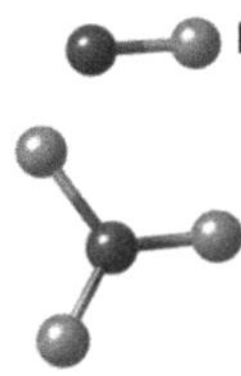 NO, NO_3	3.5	−4.6
(4) $N_2O_4(II) + O(s) \rightarrow N_2O_5$	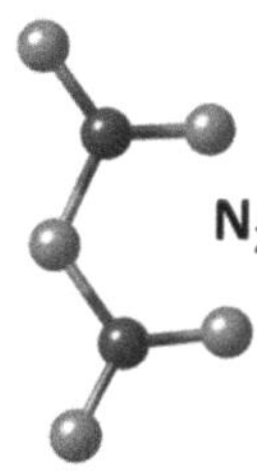 N_2O_5	2.3	3.5
(5) $NO_2 + NO \rightarrow N_2O_3$	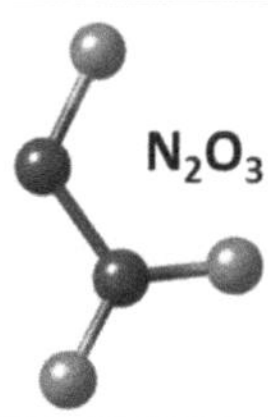 N_2O_3	−4.6	4.5
(6) $2NO \rightarrow N_2O_2$	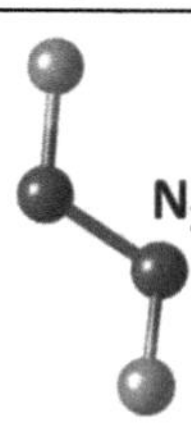 N_2O_2	4.5	2.3

trapped in the interstitial position, tends to increase the shear strain between the layers. For example, at the shear $\gamma_c = 0.1$, the crystalline DADNE slab (Fig. 3.16A) with one split-off interstitial NO_2 group relaxes to a configuration similar to that with $\gamma_c \sim 0.2$. This increased shear strain lowers the decomposition barrier for breaking the next interfacial C—NO_2 bonds, that is, facilitates further NO_2 decomposition at the interface. In other words, the shear strain reduces the activation barrier of the C–NO_2 rupture, which accelerates the decomposition reactions. An appearance of an NO_2 molecule in the confinement of the interface, in turn, increases the shear level. As the process recurs, its effect is reinforced and hence, autocatalysis.

In a perfect TATB crystal, parallel, flat, graphite-like layers can slide nearly free, that is, this structure does not produce or accumulate any shear strain (Fig. 3.16B). An interstitial NO_2 group, trapped between two TATB layers, tends to force the nearest molecules away from each other, to be exact, it disturbs the planar structure of the interfacial layers by tilting the TATB molecules out of the plane. Such a molecular orientation disorder at the interface increases while the concentration of interstitial NO_2 molecules grows. Undoubtedly, this disorder will affect chemistry. However, the number of detached NO_2 groups, confined between molecular layers in TATB is greater than in DADNE. A simple estimate shows that the critical number of free interstitial NO_2 groups is about four times higher in TATB than in DADNE. The high concentration of confined NO_2 groups in the TATB structure will cause the molecules to be significantly disoriented and the structure of the layer will be far from the initial planar structure.

Further, interactions between the interstitial NO_2 groups yield primary products, isomers of N_2O_4, and secondary products, N_2O_x and NO_y. The corresponding energies of reacting interstitial NO_2 molecules are slightly different in the two materials (Table 3.5). The formation of nitrogen oxides with larger molecules (N_2O_4) is more exothermic in the DADNE structure. The secondary formation of oxides (N_2O_5 or NO_3 and NO) is more exothermic in TATB. Although the transformation reactions of nitrogen oxides are only slightly exothermic, they demonstrate that the exothermic chemistry in DADNE and TATB crystals may be triggered early on by the products of very initial endothermic reactions due to the autocatalytic effect. In addition, these results help provide a consistent interpretation to experiment. In particular, although gaseous nitrogen oxide molecules (NO_x and N_2O_y) are always observed in decomposition experiments of nitro compounds, NO_2 is found in much lower concentrations than it would be expected due to splitting off nitro groups as a primary decomposition reaction. Conclusions obtained here may explain this seeming contradiction by the autocatalytic character of the condensed phase decomposition.

5.3. Concept of hot spots: Revisited

Among the least understood are low velocity impact ignition mechanisms in which an energetic solid subjected to a mild dynamic mechanical loading; they likely to occur in a variety of accident scenarios. As the key to initiation of chemistry is proposed to be the formation of localized hot spots,[210] factors, which influence the dissipation and localization of the mechanical energy and its transfer into the thermal energy are strong functions of material properties. It is widely accepted, for example, that sensitivity to ignition correlates with the thermal decomposition energy,[142] (also see above). Relative roles of void collapse, friction, shear, dislocation pile-ups, and other postulated hot spot formation mechanisms need to be better determined by both experiment and theory.

An examination of a crystalline arrangement of TATB in comparison with DADNE demonstrates that no specific in-plane slips are allowed in the *ab*-plane of TATB. Such rigidity is determined by the high symmetry of the TATB molecule, which results in strong inter- and intramolecular hydrogen bonds in the *ab*-plane. The shear between these layers, however, is not energetically costly because the interlayer overlapping of electron functions is minimal. This slip motion does not cause any effects on dissociation energies of TATB molecules (and barely affects the electronic structure).

DADNE, on the other hand, lacks molecular symmetry and is built out of polar molecules linked with each other by cohesion of NO_2 and NH_2 groups located at adjacent molecules. This also creates strong hydrogen bonding in corrugated chains of molecules along the *c*-axis but not in all directions (this is why the shear strain along the *a*-direction does not produce much of the structural damage). Besides, unlike TATB, several possible motions of DADNE molecules are allowed by symmetry. This results in an appearance of polar surfaces in the structure and allows a richer variety of local structural defects to occur, for example, different conformers, or a "flipped" molecule. A major consequence of this defect-induced disorder is attributed to a decomposition barrier decrease and, in turn, to a higher sensitivity of DADNE to initiation as compared to TATB.

We will analyze now an assumption that the formation of hot spots is totally attributed to the shear strain due to the mechanical loading of the material. Several important observations based on our first principles calculations would immediately follow from this proposition. We found that the shear strain results in the reduction of the decomposition barrier[5–8] and the reduction of the band gap in DADNE.[9] Those both are dynamic effects, which are crucial within a short time after the shear strain was generated by, for example, an impact or a shock wave.[9] We conclude that the reduced NO_2 detachment energy barrier facilitates the decomposition of the interfacial DADNE molecules before the system relaxes. In addition, the band gap reduction of DADNE can trigger sensitivity to optical and/or thermal stimuli via electronic excitations,[17] which can lead to either direct bond breaking[108] or high population of vibronically hot states.[211] The mechanical relaxation of the system will cause local heat of the material in the vicinity of the deformation; this can be associated with the temperature rise in the hot spots.

After the relaxation occurred, the decomposition barrier regains its original value as in an ideal crystal, and the decomposition rate may decrease. If the chemical reaction did not spread around sufficiently far by this moment, the initiation will die out. If the chemical reaction propagated to involve other hot spots and initiated chemistry in them, the process most likely will be transferred into a chain reaction due to multiplication of excitations eventually leading to an explosion.

Comparison between TATB and DADNE illustrates that hot spots are likely more complex than just regions with high density of shear-strain-induced defects or deformations. However, if shear strain is taken into account alone, DADNE is predicted to be more sensitive to mechanical loading than TATB, contrary to earlier predictions.

This study only gives us a foretaste of the possible molecular mechanisms of shear-strain sensitivity of molecular energetic materials in the course of detonation initiation. In our limited (and fairly simple minded) consideration of a single ground state decomposition mechanism, the $C—NO_2$ bond dissociation in DADNE and TATB, we aimed at illustrating how the interplay among mechanical and chemical processes in energetic materials may be correlated with their sensitivity and relevant details of the initiation processes. We showed that initial products of the chemical reactions interact with the host molecular crystal and may enhance the shear strain (DADNE) or introduce an orientation disorder in interfacial crystalline layers (TATB). The scenario(s) of mechanochemical autocatalysis are material dependent and likely to be manifested in significant variations in behavior of energetic crystals (e.g., an appearance of melting in experiments). While the C–NO_2 decomposition reactions are endothermic, the followed up C–NO_2 dissociations in the interlayer confinement require progressively less energy. Exothermic reactions, which are triggered by interactions of detached interstitial NO_2 groups, start early on in both DADNE and TATB, however, one needs to pump up a greater energy into TATB (and accumulate higher concentration of interstitial NO_2 groups) before the exothermic chemistry is ignited. A further detailed investigation is needed to uncover all the complexity of autocatalytic aspects of the sensitivity to initiation of energetic materials.

Perhaps, the most important conclusion that follows from this illustration is that, while the concept of hot spots still represents an appealing idea, the view on the notion should be significantly broaden to include other initiation mechanisms beyond simple thermal initiation.

5.4. Effect of defects: Vacancies, voids, and surfaces

5.4.1 Decomposition of the (101) and (110) crystal surfaces in PETN

In turning to the decomposition of solid-state PETN, we explore only two low energy feasible mechanisms, the O–NO_2 homolysis and HONO elimination pathways.[109] With the understanding that the condensed phase processes may develop differently due to interactions with the crystalline field, we realize that the dissociation barriers be affected in solids by defects,[13] shear-strain,[5–9] surfaces[14,15], charged or excited states,[123] and other imperfections.[16,17] However, we note that the most significant change (e.g., the reduction of the activation barrier or the reaction energy[123]) will impact the two low energy pathways identified above, and hence it is reasonable to assume that they would remain the major reactions of interest in PETN crystals. In addition, the structures of the transition states point to large activation volumes of Eq. (3.3)–(3.7) (see Section 3.2),[109] which serve as a strong signal towards the corresponding activation barriers being even higher in the *ideal* bulk crystal than in the gas phase. Subsequently, we propose that the choice of the O—NO_2 bond cleavage and the HONO-isomerization mechanisms for solid-state modeling is well justified.

Before starting periodic calculations of chemical reactions, which are based here solely on PBE functional, as implemented in the VASP code, we performed test simulations of the molecular reactions, to validate our conclusions and to establish a reliable reference point. Table 3.6 shows that PBE functional by VASP finds the decomposition barriers of 40.9 kcal/mol for the NO_2 loss in

Table 3.6 The calculated activation barriers E_a (kcal/mol), zero-point energy corrected barriers E_{ZPE} (kcal/mol), and preexponential factors log A (s^{-1}) of the PETN decomposition reactions in an isolated molecule, the ideal bulk crystal, and on the (101) and (110) PETN surfaces

	Decomposition reaction					
	NO_2 loss			HONO elimination		
PETN	E_a[a]	E_{ZPE}	log A	E_a[a]	E_{ZPE}	log A
Isolated PETN molecule[b]	40.9	–	–	34.2 [−10.8][c]	–	–
Ideal bulk	46.2	38.4	16.3	40.7 [−5.0]	36.6	11.1
(101) surface	40.0	35.3	17.8	34.6 [−10.4]	29.5	14.2
(110) surface	41.3	35.7	17.9	35.8 [−12.0]	31.2	13.9

[a]Obtained from VASP code with PBE functional.
[b]The energies for the molecular calculations are given for comparison.
[c]The corresponding reaction energy is given in brackets.

a PETN molecule (to compare with 41.8 kcal/mol obtained with PBE by GAUSSIAN, Table 3.2) and 34.2 kcal/mol for the HONO elimination (to compare with 33.6 kcal/mol obtained with PBE by GAUSSIAN, Table 3.2). The observed accuracy is within 1 kcal/mol; hence the corresponding energies ofboth reactions are in agreement and suggest that conclusions obtained from crystal modeling would be sufficiently reliable.

The activation barriers of the O—NO_2 bond fission and the HONO elimination calculated for bulk PETN are consistent with the gas-phase calculations. The homolytic scission of the O—NO_2 bond requires 38.4 kcal/mol (Table 3.6), a little bit higher than that of the isolated PETN molecule (Table 3.2). Taking into account the PBE tendency to underestimate the HONO barrier, 36.6 kcal/mol appears consistent and reasonable. As expected, the HONO mechanism develops significantly slower than the O–NO_2 homolysis. A comparison of the obtained barriers (Table 3.6) and the reaction rates (Fig. 3.17) finds the NO_2 loss to be the dominating decomposition mechanism in the ideal bulk, similarly to the gas-phase process. The trend of the ideal crystal reactions to exhibit higher energies than the equivalent pathways in the isolated molecules is reflective of the previously observed inclination in other relevant energetic materials, RDX, HMX, DADNE, and TATB.

In was experimentally demonstrated that PETN crystals primarily expose the (101) and (110) facets[212] therefore those two surfaces were chosen for our modeling (Fig. 3.18). Table 3.6 shows that the O—NO_2 bond cleavage on both (101) and (110) surfaces requires significantly lower energies than

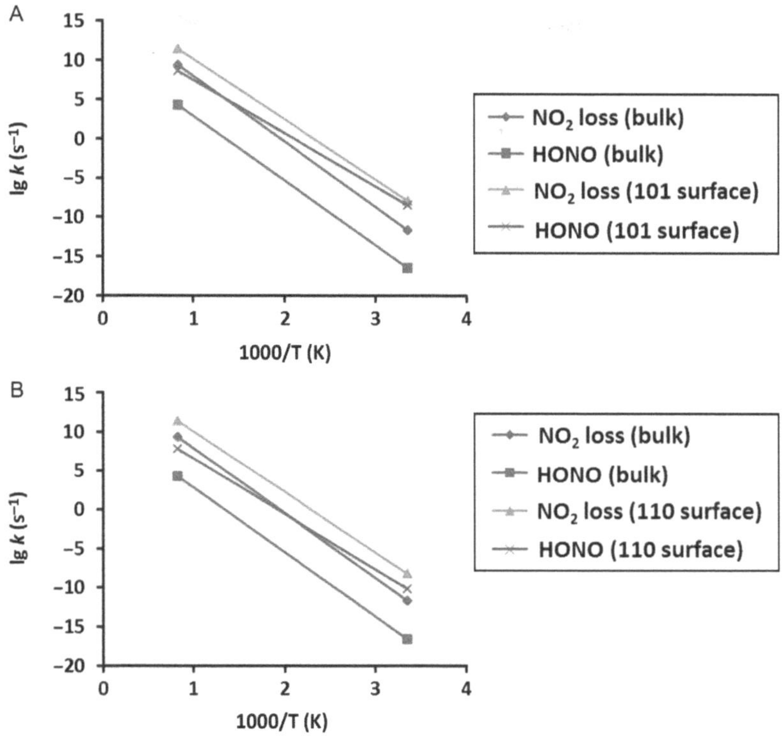

Figure 3.17 Reaction rates for the decomposition channels on PETN (A) (101) surface compared to the gas phase and ideal bulk, and (B) (110) surface compared to the gas phase and ideal bulk. (For the color version of this figure, the reader is referred to the online version of this chapter.)

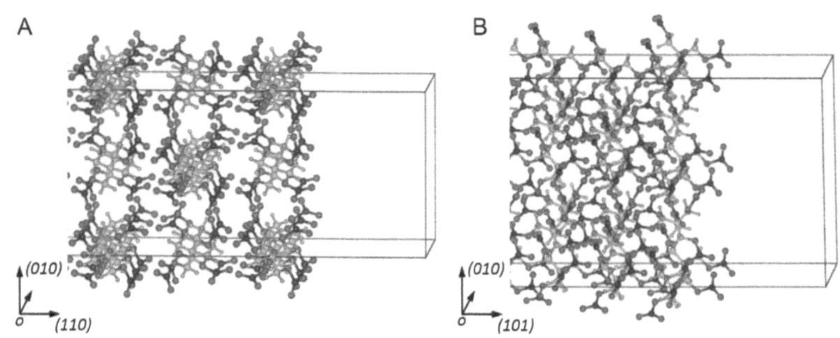

Figure 3.18 Structures of a model supercell fragment of (A) the (110) surface and (B) the (101) surfaces in PETN. (For the color version of this figure, the reader is referred to the online version of this chapter.)

those in the ideal bulk and slightly lower than those in the gas phase (Table 3.2). Similarly, the activation barriers of the HONO elimination on both facets also need lower energies than those in bulk. In addition, the HONO pathway remains exothermic while taking place on the (101) and (110) facets (Table 3.6). As it has been already observed for reactions in the gaseous phase and ideal bulk, the HONO elimination appears to require a lower energy than the NO_2 loss due to the underestimate of PBE. Nevertheless, the homolytic O—NO_2 bond cleavage remains the fastest decomposition channel due to high preexponential factors (Tables 3.2 and 3.6) and the reaction rates plotted in Fig. 3.17.

Results of our modeling of chemical decomposition reactions in periodic supercells confirm that the O—NO_2 bond homolysis is the predominant pathway of the thermal decomposition of PETN in both gaseous and solid-state phases. The obtained O—NO_2 BDE for PETN molecules placed in the ideal bulk crystals and on the free surfaces (35.3–38.4 kcal/mol, Table 3.2) fall in the measured range (32.6–47.3 kcal/mol)[97–100,105] and provide a consistent interpretation to the published experimental data. The activation barriers and reaction rates calculated for other decomposition channels demonstrate that only the HONO elimination can be considered as a coexisting complementary reaction. A pronounced exothermicity of the HONO-isomerization channel serves to accelerate the PETN decomposition in the condensed phase.

We did not observe any significant variations in response of PETN to degradation of the two surfaces (101) and (110), which were suggested to exhibit significantly different directional sensitivity to shock waves progressing across the crystals,[85,87–89] with the (110) surface being among the most sensitive facets and the (101)—among the least sensitive ones. We find this fact not really astonishing as our study was focused only on the processes triggered by the thermal perturbation rather than a mechanical shock or impact. Appropriate deformations can be simulated for PETN, for example, following the strategy applied earlier to DADNE and TATB.[5–9]

Conclusions obtained in this research make an interesting contribution towards understanding the experiments on laser initiation of PETN mixtures with light scattering additives.[108] The authors[108] proposed that PETN decomposes through a resonance photoinitiation, which proceeds via a two-stage process that is comprised of a laser-induced optical excitation of the material with an activation energy of 1.17 eV (1060 nm) followed by the thermal decomposition of molecules from their excited states with a small energy barrier of 0.4 eV. This report posed an intriguing question: how can the initiating pulse of such a low energy, 1.17 eV, trigger the photodecomposition of the samples while the optical band gap of wide-gap dielectric PETN is known to exceed 4 eV, measured experimentally[86,107] and estimated theoretically.[42]

A popular opinion shared by many researchers is that the mechanism of laser initiation is of the thermal nature, and the laser light is used as a source of heat for absorbing sites in an energetic material, which act as hot spots to originate the thermal explosive decomposition reaction.[213–216] In contrast to these ideas, an excitonic mechanism of initiation of chemistry, developed theoretically,[17,155] and laser initiation of secondary explosives, RDX[162] and PETN,[107,108,163,164] explored experimentally, suggest that the decomposition initiation may have a nonthermal photochemical nature. Our modeling of PETN decomposition here deals with the ground state chemistry only, and hence the question of a detailed mechanism of photoinitiation in PETN remains open as we did not tackle excited state processes. Nevertheless, our findings allow us to rule out a possibility of thermal initiation of PETN observed in those experiments because the theoretically obtained activation barriers (~35–36 kcal/mol) and analyzed reaction mechanisms are inconsistent with the experimental energy of laser initiation (1.17 eV=27.0 kcal/mol). Thus, the drawn conclusions lend some additional support to the notion that an excited state of a PETN molecule, created by the absorption of a photon, rather than by a thermal energy, was the primary act of initiation in laser irradiation experiments.[108,164]

5.5. Polar surfaces: Origin of surface conductivity

HMX crystallizes in four known phases labeled α, β, γ, and δ, with different sensitivities to initiate explosive decomposition reactions.[217] Among four HMX polymorphs, the β-phase, the form in which HMX is normally produced, has the highest density and is stable at room temperature.[218] When heated to temperatures above 435 K, the monoclinic β-phase converts to the hexagonal δ-HMX,[219] the most reactive phase.[217,220] The transformation involves major changes of both the HMX crystal lattice and the molecular ring conformation from β (chair) to δ (boat)[217,219,220] (Fig. 3.19). With many experimental reports declaring a high sensitivity of the δ-phase, atomistic theoretical investigations are limited,[68,74,221] and the detailed reasons behind the HMX behavior remain puzzling. The situation is further complicated by the fact that the decomposition of the δ- phase practically coincides with the β to δ phase transition and with the possible decomposition of

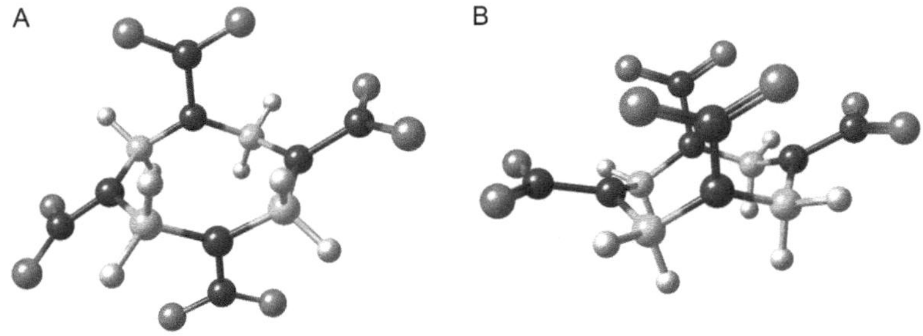

Figure 3.19 Molecular structures of (A) chair β-HMX and (B) boat δ-HMX conformations. (For the color version of this figure, the reader is referred to the online version of this chapter.)

the β-phase. It was speculated that the observed sensitivity increase of the δ-phase[219,222] is due to a significant lattice expansion and the associated density decrease from 1.90 for β to 1.78 g/cm^3 for δ, accompanying the beta-to-delta phase transition,[223] which results in an appearance of cracks, large amount of hot spots in the material, and the fast growth of reaction during shock compression.[222] Naturally, the available experimental techniques measure only global kinetics[222,224] and do not distinguish between the individual processes that have very close activation barriers.

Hope is that once the atomistic details of the δ-HMX phase decomposition mechanisms and kinetics are established and corroborated, our understanding of sensitivity will be substantially advanced. Correlations between polymorphism and chemical decomposition processes in HMX crystalline phases, when identified and understood, will enable the development of novel materials and improvements of existing energetic systems with controllable, tunable sensitivity and performance.

In this research aimed at gaining a better understanding of materials factors that control sensitivity in HMX, we explored earliest decomposition reactions in the delta phase HMX and propose a mechanistic explanation of its increased sensitivity. The N–NO_2 homolysis and concerted HONO elimination reactions are simulated for an isolated δ-HMX molecule, the molecule placed in the ideal crystal, and on a free surface. A comparison with the previously obtained results for β-HMX[14,34–36] suggests reasons that make δ-HMX so much more sensitive.

The calculated structure of the gas- and solid-phase molecule agrees well with earlier experimental and theoretical studies.[225] In simulating an ideal δ-HMX crystal with the $P6_1$ symmetry,[219] the theoretical lattice constants of the hexagonal unit cell consisting of 168 atoms are calculated as $a=8.180$ and $c=33.152$ Å, which compare to experimental[219] 7.711 and 32.553 Å within a typical accuracy of GGA-PBE functional. In calculations, the $2\times2\times1$ k-point mesh was used for the unit cell of ideal δ-HMX, and the reduced number of k-points was used for larger supercells to keep the k-point density consistent.

Gas-phase decomposition reactions were modeled by placing a single δ-HMX molecule in $15\times15\times15$ Å^3 periodic box and relaxing all atomic coordinates. The calculations were performed at G k-point only. The results of the gas-phase decomposition were additionally validated with test calculations carried out by molecular code GAUSSIAN with PBE and hybrid PBE0 functionals and the 6–31+G(2df, p) basis set. The supercell for modeling bulk reactions ($a=16.36$, $b=14.168$, $c=33.152$ Å, $\alpha=\beta=\gamma=90°$) was constructed by choosing modified lattice vectors: $\vec{a}=2\vec{a}_0+\vec{b}_0$, $\vec{b}=-\vec{a}_0+\vec{b}_0$, and $\vec{c}=\vec{c}_0$. Modeling of the solid-state decomposition reactions is especially complex due to molecular interactions[220] with the crystalline field, a manifold of local minima on the potential energy surface,[226] and an uncertainty of where to place the reaction products.

Surface reactions were simulated by using a slab model consisting of the two (100) four-molecule-thick surface layers separated by 10 Å of vacuum (Fig. 3.20) to compensate the dipole moment and eliminate a possible effect of the artificial dipole-induced electromagnetic field, leading to a so-called polarization instability. The large size of the supercells ensured that spurious

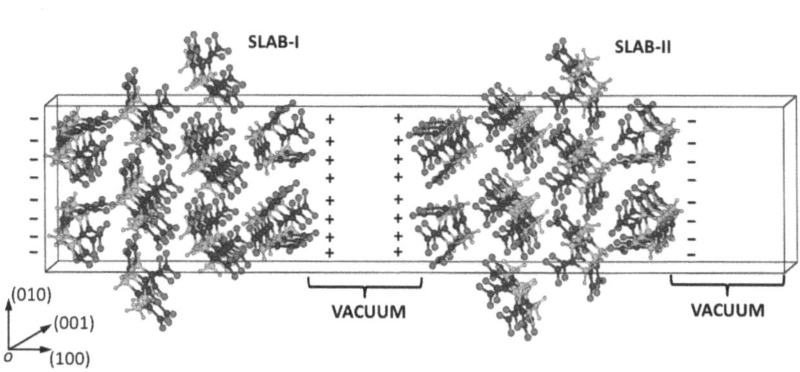

Figure 3.20 A model slab supercell of δ-HMX is constructed in such a way that the overall dipole moment is zero. (For the color version of this figure, the reader is referred to the online version of this chapter.)

Table 3.7 Activation barriers (kcal/mol) of R–NO_2 homolysis reactions of isolated molecules, ion radicals, ideal bulk crystals, and surfaces in β- and δ-HMX

HMX	β-HMX	δ-HMX
Isolated molecule	42.8[a]/44.8[b]	44.8/45.1[b]
Ion radicals [charge]	39.7[c] [+1]	39.8[c] [+1]
	16.0[c] [−1]	15.0[c] [−1]
Ideal bulk	47.9[a]	50.8[a]
Surface [charge]	40.1 [0][a]	30.5[a] [+]
		5.0[a] [−]

[a]Obtained from VASP with PBE functional.
[b]Obtained from GAUSSIAN with PBE and is given here for comparison.
[c]Obtained from GAUSSIAN with PBE0.

interactions between reactants and reaction products in neighbor cells are minimal. All atomic positions were relaxed using the quasi-Newton and conjugate gradient methods.

Out of all obtained activation barriers, only those relevant to the N–NO_2 homoysis are collected in Table 3.7. They indicate that the dissociation energies of nitro groups in the gas-phase δ-HMX molecule (44.8 kcal/mol) and barely higher than the energy of NO_2 loss in β-HMX (42.8 kcal/mol),[14,34–36] and they are in agreement with earlier calculations (~45.5 kcal/mole).[68] The activation energies in bulk of both phases are visibly higher than those in the corresponding molecules (47.9 and 50.8 kcal/mol). The ion radicals also exhibit a clear and similar trend. In particular, the ionized HMX molecules in both β- and δ-configuration show a slight reduction of the activation barrier versus the gas-phase neutral molecules (39.7 vs. 44.8 kcal/mol for the β-molecule and 39.8 vs. 45.1 kcal/mol for the δ-molecule). The negatively charged radicals require significantly lower energies (16.0 and 15.0 kcal/mol), which also happened to be quite similar to each other. This tendency resembles the concussions obtained for DADNE (see Section 4.1) and implies that an electron rich environment would facilitate decomposition in both HMX phases. Comparing the obtained energetics, we note that the decomposition scenarios in the molecules, ions, and even ideal crystals of both β- and δ-HMX are very similar. Curiously, none of the obtained energies or trends can explain differences in sensitivity of the HMX phases.

A disparity between the densely packed beta and the lower density δ-HMX structures is manifested in the presence of channels piercing the δ-HMX crystal. Connecting the large activation volume required for the $N–NO_2$ reaction to occur with the generous empty space available in the δ-HMX crystal raises an argument in favor of the reduced barriers and/or accelerated kinetics for reactions proceeding along those passages.

Our simulations, however, fall short in supporting this possibility. Placing NO_2 moiety in the middle of the crystal passage results in neither the lower activation barrier nor faster reaction rate. Therefore, the lower density of the delta phase and even the presence of the spacious empty channels in the crystal lattice do not produce any favorable conditions for the accelerated chemistry or higher sensitivity of δ-HMX. Indeed, neither the change of the molecular conformation nor the crystalline arrangement in the defect-free materials offers an explanation of a major sensitivity change from one HMX solid phase to another.

Further, we will demonstrate that the surface-induced effect on the decomposition kinetics of δ-HMX has dramatic consequences. We selected the (100) surface out of many possible HMX surfaces because it provides an opportunity to investigate both equatorial and axial $N–NO_2$ homolysis reactions of δ-HMX. Figure 3.20 shows the nonidentical configurations of the HMX molecule relative to the free surface. Figure 3.21 compares the reaction rates.

The activation barrier of the NO_2 loss reaction in β-phase (40.1 kcal/mol) is lower than in the bulk crystal and in the gas phase (Table 3.7). The difference obtained in the δ-phase is appreciably larger, with 30.5 kcal/mol on the positive surface and only 5 kcal/mol on the

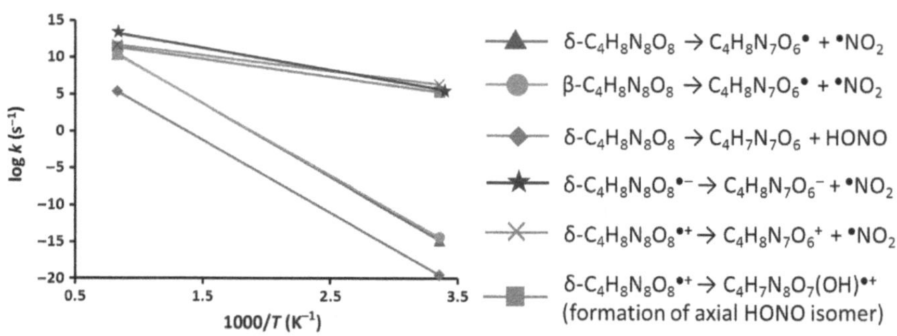

Figure 3.21 Reaction rates for the decomposition channels of HMX. (For the color version of this figure, the reader is referred to the online version of this chapter.)

negative surface (vs. 44.8 in the gas phase). This is a direct result of molecular dipole–surface interactions in δ-HMX, which causes charge separation and an electric conductivity on the surface. Nothing like that happens in centrosymmetric nonpolar β-phase.

Summing up, the calculated energetic and kinetic parameters analyzed and compared for β- and δ-HMX polymorphs allowed us to better understand much higher sensitivity to initiation of detonation of the solid δ-phase versus β-phase. We found that the change of the HMX molecular conformation from β (chair, C_i symmetry) to δ (boat, C_{2v} symmetry) configuration does not cause significant changes in the decomposition pathways or their kinetics. The dissociation barrier for NO_2 splitting off in the gas-phase δ-HMX molecule is slightly higher than in β-HMX and the activation barrier for the HONO pathway is somewhat lower (by ~6 kcal/mol) than in the β-conformer. The similar situation is observed for the ideal (defect-free) crystals. The change of the crystalline arrangement from the β-HMX phase ($P2_1$ space group) to δ-HMX phase ($P6_1$ space group) does not induce significant differences in the decomposition chemistry. The derived activation barriers of the HONO elimination and the N–NO_2 split in δ-HMX are comparable to the corresponding reactions in β-HMX. The observed difference is small (~1–2 kcal/mol) and does not play a major in offering any justification of sensitivity.

Evidently, the surface-induced effect happened to be much more pronounced in δ-HMX than in β-HMX. While the activation barrier for the surface decomposition drops ~4–8 kcal/mol in β-HMX (compare to bulk), in δ-HMX, it decreases ~30–40 kcal/mol.

We suggest that the reduction of the activation barrier is due to the polar nature of the (100) surface. The dependence of chemical reactivity on the direction of a polar axis in polar organic crystals was discussed, for example, in Ref. 227, it was argued that initiation of the reaction may be favored at a single side of the crystal and that the polar axis of the crystal directs the reaction. Speculating along those lines, we propose that the interaction between the dipole moment of the δ-HMX molecule and the polar (100) surface that induces an electronic density redistribution in the material is responsible for such a striking difference in the decomposition chemistry of δ-HMX. The most important consequence is that while both phases would exhibit much faster explosive decomposition reactions in the presence of electrons, the polar nature of δ-HMX serves to create such conditional naturally. The polarization-induced charge transfer causes the conductive (metallic) surface to trigger an early decomposition with a negligibly small activation energy barrier. The dissociation process quickly becomes autocatalytic, and that is what determines such a high sensitivity of the delta phase.

6. CONCLUSION AND FUTURE RESEARCH DIRECTIONS

Our research deals with modeling of inter- and intramolecular interactions in energetic materials and shows that quantum chemistry provides a powerful tool significantly, rigorously, and systematically improve our understanding of initiation processes at the micro-scale. Much more needs to be done. Our current results reveal that defects, deformations, and electronically excited or charged states play an imperative role in triggering chemical decomposition in energetics. The obtained conclusions prompt to revise the concept of hot spots to broaden it and include the most recent advances in the field. This modeling demonstrated that there are variety of ways to initiate the reaction, which means that we can manipulate chemical composition of materials and their morphology to tune sensitivity to initiation in many different combinations.

ACKNOWLEDGMENTS

I would like to thank my colleagues, students, and postdocs who contributed to this research while working in my group or collaborating with us on our projects. A special thank you goes to Craig Tarver for inspiration and many discussions that led to intriguing projects many of which are described in this manuscript. I am grateful to Roman Tsyshevsky for his help with preparation of the manuscript. This research is supported in part by ONR (Grant N00014-12-1-0529), DOE, and NSF. To perform the most recent calculations, we used NSF XSEDE resources (Grant DMR-100054) and DOE NERSC resources (Contract DE-AC02-05CH11231). I am also grateful to the Office of the Director of NSF for many years of continuous support under the IRD program. Any appearance of findings, conclusions, or recommendations expressed in this material are those of the author and do not necessarily reflect the views of NSF.

REFERENCES

1. Fried, L. E.; Manaa, M. R.; Pagoria, P. F.; Simpson, R. L. Design and Synthesis of Energetic Materials. *Annu. Rev. Mater. Res.* **2001**, *31*, 291–321.
2. Kotomin, E. A.; Merkle, R.; Mastrikov, Yu. A.; Kuklja, M. M.; Maier, J. Energy Conversion—Solid Oxide Fuel Cells: First-Principles Modeling of Elementary Processes. In *Computational Approaches to Energy Materials*; Richard, C.; Catlow, A.; Sokol, A. A.; Walsh, A. Eds.; *Computational Approaches to Energy Materials*; John Wiley & Sons: Chichester, UK, 2013; pp 149–186.
3. Kuklja, M. M.; Kotomin, E. A.; Merkle, R.; Mastrikov, Yu. A.; Maier, J. Combined Theoretical and Experimental Analysis of Processes Determining Cathode Performance in Solid Oxide Fuel Cells, Invited Perspective. *Phys. Chem. Chem. Phys.* **2013**, *15*, 5443–5471.

4. Carter, J. A.; Wang, Z.; Dlott, D. D. Ultrafast Nonlinear Coherent Vibrational Sum-Frequency Spectroscopy Methods to Study Thermal Conductance of Molecules at Interfaces. *Acc. Chem. Res.* **2009**, *42*, 1343–1351.
5. Kuklja, M. M.; Rashkeev, S. N. Self-Accelerated Mechanochemistry in Nitroarenes. *J. Phys. Chem. Lett.* **2010**, *1*, 363–367.
6. Kuklja, M. M.; Rashkeev, S. N. Molecular Mechanisms of Shear-Strain Sensitivity of Energetic Crystals DADNE and TATB. *J. Energ. Mater.* **2010**, *28*, 66–77.
7. Kuklja, M. M.; Rashkeev, S. N. Interplay of Decomposition Mechanisms at Shear-Strain Interface. *J. Phys. Chem. C (Lett.)* **2009**, *113*, 17–20.
8. Kuklja, M. M.; Rashkeev, S. N. Shear-Strain Induced Chemical Reactivity of Layered Molecular Crystals. *Appl. Phys. Lett.* **2007**, *90*, 151913.
9. Kuklja, M. M.; Rashkeev, S. N. Shear-Strain Induced Structural and Electronic Modifications of the Molecular Crystal 1,1-Diamino-2,2-Dinitroethylene: Slip-Plane Flow and Band Gap Relaxation. *Phys. Rev. B* **2007**, *75*, 104111.
10. Bolton, O.; Simke, L. R.; Pagoria, P. F.; Matzger, A. J. High Power Explosive with Good Sensitivity: A 2: 1 Cocrystal of CL-20: HMX. *Cryst. Growth Des.* **2012**, *12*, 4311–4314.
11. Millar, D. I. A.; Maynard-Casely, H. E.; Allan, D. R.; Cumming, A. S.; Lennie, A. R.; Mackay, A. J.; Oswald, I. D. H.; Tang, C. C.; Pulham, C. R. Crystal Engineering of Energetic Materials: Co-crystals of CL-20. *CrystEngComm* **2012**, *14*, 3742–3749.
12. Göbel, M.; Klapötke, T. M. Development and Testing of Energetic Materials: The Concept of High Densities Based on the Trinitroethyl Functionality. *Adv. Funct. Mater.* **2009**, *19*, 347–365.
13. Kimmel, A. V.; Sushko, P. V.; Kuklja, M. M. The Structure and Decomposition Chemistry of Isomer Defects in a Crystalline DADNE. *J. Energ. Mater.* **2010**, *28*, 128–140.
14. Sharia, O.; Kuklja, M. M. Rapid Materials Degradation Induced by Surfaces and Voids: *Ab initio* Modeling of β-Octatetramethylene Tetranitramine. *J. Am. Chem. Soc.* **2012**, *134*, 11815–11820.
15. Kuklja, M. M. Thermal Decomposition of Solid Cyclotrimethylene Trinitramine. *J. Phys. Chem. B* **2001**, *105*, 10159–10162.
16. Kuklja, M. M.; Rashkeev, S. N. Modeling of Defect Induced Phenomena in Energetic Materials. In *Static Compression of Energetic Materials*; Peiris, S.; Piermarini, G. Eds.; Springer-Verlag: Berlin, Germany, 2008; pp 322–361.
17. Kuklja, M. M. On the Initiation of Chemical Reactions by Electronic Excitations in Molecular Solids. *Appl. Phys. A* **2003**, *76*, 359–366.
18. Hohenberg, P.; Kohn, W. Inhomogeneous Electron Gas. *Phys. Rev. B* **1964**, *136*, 864–871.
19. Kohn, W.; Sham, L. J. Self-Consistent Equations Including Exchange and Correlation Effects. *Phys. Rev. A* **1965**, *140*, 1133–1138.
20. Frisch, M. J.; Trucks, G. W.; Schlegel, H. B.; Scuseria, G. E.; Robb, M. A.; Cheeseman, J. R.; Scalmani, G.; Barone, V.; Mennucci, B.; Petersson, G. A.; Nakatsuji, H.; Caricato, M.; Li, X.; Hratchian, H. P.; Izmaylov, A. F.; Bloino, J.; Zheng, G.; Sonnenberg, J. L.; Hada, M.; Ehara, M.; Toyota, K.; Fukuda, R.; Hasegawa, J.; Ishida, M.; Nakajima, T.; Honda, Y.; Kitao, O.; Nakai, H.; Vreven, T.; Montgomery, J. A., Jr.; Peralta, J. E.; Ogliaro, F.; Bearpark, M.; Heyd, J. J.; Brothers, E.; Kudin, K. N.; Staroverov, V. N.; Kobayashi, R.; Normand, J.; Raghavachari, K.; Rendell, A.; Burant, J. C.; Iyengar, S. S.; Tomasi, J.; Cossi, M.; Rega, N.; Millam, J. M.; Klene, M.; Knox, J. E.; Cross, J. B.; Bakken, V.; Adamo, C.; Jaramillo, J.; Gomperts, R.; Stratmann, R. E.; Yazyev, O.; Austin, A. J.; Cammi, R.; Pomelli, C.; Ochterski, J. W.; Martin, R. L.; Morokuma, K.; Zakrzewski, V. G.; Voth, G. A.; Salvador, P.; Dannenberg, J. J.;

Dapprich, S.; Daniels, A. D.; Farkas, Ö.; Foresman, J. B.; Ortiz, J. V.; Cioslowski, J.; Fox, D. J. *GAUSSIAN 09, Revision A.1*. Gaussian, Inc.: Wallingford, CT, 2009.

21. Perdew, J. P.; Burke, K.; Ernzerhof, M. Generalized Gradient Approximation Made Simple. *Phys. Rev. Lett.* **1996**, 77, 3865–3868.
22. Becke, A. D. Density-Functional Thermochemistry. III. The Role of Exact Exchange. *J. Chem. Phys.* **1993**, *98*, 5648–5652.
23. Lee, C.; Yang, W.; Parr, R. G. Development of the Colle-Salvetti Correlation-Energy Formula into a Functional of the Electron Density. *Phys. Rev. B* **1998**, *37*, 785–789.
24. Adamo, C.; Barone, V. Toward Reliable Density Functional Methods Without Adjustable Parameters: The PBE0 Model. *J. Chem. Phys.* **1999**, *110*, 6158–6169.
25. Zhao, Y.; Truhlar, D. G. The M06 Suite of Density Functionals for Main Group Thermochemistry, Thermochemical Kinetics, Noncovalent Interactions, Excited States, and Transition Elements: Two New Functionals and Systematic Testing of Four M06-Class Functionals and 12 Other Functionals. *Theor. Chem. Acc.* **2008**, *120*, 215–241.
26. Chai, J.-D.; Head-Gordon, M. Long-Range Corrected Hybrid Density Functionals with Damped Atom–Atom Dispersion Corrections. *Phys. Chem. Chem. Phys.* **2008**, *10*, 6615–6620.
27. Dunning, T. H., Jr. Gaussian Basis Sets for Use in Correlated Molecular Calculations. I. The Atoms Boron Through Neon and Hydrogen. *J. Chem. Phys.* **1989**, *90*, 1007–1023.
28. Zhao, Y.; Truhlar, D. G. Density Functionals with Broad Applicability in Chemistry. *Acc. Chem. Res.* **2008**, *41*, 157–167.
29. Hratchian, H. P.; Schlegel, H. B. Accurate Reaction Paths Using a Hessian Based Predictor-Corrector Integrator. *J. Chem. Phys.* **2004**, *120*, 9918–9924.
30. Hratchian, H. P.; Schlegel, H. B. Using Hessian Updating to Increase the Efficiency of a Hessian Based Predictor-Corrector Reaction Path Following Method. *J. Chem. Theory Comput.* **2005**, *1*, 61–69.
31. Eyring, H. The Activated Complex and the Absolute Rate of Chemical Reactions. *Chem. Rev.* **1935**, *17*, 65–77.
32. Hanggi, P.; Talkner, P.; Borkovec, M. Reaction-Rate Theory: Fifty Years After Kramers. *Rev. Mod. Phys.* **1990**, *62*, 251–341.
33. Truhlar, D. G.; Garrett, B. C. Variational Transition State Theory. *Ann. Rev. Phys. Chem.* **1984**, *35*, 159–189.
34. Sharia, O.; Kuklja, M. M. Ab Initio Kinetics of Gas Phase Decomposition Reactions. *J. Phys. Chem. A* **2010**, *114*, 12656–12661.
35. Sharia, O.; Kuklja, M. M. Modeling Thermal Decomposition Mechanisms in Gaseous and Crystalline Molecular Materials: Application to β-HMX. *J. Phys. Chem. B* **2011**, *115*, 12677–12686.
36. Sharia, O.; Kuklja, M. M. Surface-Enhanced Decomposition Kinetics of Molecular Materials Illustrated with Cyclotetramethylene-Tetranitramine. *J. Phys. Chem. C* **2012**, *116*, 11077–11081.
37. Blöchl, P. E. Projector Augmented-Wave Method. *Phys. Rev. B* **1994**, *50*, 17953–17979.
38. Kresse, G.; Furthmüller, J. Efficiency of Ab-Initio Total Energy Calculations for Metals and Semiconductors Using a Plane-Wave Basis Set. *Comput. Mater. Sci.* **1996**, *6*, 15–50.
39. Kresse, G.; Furthmüller, J. Efficient Iterative Schemes for Ab Initio Total-Energy Calculations Using a Plane-Wave Basis Set. *Phys. Rev. B* **1996**, *54*, 11169–11186.
40. Kresse, G. Hafner, Ab Initio Molecular Dynamics for Liquid Metals. *J. Phys. Rev. B* **1993**, *47*, RC558–RC561.
41. Cady, H. H.; Larson, A. C. Pentaerythritol Tetranitrate II: Its Crystal Structure and Transformation to PETN I; an Algorithm for Refinement of Crystal Structures with Poor Data. *Acta Cryst.* **1975**, *B31*, 1864–1869.

42. Conroy, M. W.; Oleynik, I. I.; Zybin, S. V.; White, C. T. First-Principles Investigation of Anisotropic Constitutive Relationships in Pentaerythritol Tetranitrate. *Phys. Rev. B* **2008**, 77, 094107.
43. Sorescu, D. C.; Rice, B. M.; Thompson, D. L. Theoretical Studies of the Hydrostatic Compression of RDX, HMX, HNIW, and PETN Crystals. *J. Phys. Chem. B* **1999**, *103*, 6783–6790.
44. Henkelman, G.; Uberuaga, B. P.; Jónsson, H. A Climbing Image Nudged Elastic Band Method for Finding Saddle Points and Minimum Energy Paths. *J. Chem. Phys.* **2000**, *113*, 9901–9904.
45. Dovesi, R.; Saunders, V. R.; Roetti, C.; Orlando, R.; Zicovich-Wilson, C. M.; Pascale, F.; Civalleri, B.; Doll, K.; Harrison, N. M.; Bush, I. J.; D'Arco, Ph.; Llunell, M. *CRYSTAL 2006 User's Manual*; University of Torino: Torino, 2006.
46. Kudin, K. N.; Scuseria, G. E. Linear-Scaling Density-Functional Theory with Gaussian Orbitals and Periodic Boundary Conditions: Efficient Evaluation of Energy and Forces via the Fast Multipole Method. *Phys. Rev. B* **2000**, *61*, 16440–16453.
47. Grimes, R. W.; Catlow, C. R. A.; Shluger, A. L. Eds.; *Quantum Mechanical Cluster Calculations in Solid State Studies*; World Scientific: Singapore, 1991; p 470.
48. Zhang, Y.; Lin, H.; Truhlar, D. G. Self-Consistent Polarization of the Boundary in the Redistributed Charge and Dipole Scheme for Combined Quantum-Mechanical and Molecular-Mechanical Calculations. *J. Chem. Theory Comput.* **2007**, *3*, 1378–1398.
49. Kimmel, A. V.; Sushko, P. V.; Shluger, A. L. Structure and Spectroscopic Properties of Trapped Holes in Silica. *J. Non-Cryst. Solids* **2007**, *353*, 599–604.
50. Ramo, D. M.; Gavartin, J. L.; Shluger, A. L.; Bersuker, G. Theoretical Prediction of Intrinsic Self-Trapping of Electrons and Holes in Monoclinic HfO_2. *Phys. Rev. Lett.* **2007**, *99*, 155504.
51. Beck, K. M.; Henyk, M.; Wang, C.; Trevisanutto, P. E.; Sushko, P. V.; Hess, W. P.; Shluger, A. L. Site-Specific Laser Modification of MgO Nanoclusters: Towards Atomic-Scale Surface Structuring. *Phys. Rev. B* **2006**, *74*, 045404.
52. Maurer, P.; Laio, A.; Hugosson, H. W.; Colombo, M. C.; Rothlisberger, U. Automated Parametrization of Biomolecular Force Fields from Quantum Mechanics/Molecular Mechanics (QM/MM) Simulations Through Force Matching. *J. Chem. Theort Comput.* **2007**, *3*(2), 628–639.
53. Senn, H. M.; Thiel, W. QM/MM Methods for Biological Systems. *Top. Curr. Chem.* **2007**, *268*, 173–290; Senn, H. M.; Thiel, W. QM/MM Methods for Biological Systems. *Angew. Chem. Int. Ed.* **2009**, *48*, 1198–1229.
54. Swerts, B.; Van Droogenbroeck, J.; Peeters, A.; Van Alsenoy, C. A Linearly Scaling QM/MM Method to Study Molecular Crystals Using BRABO/CHARMM: Application to 2-(2-Methyl-3-Chloroanilino) Nicotinic Acid. *J. Phys. Chem. A* **2002**, *106*, 4245–4250.
55. Kuklja, M. M.; Zerilli, F. J.; Sushko, P. Embedded Cluster Model: Application to Molecular Crystals. In *Symposium AA, Synthesis, Characterization and Properties of Energetic/Reactive Nanomaterials*; Armstrong, R. W.; Thadhani, N. N.; Wilson, W. H.; Gilman, J. J.; Munir, Z.; Simpson, R. L. Eds.; Invited Paper for MRS Proceedings; Vol. 800, 2004; pp 211–222.
56. Kimmel, A. V.; Sushko, P. V.; Shluger, A. L.; Kuklja, M. M. Modeling Proton Transfer and Polarons in a Molecular Crystal Diamino-Dinitroethylene. *Phys. Rev. B* **2009**, *80*, 134108.
57. Kimmel, A. V.; Sushko, P. V.; Shluger, A. L.; Kuklja, M. M. Effect of Molecular and Lattice Structure on Hydrogen Transfer in Molecular Crystals of Diamino-Dinitroethylene and Triamino-Trinitrobenzene. *J. Phys. Chem. A* **2008**, *112*(19), 4496–4500.
58. Sushko, P.; Shluger, A. L.; Catlow, C. R. A. Relative Energies of Surface and Defect States: Ab Initio Calculations for the MgO (001) Surface. *SURF. Sci.* **2000**, *450*,

153–170; Sushko, P. V.; Shluger, A. L.; Hirano, M.; Hosono, H. From Insulator to Electride: A Theoretical Model of Nanoporous Oxide 12CaO $7Al_2O_3$. *J. Am. Chem. Soc.* **2007**, *129*, 942–951.
59. Braithwaite, J. S.; Sushko, P. V.; Wright, K.; Catlow, C. R. A. Hydrogen Defects in Forsterite: A Test Case for the Embedded Cluster Method. *J. Chem. Phys.* **2002**, *116*, 2628.
60. Diwald, O.; Sterrer, M.; Knözinger, E.; Sushko, P. V.; Shluger, A. L. Wavelength Selective Excitation of Surface Oxygen Anions on Highly Dispersed MgO. *J. Chem. Phys.* **2002**, *116*, 1707.
61. Mysovsky, A. S.; Sushko, P. V.; Mukhopadhyay, S.; Edwards, A. H.; Shluger, A. L. Calibration of Embedded-Cluster Method for Defect Studies in Amorphous Silica. *Phys. Rev. B* **2004**, *69*(8), 085202.
62. Fletcher, R. A New Approach to Variable Metric Algorithms. *Comput. J.* **1970**, *13*, 317–322.
63. Goldfarb, D. A Family of Variable-Metric Methods Derived by Variational Means. *Math. Comput.* **1970**, *24*, 23–26.
64. Sushko, P. V.; Abarenkov, I. V. General Purpose Electrostatic Embedding Potential. *J. Chem. Theory Comput.* **2010**, *6*, 1323–1333.
65. Behrens, R. Thermal Decomposition of Energetic Materials: Temporal Behaviors of the Rates of Formation of the Gaseous Pyrolysis Products from Condensed-Phase Decomposition of Octahydro-1, 3, 5, 7-Tetranitro-1, 3, 5, 7-Tetrazocine. *J. Phys. Chem.* **1990**, *94*, 6706–6718.
66. Behrens, R.; Bulusu, S. Thermal Decomposition of Energetic Materials. 2. Deuterium Isotope Effects and Isotopic Scrambling in Condensed-Phase Decomposition of Octahydro-1, 3, 5, 7-Tetranitro-1, 3, 5, 7-Tetrazocine. *J. Phys. Chem.* **1991**, *95*, 5838–5845.
67. Brill, B. T.; Brush, P. G.; Kinloch, S. A.; Gray, P. Condensed Phase Chemistry of Explosives and Propellants at High Temperature: HMX, RDX and BAMO [and Discussion]. *Phil. Trans. R. Soc. Lond. A Phys. Eng. Sci.* **1992**, *339*, 377–385.
68. Chakraborty, D.; Muller, R. P.; Dasgupta, S.; Goddard, W. A., III Mechanism for Unimolecular Decomposition of HMX (1, 3, 5, 7-Tetranitro-1, 3, 5, 7-Tetrazocine), an Ab Initio Study. *J. Phys. Chem. A* **2001**, *105*, 1302–1314.
69. Lewis, J. P.; Glaesemann, K. R.; VanOpdorp, K.; Voth, G. A. Ab Initio Calculations of Reactive Pathways for α-Octahydro-1, 3, 5, 7-Tetranitro-1, 3, 5, 7-Tetrazocine (α-HMX). *J. Phys. Chem. A* **2000**, *104*, 11384–11389.
70. Cobos, C. J. DFT Study of the Thermochemistry of Gas-Phase 1, 3, 5, 7-Tetranitro-1, 3, 5, 7-Tetraazacyclooctane (β-HMX). *J. Mol. Struct. (THEOCHEM)* **2005**, *714*, 147–152.
71. Zhang, L.; Zybin, S. V.; van Duin, A. C. T.; Dasgupta, S.; Goddard, W. A., III Carbon Cluster Formation During Thermal Decomposition of Octahydro-1, 3, 5, 7-Tetranitro-1, 3, 5, 7-Tetrazocine and 1, 3, 5-Triamino-2, 4, 6-Trinitrobenzene High Explosives from ReaxFF Reactive Molecular Dynamics Simulations. *J. Phys. Chem. A* **2009**, *113*, 10619–10640.
72. Velardez, G. F.; Alavi, S.; Thompson, D. L. Theoretical Predictions of the Initial Decomposition Steps of Dimethylnitramine. *J. Chem. Phys.* **2005**, *123*, 074313.
73. Zhang, S.; Nguyen, H. N.; Truong, T. N. Theoretical Study of Mechanisms, Thermodynamics, and Kinetics of the Decomposition of Gas-Phase α-HMX (Octahydro-1, 3, 5, 7-Tetranitro-1, 3, 5, 7-Tetrazocine). *J. Phys. Chem. A* **2003**, *107*, 2981–2989, Our results can be compared with this work with caution, as α-HMX is simulated and a different methodology is used for calculations.
74. Manaa, M. R.; Fried, L. E.; Melius, C. F.; Elstner, M.; Frauenheim, Th. Decomposition of HMX at Extreme Conditions: A Molecular Dynamics Simulation. *J. Phys. Chem. A* **2002**, *106*(39), 9024–9029.

75. Brill, T. B.; Gongwer, P. E.; Williams, G. K. Thermal Decomposition of Energetic Materials. 66. Kinetic Compensation Effects in HMX, RDX, and NTO. *J. Phys. Chem.* **1994**, *98*, 12242–12247.
76. Samoilenko, N. G.; Vinokurov, A. A.; Abramov, V. G.; Merzhanov, A. G. *Russ. J. Phys. Chem.* **1970**, *44*, 22–24.
77. Campana, J. E.; Doyle, R. J., Jr. Evidence for Cluster Ion Structure and Condensed-Phase Properties in the Gas-Phase Cluster Ion Distributions from Methylenenitramine Telomers. *J. Chem. Soc. Chem. Commun.* **1985**, *2*, 45–46.
78. Kitaigorodskii, A. I. *Organic Chemical Crystallography*; Consultants Bureau: New York, NY, 1958; pp 120–124, 216–217, 263–264.
79. Olsen, C. S.; Scroggins, H. S. High-Performance Liquid Chromatographic Determination of The Nitrate Esters Isosorbide Dinitrate, Pentaerythritol Tetranitrate, and Erythrityl Tetranitrate in Various Tablet Forms. *J. Pharm. Sci.* **1984**, *73*, 1303–1304.
80. Gori, T.; Daiber, A. Non-Hemodynamic Effects of Organic Nitrates and the Distinctive Characteristics of Pentaerithrityl Tetranitrate. *Am. J. Cardiovas. Drugs* **2009**, *9*, 7–15.
81. Gruzdkov, Y. A.; Gupta, Y. M. Shock Wave Initiation of Pentaerythritol Tetranitrate Single Crystals: Mechanism of Anisotropic Sensitivity. *J. Phys. Chem. A* **2000**, *104*, 11169–11176.
82. Gruzdkov, Y. A.; Gupta, Y. M. Vibrational Properties and Structure of Pentaerythritol Tetranitrate. *J. Phys. Chem. A* **2001**, *105*, 6197–6202.
83. Zhang, G.; Weeks, B. L.; Zhang, X. Crystal Growth of Organic Energetic Materials: Pentaerythritol Tetranitrate. *Cent. Eur. J. Eng.* **2012**, *2*, 336–346.
84. Lindner, V. Explosives and Propellants. In *Kirk-Othmer Encyclopedia of Chemical Technology*, Vol. 10, Kroschwitz, J. I.; Howe-Grant, M., Eds.; 4th ed.; Wiley: New York, NY, 1993; p 22.
85. Zaoui, A.; Sekkal, W. Molecular Dynamics Study of Mechanical and Thermodynamic Properties of Pentaerythritol Tetranitrate. *Solid State Commun.* **2001**, *118*, 345–350.
86. Dreger, Z. A.; Gruzdkov, Y. A.; Gupta, Y. M.; Dick, J. J. Shock Wave Induced Decomposition Chemistry of Pentaerythritol Tetranitrate Single Crystals: Time-Resolved Emission Spectroscopy. *J. Phys. Chem. B* **2002**, *106*, 247–256.
87. Dick, J. J.; Mulford, R. N.; Spencer, W. J.; Pettit, D. R.; Garcia, E.; Shaw, D. C. Shock Response of Pentaerythritol Tetranitrate Single Crystals. *J. Appl. Phys.* **1991**, *70*, 3572–3587.
88. Dick, J. J. Effect of Crystal Orientation on Shock Initiation Sensitivity of Pentaerythritol Tetranitrate Explosive. *Appl. Phys. Lett.* **1984**, *44*, 859–861.
89. Dick, J. J. Anomalous Shock Initiation of Detonation in Pentaerythritol Tetranitrate Crystals. *J. Appl. Phys.* **1997**, *81*, 601–612.
90. Landerville, A. C.; Oleynik, I. I.; White, C. T. Reactive Molecular Dynamics of Hypervelocity Collisions of PETN Molecules. *J. Phys. Chem. A* **2009**, *113*, 12094–12104.
91. Zhurova, E. A.; Stash, A. I.; Tsirelson, V. G.; Zhurov, V. V.; Bartashevich, E. V.; Potemkin, V. A.; Pinkerton, A. A. Atoms-in-Molecules Study of Intra- and Intermolecular Bonding in the Pentaerythritol Tetranitrate Crystal. *J. Am. Chem. Soc.* **2006**, *128*, 14728–14734.
92. Klapötke, T. M.; Krumm, B.; Ilg, R.; Troegel, D.; Tacke, R. The Sila-Explosives Si(CH2N3)4 and Si(CH2ONO2)4: Silicon Analogues of the Common Explosives Pentaerythrityl Tetraazide, C(CH2N3)4, and Pentaerythritol Tetranitrate, C(CH2ONO2)4. *J. Am. Chem. Soc.* **2007**, *129*, 6908–6915.
93. Gruzdkov, Y. A.; Dreger, Z. A.; Gupta, Y. M. Experimental and Theoretical Study of Pentaerythritol Tetranitrate Conformers. *J. Phys. Chem. A* **2004**, *108*, 6216–6221.

94. Hemmi, N.; Dreger, Z. A.; Gruzdkov, Y. A.; Winey, J. M.; Gupta, Y. M. Raman Spectra of Shock Compressed Pentaerythritol Tetranitrate Single Crystals: Anisotropic Response. *J. Phys. Chem. B* **2006**, *110*, 20948–20953.
95. Ng, W. L.; Field, J. E.; Hauser, H. M. Study of the Thermal Decomposition of Pentaerythritol Tetranitrate. *J. Chem. Soc. Perkin Trans.* **1976**, *2*, 637–639.
96. Ng, W. L.; Field, J. E.; Hauser, H. M. Thermal, Fracture and Laser-Induced Decomposition of Pentaerythritol Tetranitrate. *J. Appl. Phys.* **1986**, *59*, 3945–3952.
97. Andreev, K. K.; Kaidymov, B. I. Thermal Decomposition of Nitrate Esters. II. Thermal Decomposition of Pentaerythritol Tetranitrate. *Russ. J. Phys. Chem.* **1961**, *35*, 1324–1330.
98. Chambers, D. M. *Perspectives on Pentaerythritol Tetranitrate (PETN) Decomposition*; Lawrence Livermore National Laboratory: Livermore, CA, 2002; pp 1–18, UCRLID-148956.
99. Van der Merwe, M.; Fouche, F. C.; Von Abo, M. In *17th Int. Ann. Conf. ICT*, Fraunhofer Institut fur Chemische Technologie (ICT), Karsruhe, Germany, 1986; p 62.
100. Wu, C. J.; Manaa, M. R.; Fried, L. E. Tight Binding Molecular Dynamic Simulation of PETN Decomposition at an Extreme Condition. *Mater. Res. Soc. Symp. Proc.* **2006**, *987*, 4–11.
101. Robertson, A. J. B. The Thermal Decomposition of Pentaerythritol Tetranitrate, Nitroglycerin, Ethylenediamine Dinitrate and Ammonium Nitrate. *J. Soc. Chem. Ind.* **1948**, *67*, 221–224.
102. Rogers, R. N.; Morris, E. D., Jr. On Estimating Activation Energies with a Differential Scanning Calorimeter. *Anal. Chem.* **1966**, *38*, 412–414.
103. Maycock, J. N.; Pai Verneker, V. R. Characterization of Thermal and Photosublimation of Organic Explosives by Thermobarogravimetric Techniques. *Thermochim. Acta* **1970**, *1*, 191–198.
104. Volltrauer, H. N. Real Time Low Temperature Decomposition of Explosives—PETN. *J. Haz. Mater.* **1982**, *5*, 353–357.
105. Oxley, J. C.; Smith, J. L.; Brady, J. E., IV; Brown, A. C. Characterization and Analysis of Tetranitrate Esters. *Propel. Explos. Pyrotech.* **2012**, *37*, 24–39.
106. Roos, B. D.; Brill, T. B. Thermal Decomposition of Energetic Materials 82. Correlations of Gaseous Products with the Composition of Aliphatic Nitrate Esters. *Combust. Flame* **2002**, *128*, 181–190.
107. Yu, Z.; Bernstein, E. R. Decomposition of Pentaerythritol Tetranitrate [$C(CH_2ONO_2)_4$] Following Electronic Excitation. *J. Chem. Phys.* **2011**, *135*, 154305–154305-10.
108. Aluker, E. D.; Krechetov, A. G.; Mitrofanov, A. Y.; Nurmukhametov, D. R.; Kuklja, M. M. Laser Initiation of Energetic Materials: Selective Photoinitiation Regime in Pentaerythritol Tetranitrate. *J. Phys. Chem. C* **2011**, *115*, 6893–6901.
109. Tsyshevsky, R.; Sharia, O.; Kuklja, M. Thermal Decomposition Mechanisms of Nitroesters: Ab Initio Modeling of Pentaerythritol Tetranitrate. *J. Phys. Chem. C* **2013**, *117*, 18144–18153.
110. Tsyshevsky, R. et al. Electronic Structure and Energies of Electronic Transitions of PETN Crystals, to be published, 2014.
111. Wu, C. J.; Ree, F. H.; Yoo, C.-S. A Quantum Mechanical Molecular Dynamics Study of Binary Collisions of Pentaerythritol Tetranitrate (PETN): Its Correlation to Shock Sensitivity. *Propel. Explos. Pyrotech.* **2004**, *29*, 296–303.
112. Zybin, S. V.; Goddard, W. A., III; Xu, P.; van Duin, A. C. T.; Thompson, A. P. Physical Mechanism of Anisotropic Sensitivity in Pentaerythritol Tetranitrate from Compressive-Shear Reaction Dynamics Simulations. *Appl. Phys. Lett.* **2010**, *96*, 081918–081918-3.

113. Nazin, G. M.; Manelis, G. B.; Rubtsov, Yu. I.; Strunin, V. A. *Thermal Decomposition and Combustion of Explosives and Propellants*. CRC Press: Moscow, Russia, 2003; p 376.
114. Dewar, M. J. S.; Ritchie, J. P.; Alster, J. Ground States of Molecules. 65. Thermolysis of Molecules Containing NO_2 Groups. *J. Org. Chem.* **1985**, *50*, 1031–1036.
115. Booth, R. S.; Lam, C.-S.; Butler, L. J. A Novel Mechanism for Nitric Oxide Production in Nitroalkyl Radicals that Circumvents Nitro–Nitrite Isomerization. *J. Phys. Chem. Lett.* **2013**, *4*, 547–550.
116. Brook, A. G. Molecular Rearrangements of Organosilicon Compounds. *Acc. Chem. Res.* **1974**, 7, 77–84.
117. Liu, W.-G.; Zybin, S. V.; Dasgupta, S.; Klapötke, T. M.; Goddard, W. A., III Explanation of the Colossal Detonation Sensitivity of Silicon Pentaerythritol Tetranitrate (Si-PETN) Explosive. *J. Am. Chem. Soc.* **2009**, *131*, 7490–7491.
118. Hiskey, M. A.; Brower, K. R.; Oxley, J. C. Thermal Decomposition of Nitrate Esters. *J. Phys. Chem.* **1991**, *95*, 3955–3960.
119. Khrapkovskii, G. M.; Shamsutdinov, T. F.; Chachkov, D. V.; Shamov, A. G. Energy of the O–NO_2 Bond Dissociation and the Mechanism of the Gas-Phase Monomolecular Decomposition of Aliphatic Alcohol Nitroesters. *J. Mol. Struct. (THEOCHEM)* **2004**, *686*, 185–192.
120. Sadhukhan, S.; Muñoz, D.; Adamo, C.; Scuseria, G. E. Predicting Proton Transfer Barriers with Density Functional Methods. *Chem. Phys. Lett.* **1999**, *306*, 83–87.
121. Nachimuthu, S.; Gao, J.; Truhlar, D. G. A Benchmark Test Suite for Proton Transfer Energies and Its Use to Test Electronic Structure Model Chemistries. *Chem. Phys.* **2012**, *400*, 8–12.
122. Zhang, C.; Wang, X.; Zhou, M. Isomers and Isomerization Reactions of Four Nitro Derivatives of Methane. *J. Comput. Chem.* **2011**, *32*, 1760–1768.
123. Kimmel, A. V.; Sushko, P. V.; Shluger, A. L.; Kuklja, M. M. Effect of Charged and Excited States on the Decomposition of 1,1-Diamino-2,2-Dinitroethylene Molecules. *J. Chem. Phys.* **2007**, *126*, 234711–234711-10.
124. Gindulyté, A.; Masaa, L.; Huang, L.; Karle, J. Ab Initio Study of Unimolecular Decomposition of Nitroethylene. *J. Phys. Chem. A* **1999**, *103*, 11040–11044.
125. Xu, S.; Lin, M. C. Computational Study on the Kinetics and Mechanism for the Unimolecular Decomposition of $C_6H_5NO_2$ and the Related $C_6H_5+NO_2$ and C_6H_5O+NO Reactions. *J. Phys. Chem. B* **2005**, *109*, 8367–8373.
126. Fayet, G.; Joubert, L.; Rotureau, P.; Adamo, C. Theoretical Study of the Decomposition Reactions in Substituted Nitrobenzenes. *J. Phys. Chem. A* **2008**, *112*, 4054–4059.
127. Cohen, R.; Zeiri, Y.; Wurzberg, E.; Kosloff, R. Mechanism of Thermal Unimolecular Decomposition of TNT (2,4,6-Trinitrotoluene): A DFT Study. *J. Phys. Chem. A* **2007**, *111*, 11074–11083.
128. Loebbecke, S.; Schuppler, H.; Schweikert, W. Thermal Properties of Different Substituted Energetic Furoxans. In *Energetic Materials: Synthesis, Production and Application, 33rd International Annual Conference of ICT, Karlsruhe, Federal Republic of Germany, 25-June* 28, 2002.
129. Zheng, W.; Wang, J.; Ren, X.; Zhang, L.; Zhou, Y. An Investigation on Thermal Decomposition of BNFF-CMDB Propellants. *Propel. Explos. Pyrotech.* **2007**, *32*, 520–524.
130. Kotomin, A. A.; Kozlov, A. S.; Dushenok, S. A. Detonatability of High-Energy-Density Heterocyclic Compounds. *Russ. J. Phys. Chem.* **2007**, *1*, 573–575.
131. Wang, J.; Dong, H. S.; Huang, Y. G.; Li, J. S. Properties of 3,4-Bis(nitrofurazano) furoxan. In *Energetic Materials: Insensitivity, Ageing, Monitoring, 37th International Annual Conference of ICT, Karlsruhe, Federal Republic of Germany, June 27–June 30*, 2006.

132. Stepanov, A. I.; Dashko, D. V.; Astrat'ev, A. A. 3,4-Bis(4′-nitrofurazan-3′-yl)furoxan: A Melt Cast Powerful Explosive and a Valuable Building Block in 1,2,5-Oxadiazole Chemistry. *Cent. Eur. J. Energy Mater.* **2012**, *9*, 329–342.
133. Feng-qi, Z.; Pei, C.; Rong-zu, Hu.; Yang, L.; Zhi-zhong, Z.; Yan-shui, Z.; Xu-wu, Y.; Yin, G.; Sheng-li, G.; Qi-zhen, S. Thermochemical Properties and Non-isothermal Decomposition Reaction Kinetics of 3,4-Dinitrofurazanfuroxan (DNTF). *J. Haz. Mater.* **2004**, *113*, 67–71.
134. Wei, T.; Wu, J.; Zhu, W.; Zhang, C.; Xiao, H. Characterization of Nitrogen-Bridged 1,2,4,5-Tetrazine-, Furazan-, and 1H-Tetrazole-Based Polyheterocyclic Compounds: Heats Of Formation, Thermal Stability, and Detonation Properties. *J. Mol. Model.* **2012**, *18*, 3467–3479.
135. Zhang, X.; Zhu, W.; Xiao, H. Comparative Theoretical Studies of Energetic Substituted Carbon- and Nitrogen-Bridged Difurazans. *J. Phys. Chem. A* **2010**, *114*, 603–612.
136. Zhang, X.; Zhu, W.; Xiao, H. Theoretical Studies on Heats of Formation, Detonation Properties, and Bond Dissociation Energies of Monofurazan Derivatives. *Int. J. Quant. Chem.* **2010**, *110*, 1549–1558.
137. Wang, L.; Yi, C.; Zou, H.; Liu, Y.; Li, S. Theoretical Study on the Thermal Decomposition Mechanism of 3,3'-Dinitro-4,4'-Azoxyfurazan. *Comput. Theoret. Chem.* **2011**, *963*, 135–140.
138. Guo, Y. Q.; Bhattacharya, A.; Bernstein, E. R. Excited Electronic State Decomposition of Furazan Based Energetic Materials: 3,3′-Diamino-4,4′-Azoxyfurazan and Its Model Systems, Diaminofurazan and Furazan. *J. Chem. Phys.* **2008**, *128*, 034303.
139. Bhattacharya, A.; Guo, Y. Q.; Bernstein, E. R. Nonadiabatic Reaction of Energetic Molecules. *Acc. Chem. Res.* **2010**, *43*, 1476–1485.
140. Tsyshevsky, R.; Kuklja, M. M. Decomposition Mechanisms and Kinetics of Novel Energetic Molecules BNFF-1 and ANFF-1: Quantum-Chemical Modeling. *Molecules* **2013**, *18*(7), 8500–8517, Special Issue: Computational Chemistry.
141. Brill, T. B. In *Energetic Materials: Detonation, Combustion*; Politzer, P. A.; Murray, J. S. Eds.; Elsevier, Amsterdam, 2003.
142. Brill, T. B.; James, K. J. Kinetics and Mechanisms of Thermal Decomposition of Nitroaromatlc Explosives. *Chem. Rev.* **1993**, *93*, 2667–2692.
143. Brill, T. B.; James, K. J. Thermal Decomposition of Energetic Materials. 61. Perfidy in the Amino-2,4,6-Trinitrobenzene Series of Explosives. *J. Phys. Chem.* **1993**, *97*, 8752.
144. Zhang, C. Computational Investigation of the Detonation Properties of Furazans and Furoxans. *J. Mol. Struct. (THEOCHEM)* **2006**, *765*, 77–83.
145. Korolev, V. L.; Petukhova, T. V.; Pivina, T. S.; Porollo, A. A.; Sheremetev, A. B.; Suponitskii, K. Yu.; Ivshin, V. P. Thermal Decomposition Mechanisms of Nitro-1,2,4-Triazoles: A Theoretical Study. *Russ. Chem. Bull. Int. Ed.* **2006**, *55*, 1388–1410.
146. Wu, C. J.; Fried, L. E. Ring Closure Mediated by Intramolecular Hydrogen Transfer in the Decomposition of a Push – Pull Nitroaromatic: TATB. *J. Phys. Chem. A* **2000**, *104*, 6447–6452.
147. Rice, B. M.; Sahu, S.; Owens, F. J. Density Functional Calculations of Bond Dissociation Energies for NO_2 Scission in Some Nitroaromatic Molecules. *J. Mol. Struct. (THEOCHEM)* **2002**, *583*, 69–72.
148. Shao, J.; Cheng, X.; Yang, X. The C – NO2 Bond Dissociation Energies of Some Nitroaromatic Compounds: DFT Study. *Struct. Chem.* **2006**, *17*, 547–550.
149. Politzer, P.; Concha, M. C.; Grice, M. E.; Murray, J. S.; Lane, P.; Habibollazadeh, D. Computational Investigation of the Structures and Relative Stabilities of Amino/Nitro Derivatives of Ethylene. *J. Mol. Struct. (THEOCHEM)* **1998**, *452*, 75–83.

150. Gindulytė, A.; Masaa, L.; Huang, L.; Karle, J. Proposed Mechanism of 1,1-Diamino-Dinitroethylene Decomposition: A Density Functional Theory Study. *J. Phys. Chem. A* **1999**, *103*, 11045–11051.
151. Wodtke, A. M.; Hintsa, E. J.; Lee, Y. T. Infrared Multiphoton Dissociation of Three NItroalkanes. *J. Phys. Chem.* **1986**, *90*, 3549–3558.
152. Saxon, R. P.; Yoshimine, M. Theoretical Study of Nitro-Nitrite Rearrangement of CH_3NO_2. *Can. J. Chem.* **1992**, *70*, 572–579.
153. Williams, F. *Adv. Chem. Phys.* **1971**, *21*, 289.
154. Dremin, A. N. On the Detonation Theory. *Chem. Phys. Rep.* **1995**, *14*, 1851–1870; Dremin, A. N. *Towards Detonation Theory*. Springer-Verlag: New York, Inc., 1999.
155. Kuklja, M. M.; Stefanovich, E. V.; Kunz, A. B. An Excitonic Mechanism of Detonation Initiation in Explosives. *J. Chem. Phys.* **2000**, *112*, 3417.
156. Owens, F. J.; Sharma, J. X-Ray Photoelectron Spectroscopy and Paramagnetic Resonance Evidence for Shock-Induced Intramolecular Bond Breaking in Some Energetic Solids. *J. Appl. Phys.* **1980**, *51*, 1494–1497.
157. Sharma, J.; Beard, B. C. In *Structure and Properties of Energetic Materials*; Liedenberg, D. H.; Armstrong, R. W.; Gilman, J. J. Eds.; MRS Symposium Proceedings, Vol. 296; Materials Research Society: Pittsburgh, PA, 1993, and other papers presented at this conference.
158. Duval, G. E. In *Shock Waves in Condensed Matter*; Gupta, Y. M. Ed.; Plenum: New York, NY, 1985.
159. Gilman, J. J. *Chem. Propulsion Inf. Agency* **1992**, *589*, 379.
160. Aduev, B. P.; Aluker, E. D.; Belokurov, G. M.; Krechetov, A. G. *Chem. Phys. Rep.* **1997**, *16*, 1479; Aduev, B. P.; Aluker, E. D.; Belokurov, G. M.; Krechetov, A. G. *Chem. Phys. Rep.* **1998**, *17*, 469.
161. Aluker, E. D.; Aduev, B. P.; Krechetov, A. G.; Nurmukhametov, D. R.; Pashpekin, A. S.; Tupitsyn, E. V.; Shvaiko, V. N. *J. Appl. Phys.* **2001**, *89*, 4156.
162. Kunz, A. B.; Kuklja, M. M.; Botcher, T. R.; Russel, T. P. Initiation of Chemistry in Molecular Solids by Processes Involving Electronic Excited States. *Thermochim. Acta* **2002**, *384*, 279–284.
163. Aluker, E. D.; Aluker, N. L.; Belokurov, G. M.; Krechetov, A. G.; Loboiko, B. G.; Nurmuhametov, D. R.; Tupitsyn, A. V.; Filin, V. P. *Russ. J. Phys. Chem. B* **2010**, *4*, 63.
164. Aluker, E. D.; Aluker, N. L.; Krechetov, A. G.; Mitrofanov, A. Y.; Nurmuhametov, D. R.; Shvayko, V. N. *Russ. J. Phys. Chem. B* **2011**, *5*, 67.
165. Aluker, E. D.; Krechetov, A. G.; Mitrofanov, A. Y.; Zverev, A. S.; Kuklja, M. M. *J. Phys. Chem. C* **2012**, *116*, 24482.
166. Kunz, A. B. *Phys. Rev. B* **1996**, *53*, 9733–9738.
167. Kunz, A. B. Ab Initio Investigation of the Crystalline PETN. In *Decomposition, Combustions and Detonation Chemistry of Energetic Materials*; Brill, T. B.; Russel, T. P.; Tao, W. C.; Warde, R. B. Eds.; Mat. Res. Soc. Symp. Proc.; Vol. 418, Materials Research Society: Pittsburgh, PA, 1996; pp 287–292.
168. Kuklja, M. M.; Kunz, A. B. *Ab initio* Simulation of Defects in Energetic Materials. I. Molecular Vacancy Structure in RDX Crystal. *J. Phys. Chem. Solids* **2000**, *61*(1), 35–44.
169. Kuklja, M. M.; Kunz, A. B. *Ab initio* Simulation of Defects in Energetic Materials. II. Hydrostatic Compression of Cyclotrimethylene Trinitramine. *J. Appl. Phys.* **1999**, *86*(8), 4428–4434.
170. Kuklja, M. M.; Kunz, A. B. Simulation of the Defects in Energetic Materials. III. Structure and Properties of the RDX Crystal with Vacancy Complexes. *J. Phys. Chem.* **1999**, *103*, 8427–8431.
171. Aluker, E. D.; Krechetov, A. G.; Mitrofanov, A. Y.; Zverev, A. S.; Kuklja, M. M. Topography of Photo-Initiation in Molecular Materials. *Molecules* **2013**, *18*, 14148–14160.

172. Latypov, N. V.; Bergman, J.; Langlet, A.; Wellmar, U.; Bemm, U. *Tetrahedron* **1998**, *54*, 11525.
173. Bemm, U.; Östmark, H. *Acta Crystallogr. Sect. C: Cryst. Struct. Commun.* **1999**, *C54*(1997).
174. Östmark, H.; Langlet, A.; Bergman, H.; Winborg, N.; Wellmar, U.; Bemm, U. In *Proceedings of the 11th International Detonation Symposium*, Office of Naval Research, Arlington, VA, 1998; p 807; Karlsson, S.; Östmark, H.; Eldsäter, C.; Carlsson, T.; Bergman, H.; Wallin, S.; Pettersson, A. In *12th International Symposium on Detonation Proceedings, Office of Naval Research*, Arlington, VA, 2002; pp 286–291.
175. Gao, H.-X.; Zhao, F.-Q.; Hu, R.-Z.; Pan, Q.; Wang, B.-Z.; Yang, X.-W.; Gao, Y.; Gao, S.-L. *Chin. J. Chem.* **2006**, *24*, 177.
176. Rashkeev, Tsyshesvky, Kuklja. PETN/MgO, to be published.
177. Sharia, Tsyshesvky, Kuklja. Beta-Delta-HMX, to be published.
178. Petersilka, M. G. U. J.; Gossmann, U. J.; Gross, E. K. U. Excitation Energies from Time-Dependent Density-Functional Theory. *Phys. Rev. Lett.* **1996**, *76*(8), 1212.
179. Bauernschmitt, R.; Ahlrichs, R. Treatment of Electronic Excitations Within the Adiabatic Approximation of Time Dependent Density Functional Theory. *Chem. Phys. Lett.* **1996**, *256*, 454–464.
180. Cossi, M.; Barone, V. Time-Dependent Density Functional Theory for Molecules in Liquid Solutions. *J. Chem. Phys.* **2001**, *115*, 4708.
181. Casida, M. E. *Time-Dependent Density Functional Response Theory for Molecules*; World Scientific: Singapore, 1995; Vol. 1.
182. Adamo, C.; Scuseria, G. E.; Barone, V. Accurate Excitation Energies from Time-Dependent Density Functional Theory: Assessing the PBE0 Model. *J. Chem. Phys.* **1999**, *111*, 2889.
183. Marques, M. A. L.; Gross, E. K. U. Time-Dependent Density Functional Theory. *Annu. Rev. Phys. Chem.* **2004**, *55*, 427–455.
184. Roos, B. O.; Taylor, P. R. A Complete Active Space SCF Method (CASSCF) Using a Density Matrix Formulated Super-CI Approach. *Chem. Phys.* **1980**, *48*(2), 157–173.
185. Roos, B. O. The Complete Active Space Self-Consistent Field Method and Its Applications in Electronic Structure Calculations. Advances in Chemical Physics*: Ab Initio* Methods in Quantum Chemistry Part 2; Vol. 69, 1987; pp 399–445.
186. Andersson, K.; Malmqvist, P.-K.; Roos, B. O. Second-Order Perturbation Theory with a Complete Active Space Self-Consistent Field Reference Function. *J. Chem. Phys.* **1992**, *96*, 1218.
187. Aquilante, F.; Pedersen, T. B.; Lindh, R.; Roos, B. O.; de Merás, A. S.; Koch, H. Accurate Ab Initio Density Fitting for Multiconfigurational Self-Consistent Field Methods. *J. Chem. Phys.* **2008**, *129*, 024113.
188. Fang, T.; Shen, J.; Li, S. Block Correlated Coupled Cluster Method with a Complete-Active-Space Self-Consistent-Field Reference Function: The Formula for General Active Spaces and Its Applications for Multibond Breaking Systems. *J. Chem. Phys.* **2008**, *128*, 224107.
189. Yanai, T.; et al. Accelerating Convergence in Iterative Solution for Large-Scale Complete Active Space Self-Consistent-Field Calculations. *Int. J. Quantum Chem.* **2009**, *109*, 2178–2190.
190. Ivanic, J. Direct Configuration Interaction and Multiconfigurational Self-Consistent-Field Method for Multiple Active Spaces with Variable Occupations. I. Method. *J. Chem. Phys.* **2003**, *119*, 9364.
191. Hoffmann, M. R. Third-Order Complete Active Space Self-Consistent Field Based Generalized Van Vleck Perturbation Theory. *Chem. Phys. Lett.* **1993**, *210*, 193–200.
192. Holstein, T. *Ann. Phys. (N.Y.)* **1959**, *8*, 343.
193. Emin, D. *Adv. Phys.* **1975**, *24*, 305.

194. Hannewald, K.; Stojanovic´, V. M.; Schellekens, J. M. T.; Bobbert, P. A.; Kresse, G.; Hafner, J. *Phys. Rev. B* **2004**, *69*, 075211.
195. Serra, S.; Iarlori, S.; Tosatti, E.; Scandolo, S.; Righi, M. C.; Santoro, G. E. *Chem. Phys. Lett.* **2002**, *360*, 487.
196. Pacchioni, G.; Frigoli, F.; Ricci, D.; Weil, J. A. *Phys. Rev. B* **2000**, *63*, 054102.
197. Gavartin, J. L.; Sushko, P. V.; Shluger, A. L. *Phys. Rev. B* **2003**, *67*, 035108.
198. Durand, P.; Barthelat, J. C. *Theor. Chim. Acta* **1975**, *38*, 283.
199. Sulimov, V. B.; Sushko, P. V.; Edwards, A. H.; Shluger, A. L.; Stoneham, A. M. *Phys. Rev. B* **2002**, *66*, 024108.
200. Silinsh, E. A. In *Organic Molecular Crystals: Their Electronic States*; Cardona, M.; Fulde, P.; Queisser, H.-J. Eds.; Springer Series in Solid-State Sciences; Vol. 16, Springer: Berlin, Germany, 1980.
201. Vilfan, J. *Phys. Stat. Sol. B* **1973**, *59*, 351.
202. Chojnacki, H. *Int. J. Mol. Sci.* **2003**, *4*, 408.
203. Tsyshevski, to be published.
204. Rashkeev, S. N.; Kuklja, M. M.; Zerilli, F. J. *Appl. Phys. Lett.* **2003**, *82*, 1371.
205. Kuklja, M. M.; Rashkeev, S. N.; Zerilli, F. J. *Appl. Phys. Lett.* **2006**, *89*, 071904.
206. Cady, H.; Larson, A. *Acta Cryst.* **1965**, *18*, 485.
207. (a) Yoo, C. S.; Holmes, N. C.; Souers, P. C.; Wu, C. J.; Ree, F. H.; Dick, J. J. *J. Appl. Phys.* **2000**, *88*, 70; (b) Wu, C.; Ree, F. S.; Yoo, C. S. *Propel. Explos. Pyrotech.* **2004**, *29*, 296.
208. Glascoe, E. A.; Zaug, J. M.; Armstrong, M. R.; Crowhurst, J. C.; Grant, C. D.; Fried, L. E. Nanosecond Time-Resolved and Steady-State Infrared Studies of Photo-induced Decomposition of TATB at Ambient and Elevated Pressure. *J. Phys. Chem. A* **2009**, *113*, 5881–5887.
209. Kuo, K. K., et al. *Challenges in Propellants and Combustion: 100 Years after Nobel*; Mega Energy Technology Vaesterhaninge (Sweden), 1997.
210. Bowden, F. P.; Yoffe, Y. D. *Initiation and Growth of Explosion in Liquids and Solids*; Cambridge University Press: London, 1952; pp 64–65.
211. Tarver, C. M.; et al. *J. Phys. Chem.* **1996**, *100*, 5794–5799.
212. Zepeda-Ruiz, L. A.; Maiti, A.; Gee, R.; Gilmer, G. H.; Weeks, B. L. Size and Habit Evolution of PETN Crystals—A Lattice Monte Carlo Study. *J. Cryst. Growth* **2006**, *291*, 461–467.
213. Bourne, N. K. On the Laser Ignition and Initiation of Explosives. *Proc. R. Soc. Lond. A* **2001**, *457*, 1401–1426.
214. Rubenchik, A. M. On the Initiation of High Explosives by Laser Radiation. *Propel. Explos. Pyrotech.* **2007**, *32*, 296–300.
215. Tarzhanov, V. I. Ed.; *Fast Initiation of Explosives. Special Regimes of Detonation: A Collection of Articles*; RFYaTs VNIITF: Snezhinsk, 1998 (in Russian).
216. Danilenko, V. V. *Explosion: Physics, Engineering, Technology*; Mashinostroenie: Moscow, Russia, 2010 (in Russian).
217. Brill, T. B.; Reese, C. O. *J. Phys. Chem.* **1980**, *84*, 1376–1380.
218. Choi, C. S.; Boutin, H. P. *Acta Crystallogr. B* **1970**, *26*, 1235–1240.
219. (a) Cady, H. H.; Larson, A. C.; Cromer, D. T. *Acta Crystallogr. B* **1963**, *16*, 617; (b) Cobbledick, R. E.; Small, R. W. H. *Acta Crystallogr.* **1974**, *B30*, 1918–1922; (c) Herrmann, M.; Engel, W.; Eisenreich, N. *Z. Krist.* **1993**, *204*, 129–142.
220. Brill, T. B.; Karpowicz, R. J. *J. Phys. Chem.* **1982**, *86*, 4260–4265.
221. Lewis, J. P. *Chem. Phys. Lett.* **2003**, *371*, 588–593.
222. Urtiew, P. A.; Forbes, J. W.; Tarver, C. M.; Vandersall, K. S.; Garcia, F.; Greenwood, D. W.; Hsu, P. C.; Maienschein, J. L. In *AIP Conf. Proc. 706*, American Institute of Physics, 2004; pp 1053–1056.

223. Henson, B. F.; Asay, B. W.; Sander, R. K.; Son, S. F.; Robinson, J. M.; Dickson, P. M. *Phys. Rev. Lett.* **1999**, *82*, 1213–1216.
224. (a) Whelan, D. J.; Mark R. F. A Reassessment of the Kinetics of the Thermal Decomposition of the High Explosive, Delta-HMX, in the Range 508 K to 524 K, as Studied by Isothermal Gravimetry. NASA 19980218798, 1998; (b) Bulusu, S.; Weinstein, D. I.; Autera, J. R.; Velicky, R. W. *J. Phys. Chem.* **1986**, *90*, 4121–4126.
225. Brand, H. V.; Rabie, R. L.; Funk, D. J.; Diaz-Acosta, I.; Pulay, P.; Lippert, T. K. *J. Phys. Chem. B* **2002**, *106*, 10594–10604.
226. (a) Zerilli, F. J.; Kuklja, M. M. *J. Phys. Chem. A* **2010**, *114*, 5372–5376; (b) Zerilli, F. J.; Kuklja, M. M. *J. Phys. Chem. A* **2006**, *110*, 5173–5179.
227. Curtin, D. Y.; Paul, I. C. *Chem. Rev.* **1981**, *81*, 528–541.

CHAPTER FOUR

Geometric Metastability in Molecules as a Way to Enhance Energy Storage

Ajith Perera, Rodney J. Bartlett[1]
Quantum Theory Project, University of Florida, Gainesville, Florida, USA
[1]Corresponding author: e-mail address: bartlett@qtp.ufl.edu

Contents

Abstract

This chapter traverses through nearly 20 years of theoretical developments and applications devoted to investigating the geometric metastability of molecules that are speculative, but if they could indeed be made and if they were to scale up to practical applications, they could revolutionize the applications in defense, space travel, etc., where such materials are in high demand. The underlying theme of this research has been the development of new predictive level theory—theory that is accurate and reliable to use in the absence of experiment. Without such assurance from theory, application to "speculative molecules" that are discussed here would have been without merit.

1. INTRODUCTION

New high-energy density materials (HEDM) that can store significantly more chemical energy than those that are currently being used and deliver them safely with the least impact on the environment are being

Advances in Quantum Chemistry, Volume 69
ISSN 0065-3276
http://dx.doi.org/10.1016/B978-0-12-800345-9.00004-0

sought continually.[1] In 1989, we submitted a proposal to the United States Air Force Office of Scientific Research (AFOSR) on the basic research topic of investigating "metastability" in molecules, meaning states whose energies are far above the norm even for highly endothermic molecules. One consequence of such an investigation might be the discovery of ways to enhance the energy content of potential rocket fuels. The avowed objective of the initial HEDM program was to explore ways to make potentially revolutionary improvements in thrust to provide significant gains in payload. Obviously, the applications are not limited to rocket fuels as the same concepts pertain to explosives. Metastability can occur because of the long-lived excited states of a molecule, but no solution to trapping such states is apparent. The other alternative is what we termed "geometric metastability," which means to put atoms together into a molecule in unusual ways and depend upon the bonds formed to store the energy. An additional condition was that the prospective fuels had to be lightweight yet have more energy per mass unit than known bipropellant species, such as LOX + H_2 or a molecule such as hydrazine, as a monopropellant. Since the molecules to be proposed are not known experimentally, and since synthesis and characterization of candidate systems would be very expensive and difficult, *only predictive quantum chemical methods make it possible to assess the energy content and kinetic stability (barrier to dissociation) of such structures, to identify the best prospects for success prior to any synthetic effort.* Modern quantum chemistry also provides very accurate spectroscopic signatures to identify such energetic molecules when they are formed, albeit in very small quantities. In this chapter, we review the developments that have led to new methods (and the accompanying software), which allow computing nearly all molecular properties with predictive accuracy and are indispensable in the tool kit of today's computational chemist, and the applications of those new methods to molecules that will draw the interest of synthetic chemists in the years to come.

The conventional carbon-, hydrogen-, nitrogen-, and oxygen (C, H, N, and O)-based materials are at the low end of the mass-based energy spectrum (10^3 cal/gm).[2–4] To further clarify this point, Fig. 4.1 shows the C-, H-, N-, and O-based explosives and their detonation pressure mapped to the year that they were used on a large scale. Notice that the nitramines RDX, HMX, and CL-20 are nearing a plateau and that the specific impulse (I_{sp}) of most widely used propellants is about 200–300 s. So, the question is whether novel materials (or processes) that can yield a significant increase in the energy density and/or I_{sp} can be found. However, energy content is not the only criteria that matters. It is important to keep in mind that

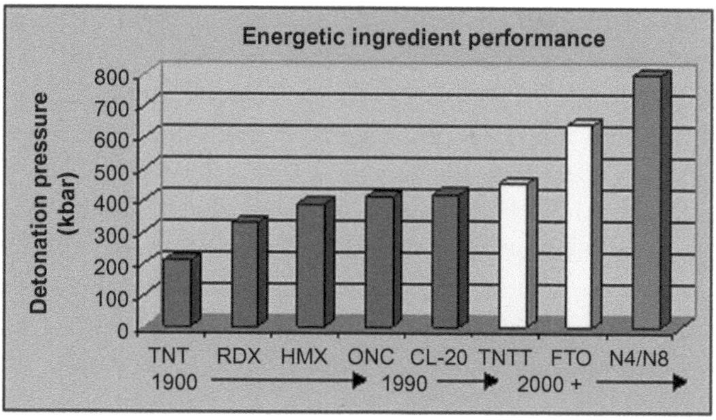

Figure 4.1 Energetic ingredient performance. (For color version of this figure, the reader is referred to the online version of this chapter.)

for practical applications, the HEDM must be amenable to safe handling and storage. Furthermore, the environmental impact upon their use in peaceful as well as military applications is another important factor.

The process of searching for novel energetic materials requires first the identification of energy storage mechanisms. Energy can be stored in the form of chemical bonds, strain energy, the metastable state of a molecule or ion, and the lattice energy of an ionic crystal. Chemical bond energy or strain (geometric metastability) is the most common form of the energy storage mechanisms. In this regard, the first-row atoms come into focus because of their smaller size and mass, multiple bonding capabilities, and their ability to form rings, cages, tubes, etc. Therefore, it is not a surprise that nearly all the HEDMs currently in use are based on C, H, N, and O. These atoms are constituents of small "energetic units" that are combined to yield larger, more powerful HEDMs. This is the concept behind the RDX, HMX, and CL-20, all of which are based on the nitramine (H_2NNO_2) unit. Other units have also been proposed, for example, $CNNO_2$. Another concept that is being explored is building rings or cage structures of C or N atoms. From this view point, cubane and substituted cubanes have been synthesized and are at the forefront of current HEDM research. Polynitrogen species similar to carbon analogues have fascinated the community for a long time, but except for N_2, N_3^-, and the recently synthesized N_5^+, other polynitrogens have not been produce in gram quantities.

So, the question is how we optimize the search for new HEDMs. The new "energetic units" or structures must be identified. These energetic units must be combined in various possible ways, and the viability of the resulting structures, as HEAD and to synthesis, must be assessed. This is a time-consuming and costly process since the synthesis of HEDM requires "specialized" facilities and personnel who are "experts" in handling potentially explosive materials. It is generally accepted that the development cycle for a new HEDM is more than 20 years. TNT is a good case in point: Until the1900s, TNT had not been widely used even though it had been known long before. Can this process be accelerated? Are there other areas of science that share similar constraints and can we learn from their experience? The pharmaceutical industry comes to mind and the automated processes in practice today in rational drug design are designed to cut the cost and the length of the development cycle. One element of this automated process is computer simulation. Computer simulations can range from multiscale simulations to high-accuracy quantum mechanical calculations of isolated molecules in the gas phase. There are many reasons why computer simulations are preferred. The accurate quantum chemical design and characterization of experimentally unknown molecules open the door to predicting molecular structures that will enable larger amounts of energy to be stored. Computational studies are far more cost effective than the often expensive experiments required to measure some essential quantity. Furthermore, they can be applied in extreme conditions where experiments might be impossible, and to transient molecules such as reaction intermediates whose experimental observation is difficult. Related examples include cations and anions that typically will not exist except as transients in reactions or in interstellar space. Short-lived free radicals are another category of molecules about whom essential information might be desired, but hard to obtain without predictive quantum chemical methods. Obviously, there are limitations. The most accurate quantum mechanical treatments are mostly limited to isolated molecules in the gas phase. However, the actual material used in practice is far from this ideal case and a simulation of the behavior of the actual material requires merging theoretical models that span various length- and time scales. These are the research topics that continue to make advances and constitute the subject matter of several chapters in this volume.

As mentioned earlier, our group's foray into this arena of research goes back to the early 1990s. Then, as now, we were motivated by assessments of future military requirements for significantly smaller, lighter, more versatile weapon systems, and the ceaseless quest for more efficient and versatile propellants or explosives with the least impact on the environment for space exploration and other industrial applications. We understood that the number of existing technical approaches can be systematically combined to make marginal improvements in these applications; however, significantly new capabilities in stored energy density and energy release rates are required for orders-of-magnitude improvements. Our effort addressed the question of how much energy could be stored in a molecule by exploiting the predictive quantum chemical methods we had developed, coupled-cluster (CC) and many-body perturbation theory, to assess how to increase the energy content of molecules while ensuring that they have sufficient barriers to unimolecular dissociation to enable them to have adequate lifetimes to be used. In this context, we outline the theoretical developments we pursued in order to accurately identify the best possible HEDM candidates and highlight several important applications that we made in the pursuit of new HEDMs.

2. THEORY DEVELOPMENTS

The development and application of electronic structure theory to atoms, molecules, solids, materials, and biosystems permeates all of science as matter is composed of electrons and positively charged nuclei, of which the latter mostly have a passive role since the nuclei move much more slowly than do the electrons. If we can solve the quantum mechanical equations that describe the electrons in matter to sufficient accuracy, then we can know virtually all there is to know about a molecule's structure, its energetics, all its spectral characteristics (IR, Raman, NMR, ESR, UV–vis, PES), and for the more dynamical aspects of chemistry, its reaction and decomposition paths. Cross-sections that measure the probability of certain phenomena occurring, as in collision-induced vibrational emission are quite important in plume detection and identification.[5] Many of the computed properties are summarized in Fig. 4.2 for an arbitrary molecule, *ABC*, along with some expected levels of accuracy determined from a large number of molecules obtained with today's reference method, CC theory.[6–9]

As the figure shows, molecular structure is determined for the adiabatic potential energy surfaces by the places where all unique bond lengths R and angles, Θ, assume an energy minimum (R_{eq}), meaning that $\partial E/\partial X_\alpha = 0$. The Hessian matrix $\partial^2 E/\partial X_\alpha \partial X_\beta > 0$ if the molecule's geometry is a minimum. The Hessian matrix provides the harmonic approximation for vibrational frequencies ω_{vib} essential to IR and Raman spectra, while derivatives of the dipole moment and polarizabilities give the IR and Raman intensities, respectively. Transition states (TS) occur where there is a saddle-point, meaning that the first derivatives vanish, but the Hessian matrix is

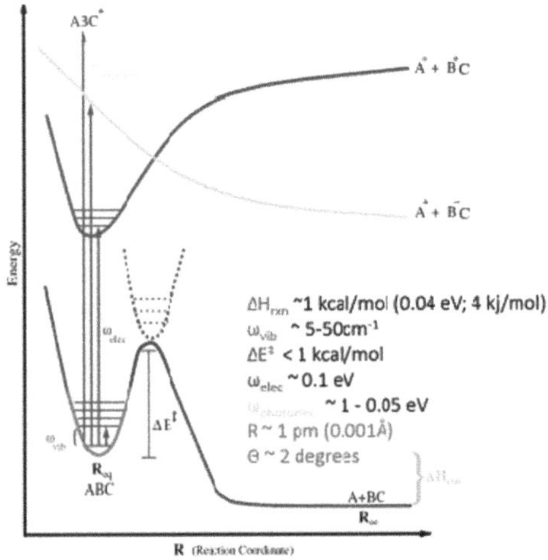

Figure 4.2 Molecular structure and expected accuracies of the high-level CC methods. (For color version of this figure, the reader is referred to the online version of this chapter.)

positive in all degrees of freedom but the one that defines the reaction path. The TS indicates a decomposition to products, $ABC \rightarrow A + BC$. The activation barrier for the process is $\Delta E^{\ddagger}$. Note that it is essential in all applications of theory to be able to readily evaluate the first and second derivatives to verify the nature of the critical points and provide the spectra. This was a major accomplishment for the nonvariational, size-extensive CC theory.[10]

The applications of quantum chemistry techniques also have to include excited states, described by the equation-of-motion (EOM-CCSD or EOM-CCSD(T)) method,[11] as a transition to a one bound and one unbound state. Such electronic excitations normally occur in the UV–vis range. The intensity of the transition comes from the computed dipole strength. For dissociative states, in particular, this information provides cross-sections that are critical to assessing the sensitivity of a HEDM.[12]

Additional excitations to ions ABC^{+} are also shown. They correspond to photoelectron spectra, and they can be accompanied by further "shake-ups" or Auger effects where other electrons move to now-vacated orbital locations. The energy of the bound ground state of the molecule is given by its reaction enthalpy, ΔH_{rxn}. Its free energy can be obtained from the enthalpy and the vibrational and rotational entropy contributions evaluated from their computed energy levels.

The solution of the electronic structure problem provides the starting point for molecular dynamics, which determine the time evolution of a system.[13] Furthermore, properties that depend on ensembles of units as described by statistical mechanics are obtained from the electronic properties of the units. So, developments of accurate methods for solving the Schrödinger equation for the electrons in matter (or in the case of heavy elements where relativistic effects are important, the Dirac equation) is the all-important first step toward a direct, computational solution for many problems, or a way to assist experimental interpretations for systems of interest to the community. The most critical element in the success of this work, however, is that the theory must be "predictive." Achieving this goal demands further development of the theory. Such extensions over time not only eliminate the failings encountered in commonly applied methods, but also outstrip the advances made in computer hardware, making more and more accurate applications possible.

Currently, there are two dominant approaches to the electronic structure theory of molecules. One derives from wavefunction (WFN) theory; the other from density functional theory (DFT).[14,15] It is well known that because of the two-particle interactions we have to describe in molecules, any quantum chemical method that fails to properly include the instantaneous interactions of electrons that correlate their motions cannot be "predictive." The most satisfactory realization of the former is CC theory, primarily developed in our laboratory with support from DOD. The CC WFN is $\Psi_{CC}=\exp(T)\Phi_0$, where Φ_0 is some reference function such as Hartree–Fock, and $T=T_1+T_2+T_3+\cdots$ is an excitation operator composed of single, double, triple, and higher excitations of Φ_0. By virtue of the exponential, all products of the T_n operators are naturally included into the WFN making it highly correlated and size-extensive,[16] meaning that it scales properly with the number of electrons in the system. This is contrary to truncated CI methods and largely accounts for CC's superior performance. CCSD[17] means all T_1 and T_2 operators, while T_3 introduces triple excitations either approximately, CCSD(T),[18–20] or fully, CCSDT.[21] CC theory provides the most accurate answers for the largest number of problems encountered for molecules.[7] The average accuracy with good basis sets (triple zeta or better) for CCSD(T) is shown in Fig. 4.2. The primary advantage of correlated WFN methods is that they have to converge to the right answer in the limit of electron correlation and basis set. This makes such approaches "predictive," and with the development of the massively parallel program ACES III,[22] they can be applied to (~10–100 atoms). However, such calculations demand substantial computational resources for larger molecules.

The DFT approach to electron correlation attempts to bypass the complications of a direct treatment of the two-particle interactions, as in correlated WFN methods, by replacing them with an effective correlation functional that attempts to describe such two-particle interactions analytically. Then, this analytical description, usually written as an exchange-correlation functional of the density, $E_{XC}[\rho(1)]$, leads to a set of one-particle Kohn–Sham equations[14] whose solution can be obtained self-consistently. This is operationally like the uncorrelated HF theory, but now it has some treatment of correlation. Because of their one-particle structure, DFTs can be applied to much larger systems than can *ab initio* correlated WFN methods such as CC theory. However, the correlated WFN methods have to converge to the exact answer in the limit of basis and correlation, while DFT methods use an enormous number of different choices for the $E_{XC}[\rho(1)]$ functional, and it is not clear when one is better than another; there is no route toward converging on the exact answer. Instead, for a given choice of $E_{XC}[\rho(1)]$, or gradient corrected functionals[23–26] ($E_{XC}[\rho(1), \Delta\rho(1)]$, or hybrid functionals[14] where some amount of HF exchange is added, or even double hybrid functionals that mix some second-order wavefunction correlation and HF exchange with some generalized gradient approximation corrected (GGA) exchange and correlation functionals[15]; there is no guarantee that the answers will improve. In fact, it is commonly observed that the simplest local density approximation (LDA)[27,28] is usually better for solids than the GGA or hybrid approximations that are meant to be improvements. For molecules, however, LDA is a poor approximation. So there is a dilemma in the field. On the one hand, we have a computationally attractive, often fairly good but nonpredictive approach, which has almost as many choices for E_{XC} as there are investigators. On the other is a predictive method that is computationally demanding, but entirely rigorous.

3. PREDICTIVE THEORY IN SEARCH OF HEDM

Boron and Carbon are known to have their own chemistries; they both have prolific affinities to react and to form molecules with diverse bonding characteristics and structures. On the other hand, nitrogen, despite being next to carbon in the periodic table, was once considered so inert that Laviosier[29] named it azote, meaning "without life." Is it really without life, or is it awaiting the discovery of it mysteries?

We sought to answer to this question by predictive theory. The prototypes are the tetrahedral form of N_4 (Fig. 4.3A) and the cubane-like

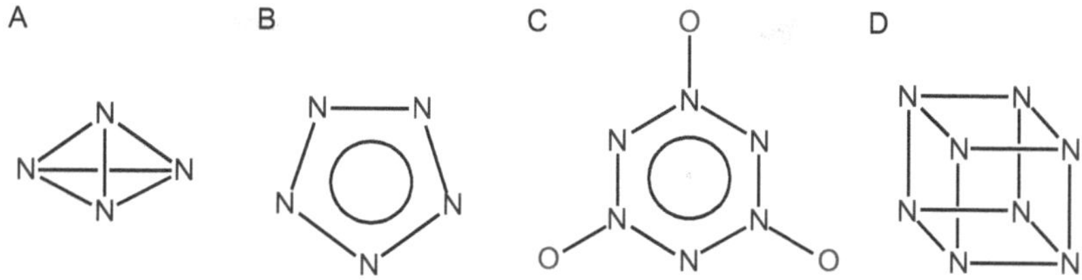

Figure 4.3 (A) Tetrahedral N_4, (B) pentazole anion (N_5^-), (C) $N_6O_3^{(3-)}$, and (D) octatetrane (N_8). (For color version of this figure, the reader is referred to the online version of this chapter.)

octahedral form, N_8 (Fig. 4.3D). In addition, we proposed that the pentazole anion, N_5^- (Fig. 4.3B) should exist along with several other highly energetic systems. The beauty of these structures is that even freshman-level valence bond arguments predict that they should exist. One of us fondly refers to them as "molecules that should exist but don't."[30].

Before we discuss high-level theoretical calculations of these molecules, let us discuss some interesting qualitative data pertaining to bonding of nitrogen and carbon. The most stable form of nitrogen is N_2 with a triple bond. The bond enthalpies of nitrogen single, double, and triple bonds are 38, 100, and 226 kcal/mol, while the bond enthalpies of the C analogues are 85, 143, and 194 kcal/mol, respectively.[31] Therefore, purely nitrogen molecules that are made of N—N single and double bonds have the capacity to store very large amounts of energy relative to their detonation product of N_2, particularly if more single bonds are formed. For example, Fig. 4.1 shows an estimate of the detonation pressure of N_4 and N_8 compared to other known explosives. This shows that polynitrogens could increase the performance by a factor of 2. Also, the fuels and explosives based on them would have no signature and be safe for the environment. So, what is it that makes all carbon structures so ubiquitous but not all nitrogen structures? There are two competing factors at work. Nitrogen atoms have lone pairs of electrons that repel each other, so, naively, their single bonds are likely to be much weaker than, for example, carbon–carbon single bonds. Nitrogen also forms relatively weak double bonds. However, in addition, at a very basic level, a molecule exists only if the barrier to dissociation is enough to keep it from falling apart to its dissociation products. So, if there is a sufficient barrier, it should be possible to create and conceivably trap the molecule. Since the dissociation occurs via bond breaking, to a good approximation, an activation energy is the bond enthalpy of its weakest link. In the case of polynitrogens, it is the N—N single bond, which is less stable than the C—C single bond. So, despite the fact that the N—N single bond stores more energy, the design of a polynitrogen might need to introduce double bond character through delocalization to ensure a sufficient barrier to dissociation.

3.1. The tetrahedral N_4

The tetrahedral form of N_4 (Tetrazete) is a highly sought after HEDM.[32–35] Even though the possibility of its existence has been predicted theoretically for a long time, despite numerous attempts, no experiments have yet been able to conclusively prove that it can be made in any medium. The isoelectronic N_3O^+ has been observed in mass spectrometry. The paper titled, "Tetrazete (N_4)—Can it be prepared and observed?" by Radziszewski and coworkers[36] reports on the potential preparation of N_4 in the gas phase. N_4 has only one IR active mode and the best currently available theoretical frequency (CCSD(T)/aug-cc-pVTZ) is 936 and 904 cm^{-1} for $^{14}N_4$ and $^{15}N_4$, respectively.[37] Radziszewski et al. report observing a fundamental IR transition at 936.7 and 900 cm^{-1} for $^{14}N_4$ and $^{15}N_4$, respectively. Considering the fact that the computed frequencies are harmonic and have at least ± 5 cm^{-1} error bars, the agreement between the computed and theoretical $^{14}N_4$ frequency is remarkable. However, the computed and measured isotopic shifts differ by about 5 cm^{-1}, which is larger than expected since computed isotopic shifts are considered to be more accurate because of error cancellations. This contradictory observation and the lack of further evidence such as Raman in the experiment by Radziszewski et al. because of low column density, has left the title question, "Tetrazete (N_4)—Can it be prepared and observed?" remains unanswered.

From the theoretical standpoint, the quality of the calculations must also be further improved to obtain experimentally accurate spectroscopic and energetic data so that if the material is indeed made, then it can be unambiguously identified. Therefore, we have recently undertaken basis set limit calculations of anharmonic vibrational frequencies and intensities, and the Raman intensities. The excitation energies, ionization potentials, electron affinities, and other relevant data, such as activation energy for decomposition to N_2, dissociation enthalpies, are also obtained with the current best theoretical models. In Table 4.1, we report the basis set limit geometries and harmonic frequencies of N^{14} and N^{15}.

As we can see from the theoretical IR spectrum, among the three vibrational modes, only one is IR active. The others are Raman active. Therefore, we concluded that a conclusive identification would require matching

Table 4.1 Basis set limit bond lengths (in Å) and harmonic IR frequencies (in cm^{-1}) of N^{14} and N^{15}; IR intensities given in parenthesis (in km/mol) obtained at CCSD(T)

		CCSD(T)/aug-cc-pCVTZ (frozen core)		**CCSD(T)/aug-cc-pCVTZ (all MOs)**		**CCSD(T)/NASA-AMES-ANO**		**CCSD(T)/aug-cc-pCVQZ**	
R(NN)		**1.4605**		**1.4577**		**1.4434**		**1.4524**	
	Symm.	**N^{14}**	**N^{15}**	**N^{14}**	**N^{15}**	**N^{14}**	**N^{15}**	**N^{14}**	**N^{15}**
Harmonic vib. freqs. and intensities	E	725.9	701.4	728.6	704.0	732.6	707.8	732.3	707.6
	T_2	937.6(13)	905.9(13)	942.0(3)	909.7(3)	950.9(4)	918.7(4)	950.6(3)	918.3 (3)
	A_1	1296.7	1252.9	1301.6	1257.6	1325.2	1280.4	1316.2	1271.7

the other two modes with the computed Raman intensities, developed necessary extensions of CC theory to obtain predictive quality Raman intensities, and reported the first calculation of CCSD Raman intensities of N_4 and N_5^- (Ref. 37). As stated earlier, we have been extending our earlier calculations on N_4 to obtain the most definitive spectroscopic properties. The CCSD Raman intensities obtained with the aug-cc-pCVTZ basis set are 3.8 (0.75), 8.9 (0.75), and 39.1 (0.0) (in $Å^4$/a.m.u.) for E, T_2, and A_1 vibrational modes, respectively (the depolarization ratios are given in parenthesis).* It should be noted that a complete list is published elsewhere.

3.2. The pentazole anion (N_5^-)

The pentazole anion[38–40] (Fig. 4.1B) has a rich, though controversial, history. Its first observation was reported in 1915[38] although that proved to be false.[39] It is isoelectronic with the cyclopentadienyl anion, aromatic, and has D_{5h} symmetry, and state-of-the-art theoretical calculations[40] predict that it is a minimum on its potential energy surface with an activation barrier of 27 kcal/mol to prevent auto-decomposition. Despite all the available theoretical and conceptual data pointing to its existence, N_5^- remained elusive until very recently; there are reports that it has now been observed in negative ion mass spectra,[42,43] followed by a potential NMR observation in solution.[44]

However, the initial NMR results have been controversial. They were disputed by Schroer et al.[45] Subsequent work by Butler et al.[46] concluded that their earlier paper[44] had not observed the pentazole anion because it was too short-lived. However, Butler et al.[46] argue that it was made in the experiment as proven by its unique by-product of an isotopically labeled azide anion. The latter could only appear from the terminally labeled *para*-methoxyphenylpentazole (4-$MeOC_6H_4N_5$) precursor. We proposed to employ predictive theory as a third voice in resolving the differences in opinion regarding the different NMR assignments. Our proposal was to compute the full NMR spectrum (chemical shifts[47] and spin–spin coupling constants) of all the relevant species, in particular NO_3^-, N_3^-, and 4-$MeOC_6H_4N_5$, and they are shown in Tables 4.2 and 4.3. A detailed account of this work is published elsewhere.[48,49] Here, we only summarize those findings to emphasize that our theoretical methods can compute properties such as NMR spin–spin coupling constants, as originally developed in

*We note that in our previous paper on Raman intensities,[37] the intensities of E and T_2 vibrational modes are reported as integer multiples of the degeneracy of the mode.

Table 4.2 The computed and measured (experimental gas phase data shown in bold) N^{15} chemical shifts of the molecules that are of interest in the current context (in ppm with respect to CH_3NO_2 internal standard. The unique atoms of N_5^+, N_5H, $CH_3OC_6H_5N$, and $CH_3OC_6H_5N_5$ are labeled as N_1, N_2, and N_3 in accordance with the literature)

	cc-pVTZ	cc-pVQZ	PZ3D2F	CP corrected	Experiment
N_2	−85.8	−80.1	−77.5	−70.5	**−75.3**[a], −70.2[a]
N_tN_cO	−251.4(N_t)	−246.3	−244.9	−227.4	**−232.3**[a], −225.0[b]
	−159.1(N_c)	−152.7	−149.9	−143.9	**−148.0**[a], −140.0[b]
NH_3	−410.7	−408.7	−407.2	−389.7	**−400.1**[a], −380.2[a–c]
CH_3CN	−137.6	−134.4	−132.3	−137.3	**−126.7**[a], −137.1[b]
ClCN	−150.3	−147.0	−145.0	−139.0	−144.0[a]
HCN	−127.3	−125.2	−123.0	−128.0	−127.5[a]
NO_3^-	−11.7	−2.6	1.5	−5.5	−11.5[d], −4.0[a]
$N_t(N_cN)^-$	−310.6(N_t)	−309.0	−309.1	−291.6	−283[e], −282.2[d]
	−145.1(N_c)	−136.2	−131.9	−126.9	−147[e], −133.6[d], −130[b]
NH_4^+	−380.7	−382.4	−381.3	−363.8	−359.8[d], −359.6[a]
N_5^+	−264.1(N_1)	−262.4	−260.8	−243.3	−237.3[f]
	−179.0(N_2)	−174.2	−171.7	−165.7	−165.3[f]
	−111.7(N_3)	−105.7	−103.1	−98.1	−100.4[f]
N_5^-	−22.8	−16.5	−13.5	−6.5	
N_5H	−136.7(N_1)	−131.3	−128.5	−121.5	
	−44.0(N_2)	−37.6	−34.6	−27.6	
	−8.7(N_3)	−0.5	3.3	10.2	
$MeOC_6H_4N_3$	−295.9(N_1)			−278.4	
	−139.5(N_2)			−133.5	−135.5[d]
	−137.2(N_3)			−131.2	−148.2[d]
$MeOC_6H_4N_5$	−87.8(N_1)			−92.8	
	−27.0(N_2)			−34.0	−26.7[e]
	13.9(N_3)			6.9	5.6[e]

[a]Reference 50.
[b]Reference 51.
[c]Reference 52.
[d]Reference 53.
[e]Reference 43.
[f]Reference 54.

Table 4.3 The $^1J(^{15}N^{17}O)$ and $^1J(^{15}N^{15}N)$ (in Hz) spin-spin coupling constants of NO_3^-, N_3^- and N_5H (FC, PSO, SD and DSO are Fermi-contact, paramagnetic spin-orbit, spin-dipole and diamagnetic spin-orbit contributions). The only couplings that are > 1 Hz are shown

	Coupling	FC	PSO	SD	DSO	Total
NO_3^-	$^1J(^{15}N^{17}O)$	−34.4	−7.6	−0.2	0.1	−42.1
		−32.5	−7.7	−0.3	0.1	−40.4
	$^1J(^{17}O^{17}O)$	1.6	−7.3	1.7	0.0	−4.0
		1.5	−7.4	1.7	−0.1	−4.1
$N_tN_cN^-$	$^1J(^{15}N_t{}^{15}N_c)$	−12.3	−2.0	0.6	0.0	−13.7
		−11.2	−2.0	0.6	0.0	−12.6
	$^2J(^{15}N_t{}^{15}N_t)$	−3.6	2.7	3.4	−0.1	2.4
		−3.8	2.9	3.5	−0.1	2.7
N_5H[a]	$^1J(^{15}N_1{}^1H)$	−113.1	−0.7	−0.2	−0.5	−114.4
	$^1J(^{15}N_1{}^{15}N_2)$	−13.3	−3.3	0.1	0.0	−16.5
	$^2J(^{15}N_2{}^1H)$	−10.3	0.7	0.0	0.3	−9.3
	$^1J(^{15}N_2{}^{15}N_3)$	−13.2	−4.8	0.5	0.0	−17.5
	$^1J(^{15}N_3{}^{15}N_4)$	−15.9	−4.1	0.3	0.0	−19.7
$C_6H_5N_3$[b]	$^1J(^{15}N_1{}^{15}N_2)$	−12.5	−1.8	1.8	0.0	−12.4
	$^1J(^{15}N_2{}^{15}N_3)$	−14.1	−2.3	0.0	0.1	−11.7
$MeOCH_4N_3$[b]	$^1J(^{15}N_1{}^{15}N_2)$	−12.8	−1.9	0.5	0.0	−14.1
	$^1J(^{15}N_2{}^{15}N_3)$	−12.6	−1.8	1.0	0.0	−13.4

[a]The N_5H coupling constants are obtained with the qz2p basis set[52] only.
[b]See footnote of Table 4.2 of Reference 51.

this group. These new developments have helped to offer solutions to otherwise intractable controversies (Tables 4.2 and 4.3).

Let us first consider the computed shifts for N_5^- and NO_3^- and the conflicting assignments proposed. Both groups observed a ^{15}N NMR signal in the region of −10 ppm. We observed that the computed shift for NO_3^- can vary greatly with polarity; experimentally from −11.5 to −4.0 ppm, but except for the pentazole itself, only NO_3^- falls in this range. On the basis of the agreement shown by our corrected shifts of N_5^+ and $MeOC_6H_4N_5$ with experiment, we expected a similar agreement of computed and observed N_5^- shifts. This led us to conclude that the observed NMR signal in the -10 ppm region is indeed due to NO_3^- rather than N_5^-. As noted earlier, the chemical shifts for N_c and N_t of N_3^- are close to what is reported for its chemical shifts by both groups. The ^{15}N chemical shifts of $MeOC_6H_4N_3$ are also in the same region as N_3^-, further complicating the assignments. Nevertheless, Schroer et al.[45] assigned the shifts in the first experiment to $MeOC_6H_4N_3$, and concluded that the N_3^- shifts should be -133.6 ppm (N_c) and -282.2 ppm (N_t). Schroer et al.[45] also questioned the presence of N_3^- in the highly oxidative medium. Despite these coincidences, the presence of the centrally ^{15}N-labeled N_3^- is a key piece of evidence that supports the presence of N_5^-, though it is of too short a lifetime to be seen in NMR (Tables 4.2 and 4.3).

To resolve this remaining question, we suggest an analysis of the scalar NMR spin–spin coupling constants. The NMR spin–spin coupling constants that would help unambiguously resolve the disputed evidence and establish whether the N_5^- is formed are $^1J(^{15}N^{17}O)$ of NO_3^- and $^1J(^{15}N_t^{15}N_c)$. The computed values for the couplings for qz2p and pz3d2f basis[52,55] sets are shown in Table 4.3. Both basis sets give the same results, indicating that they are essentially converged. (The qz2p basis has been previously calibrated and shown to be well suited for NMR coupling constants.[53] The pz3d2f basis set is used to maintain consistency with shift calculations.) Unlike chemical shifts, the spin–spin couplings are less sensitive to the medium effects, and it is our experience that their computed EOM-CCSD values are accurate to within ± 5 Hz. A new experiment to measure the relevant coupling constants would resolve the disputed assignments and conclusively establish whether the N_5^- had been made in solution. (Upon request from us, Prof. Butler extracted the observed spin–spin coupling constants from their spectra.[51] They are 11.4 ± 0.5 and 13 ± 0.5 Hz in CD_3OD–D_2O, with and without Ce^{3+}, respectively, which is in near-perfect agreement with our prediction. This is substantial evidence to support the presence of centrally ^{15}N-labeled N_3^-, which could only be possible had N_5^- also been present.)

3.3. Salts of N_5^+ with N_3^- and N_5^- ($N_5^+N_3^-$, $N_5^+N_5^-$)

Since N_5^+ is stable at room temperature (as salts of AsF_6^-),[56] it is conceivable to combine them with a suitable anion. What counter anion is suitable? To a first approximation, it is determined by the electron affinity (EA) of the anion and the ionization potential (EA) of the N_5^+. It is obvious that the ideal anions should have IPs whose magnitude is similar to that of the EA of N_5^+. Both IPs and EAs can be very accurately computed by the IP-EOM and EA-EOM methods that we developed in our laboratory.

These concepts are being explored in possible N_8 and N_{10} species by combining N_5^+ with N_3^-, which is well known, and N_5^-, which has been observed in negative ion mass spectroscopy, as possible anion.[42,57] The EA of N_5^+ and IP of N_3^- and N_5^- computed at the IP/EA-EOM-CCSD/POL1 level are shown in Table 4.4.

A closer look at the EAs and IPs suggests that N_5^+ should be much more ionic and, consequently, more stable with N_5^- than with N_3^-. In fact, most simple anions (as opposed to AsF_6^- and others) cannot be combined with N_5^+, because of its exceptional vertical EA (6.0 eV). However, N_5^- appears to be a potential exception because of the balance between its vertical IP (5.6 eV) and the EA of N_5^+. Nevertheless, as part of a study of N_8, we first considered an $N_5^+N_3^-$ ion pair with fragment charges of ± 0.3 and demonstrated that they are not very stable. The barriers toward dissociation or rearrangement to diazidyldiazenes are less than 5 kcal/mol at B3LYP526/aug-cc-pVDZ and just a few kcal/mol larger at MBPT(2)/aug-cc-pVDZ. In dissociative TS, the already large distance between the N_3 and N_5 fragments increases slightly, while the N_5 fragment splits into N_2 and N_3. We also concluded that the isolation of covalent N_8 (diazidyldiazene) is unlikely because of a small (18 kcal/mol) decomposition barrier and may not even form because one of the approach pathways has great potential for mutual neutralization and fragmentation.

Our exploration of the ion pair $N_5^+N_5^-$ has shown that it is a minimum on the N_{10} surface. There are lower energy structures, including an N_{10} analogue to azidylpentazole and the bipentazole, initially proposed in 1992. The TS between these two structures has not been located. Gas-phase

Table 4.4 Vertical EA and IPs[a] in eV at EOM-CCSD/POL1/B3LYP/aug-cc-pVDZ

N_5^+, EA			N_3^-, IP			N_5^-, IP		
C_{2v}	C_s		$D_{\infty h}$	C_s[b]		D_{5h}	C_s[b]	
B_1	A'	6.04	Π_g	π (nb)	2.53	E_1'	A′	5.58
A_2	A''	5.63	Σ_u	σ (LP)	7.21	E_1''	A″	6.65
A_1	A'	5.53	Σ_u	σ (LP)	8.68	E_2'	A′	6.14
A_1	A'	3.93	Π_u	π (nb)	8.82	A_1'	A′	9.98

nb, nonbonding; LP, lone pair.
[a]Molecular plane of the complexes.
[b]Molecular plane of the complexes.

$N_5^+N_5^-$ has enough charge separation to be called an ion pair. This is in contrast to the corresponding $N_5^+N_3^-$ structure, which has so much electron transfer that it should be called a complex. As discussed earlier, the relatively ionic structure of $N_5^+N_5^-$ is due to the unusually large ionization potential of N_5^-. In the gas phase, $N_5^+N_5^-$ has a barrier of 25 kcal/mol toward loss of N_3 from the N_5^+ fragment, twice as high as the corresponding barrier in $N_5^+N_3^-$. A correlation between dissociation barriers and charge of the N_5 fragment at the equilibrium geometry suggests that a larger positive charge of N_5^+ must be accompanied by greater kinetic stability. Quasiperiodic cluster calculations show increased partial charges of the N_5^+ fragment. This would appear to move the dissociation barrier into a practically useful range. The first-principle prediction of crystalline structures and their stability is not yet possible at high levels of theory, as it would be for isolated molecules. Studies at the DFT level can be made to a degree, but studies with *ab initio* methods such as CC theory are at the forefront of new theory developments. This is a research topic that will be active in the next few years.

3.4. Other candidates

There is little reason to limit the theoretical design and characterization of such molecules to just nitrogen. Although nitrogen has the capacity to store a great deal of energy, the judicious use of other elements can provide better stability and performance. The vast majority of energetic molecules combine N with C and O, but this continues the well-worn path toward C, H, N, O chemistry that is not likely to offer any revolutionary improvements. However, it is well known that adding metals to explosives can greatly enhance their performance. One way to achieve this is to conceive of making new species by adding Li in place of H to known propellants or explosives.

Hydrazine, $H_2N{-}NH_2$, is an extensively used monopropellant. When substituted by Li, Li_2NNLi_2, it would offer major improvements in its theoretical performance. Hydrazine's heat of formation is 15.7 kcal/mol with a specific thrust of 185.7 s. Replacing H with Li gives a heat of formation of 79.8 and an I_{sp} of 319, suggesting a revolutionary improvement. This unknown molecule's particulars from our calculations are shown in Fig. 4.4 and its harmonic frequencies, and IR and Raman intensities are given in Table 4.5.

Another quite interesting molecule is N_3Li_3. The D_{3h} structure shown in Fig. 4.4 is a minimum at the theoretical level that was employed and has a

Table 4.5 Harmonic frequencies, and IR and Raman intensities of Li_4N_2[a]

No.	Symm.	Frequency (cm^{-1})	IR intensity (km/mol)	Raman intensity Å^4/(a.m.u.)	Depolarization ratio
1	A_u	219.1	0.0	0.0	0.41
2	A_u	375.3	40.7	0.0	0.46
3	B_u	376.0	40.8	0.0	0.75
4	A_g	472.8	0.0	973.4	0.00
5	B_g	484.5	0.0	196.0	0.75
6	A_g	485.3	0.0	149.2	0.75
7	A_g	486.3	0.0	58.0	0.73
8	B_u	514.5	299.0	0.0	0.75
9	B_g	597.6	0.0	143.9	0.75
10	A_u	727.6	214.8	0.0	0.04
11	B_u	728.0	214.9	0.0	0.62
12	A_g	735.1	0.0	325.9	0.03

[a]Mode nos. 2, 3, 7, 10, 11, and 12 correspond to Li–Li–Li–Li ring opening, Li–Li–Li–Li ring opening, Li ring rupture (two bonds breaking), Li—N bond breaking (two bonds breaking), Li—N bond breaking (four bonds breaking), and dissociation into $2Li+N_2Li_2$, respectively.

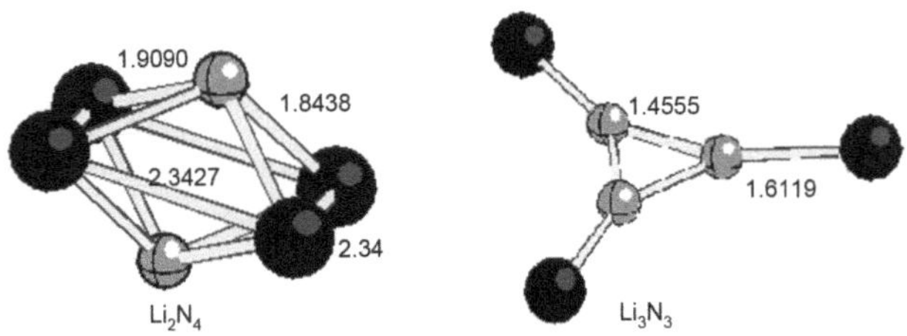

Figure 4.4 Structures of Li-substituted hydrazine, Li_2N_4, and Li_3N_3. (For color version of this figure, the reader is referred to the online version of this chapter.)

heat of formation of 173.50 kcal/mol and $I_{sp}=440.3$ s. Notice its similarities to $C_3H_3^+$, another molecule that we considered purely on the basis of its academic interest (A. Perera, Ben Mccall and R. J. Bartlett). The harmonic frequencies, and IR and Raman intensities are shown in Table 4.6.

The vibrational frequencies demonstrate that the molecule is a true minimum on its potential energy surface, implying that it can be made and used. Many other logical but experimentally unknown molecules can be similarly

Table 4.6 Harmonic frequencies, and IR and Raman intensities of Li_3N_3[a]

No.	Symm.	Frequency (cm^{-1})	IR intensity (km/mol)	Raman intensity $Å^4$/(a.m.u.)	Depolarization ratio
1	E'	83.4	58.1	768.5	0.72
2	A_2'	123.0	0.0	~0.0	0.27
3	A_2''	173.4	106.9	~0.0	0.75
4	E''	299.3	0.0	>1000	0.75
5	E'	587.7	45.8	>1000	0.67
6	A_1'	637.1	0.0	>1000	0.25
7	E'	868.4	45.1	>1000	0.69
8	A'	1338.4	0.0	>1000	0.24

[a]Mode nos. 5, 7, and 8 represent ring opening, dissociation into $Li+N_3Li_2$, and dissociation into $3N+3Li$ (three bonds breaking), respectively.

formed by substituting Li for H. Only predictive electronic structure theory can readily survey a wide number of such prospective molecules for their existence and spectroscopic signatures. A complete survey of various polynitrogens, hydrazines, and Li-substituted hydrazine is cataloged.[58]

All the numerical results reported in this chapter have been obtained with the ACES II program system[59] developed under the auspices of the HEDM program. With the recent development of ACES III,[22] the massively parallel program system that scales to tens of thousands of processors, predictive quantum chemistry can be applied to quite large potential molecular structures. Besides the predictive quantum chemical calculations role as a screening tool for potential HEDM candidates, this means that a large number of possible reactions paths can also be accurately computed to elucidate the critical steps in a given synthetic procedure or a decomposition mechanism as well. Although they are not in the context of this review, which focuses on speculative HEDM, we would like to mention the recent applications of the ACES III program to study isomers, vibrational spectra, and the excited states of RDX, HMX, and CL-20 that are currently being used in practical applications.[60–62]

4. FUTURE PROSPECTS

No conventional synthesis has yet emerged for making these systems, but another route is to use high-pressure, anvil cell experiments to create solids of highly energetic N, as in the independent efforts of Hemley

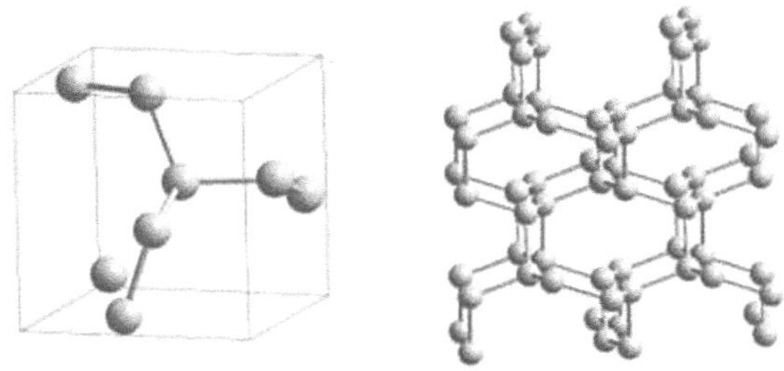

Figure 4.5 cg-N structure. Each atom of nitrogen is connected to three neighbors with three single covalent bonds. The primitive cell is shown on the left, and an extended structure of the polymeric nitrogen is shown on the right.

et al.[63] and Eremets et al.[64,65] This novel alternative has made it possible to "synthesize" polynitrogen species with single bonds that we had predicted would be highly energetic. The structure found by Eremets et al.[64] is shown in Fig. 4.5, with its unit cell on the left.

This is a proof of principle as under extreme conditions, many polynitrogen forms can be created, and depending upon the comparative amounts of the different forms of N—N bonds in the species, they could have very large heats of formation. The full characterization of such species, including whether they can be expected to have an adequate lifetime for application, requires the combined efforts of experiment and theory, as predictive theory can address the structure of the species, their spectra, and the low-lying excited states that can potentially provide alternative decomposition pathways that could cause energy loss.

HEDMs must have reduced sensitivity and should meet the minimum thresholds of the effects of their widespread use on the environment to comply with regulatory requirements. These requirements add considerably to the cost and the length of the development cycle. Predictive theory, of course, is not limited to applications to search for new molecules. It can be used to map all possible decomposition pathways and conclusively identify all possible decomposition products. Prior knowledge of the decomposition products helps to assess their impact on the environment. It has been proposed that the low-lying excited states play a key role in the initiation of a cascade of reactions that lead to detonation. The excitation energies and the excited-state structures can be computed accurately by EOM-CC methods to assist in the modeling of shock sensitivity. In closing, we note that computational modeling has the potential to expedite the development cycle, if the modeling is sufficiently accurate. Multiscale modeling, where the atomistic scale processes are described by predictive level quantum mechanical methods, seems to hold promise for the future.

ACKNOWLEDGMENTS

The United States Air Force Office of Scientific Research (AFOSR) HEDM program under the sponsorship of Dr. Michael Berman is instrumental in supporting the basic research described in this chapter. It would not be an overstatement to say that the coupled-cluster theory would not enjoy the success it enjoys today without this support. We also thank the Defense Advancement Research Agency (DARPA) and the sponsorship of the United States Army Research Office (ARO) through Dr. James Parker. Over nearly 20 years of involvement with HEDM, many postdoctoral associates, visiting scientists, and graduate students have contributed to the material covered in this survey. In particular, we would like to acknowledge Drs. John Stanton, Jurgen Gauss, John Watts, Walter Lauderdale, David Bernholdt, Kim Ferris, Ken Wilson, Stefan Fau, Motoi Tobita, Adriana Gregusova, Robert Molt, Thomas Watson, and Mr. Alexandre Bazante. We also thank Dr. William Kopess at NSWC-IH for allowing us to use Fig. 4.1.

REFERENCES

1. National Research Council. *Advanced Energetic Materials*; National Academies Press: Washington, DC, 2004; Vol. 32.
2. *Materials Research to Meet 21st Century Needs*; The National Research Council of the National Academies of the United States, National Academies Press: Washington, DC, 2001.
3. Klapötke, T. M. Ed.; *High Energy Density Materials*; Series: Structure and Bonding, Vol. 125; Springer-Verlag: Heidelberg, 2007.
4. Miller, R. S. Three Phase Combustion Modeling; Frozen Ozone, a Prototype System. In *Decomposition, Combustion and Detonation Chemistry of Energetic Materials*; Brill, T. B.; Russell, T. P.; Tao, W. C.; Wardle, R. B. Eds.; Materials Research Society Symposium, Proceedings, Vol. 418; Materials Research Society: Pittsburgh, PA, 1995.
5. Redmon, M. J.; Bartlett, R. J.; Garrett, B. C.; Purvis, G. D., III; Saatzer, P. M.; Schatz, G. C.; Shavitt, I. Collisional Excitation of H_2O by o-Atom Impact-Classical Dynamics on an Accurate Ab Initio Potential-Energy Surface. In *Potential Energy Surfaces and Dynamics Calculations*; Truhlar, D. G. Ed.; Plenum: New York, 1981; pp 771–803.
6. Shavitt, I.; Bartlett, R. J. MBPT and Coupled-Cluster Theory. *Many-Body Methods in Chemistry and Physics*; Cambridge University Press: New York, 2009.
7. Bartlett, R. J.; Musial, M. Coupled Cluster Theory in Quantum Chemistry. *Rev. Mod. Phys.* **2007**, *79*, 291.
8. Bartlett, R. J. Coupled Cluster Theory: An Overview of Recent Developments. In *Modern Electronic Structure Theory Part II*; Yarkony, D. R. Ed.; Vol. 2, World Scientific: Singapore, 1995; pp 1047–1131.
9. Crawford, T. D.; Schaefer, H. F. An Introduction to Coupled Cluster Theory for Computational Chemists. In *Reviews in Computational Chemistry*; Lipkowitz, K. B.; Boyd, D. B. Eds.; Vol. 14, VCH: New York, 2000; pp 33–115.
10. Salter, E. A.; Trucks, G. W.; Bartlett, R. J. Analytic Energy Derivatives in Many Body Methods. I. 1st Derivatives. *J. Chem. Phys.* **1989**, *90*, 1752.
11. Stanton, J. F.; Bartlett, R. J. The Equation of Motion Coupled-Cluster Method: A Systematic By orthogonal Approach to Molecular Excitation Energies, Transition-Probabilities and Excited-State Properties. *J. Chem. Phys.* **1993**, *98*, 7029.
12. Melnichuk, A.; Perera, A.; Bartlett, R. J. Ab-Initio Simulation of UV/VIS Spectra for Atmospheric Modeling: Method Designed for Medium Size Molecules. *J. Phys. Chem. Chem. Phys.* **2010**, *12*, 9726.
13. Heile, J. M. *Molecular Dynamics Simulations: Elementary Methods*; John Wiley: New York, 1992.

14. Parr, R. G.; Yang, W. T. *Density-Functional Theory of Atoms and Molecules*; Oxford University Press: New York, 1989.
15. Engel, E.; Dreizler, R. M. Density Functional Theory: An Advanced Course. In *Theoretical and Mathematical Physics*; Springer: Berlin, Heidelberg, 2011.
16. Bartlett, R. J.; Purvis, G. D. Many-Body Perturbation Theory, Coupled-Pair Many Electron Theory, and Importance of Quadruple Excitations for Correlation Problem. *Proceedings of the American Theoretical Chemistry Conference, Boulder, Colorado. Int. J. Quantum Chem.* **1978**, *14*, ; 1978561.
17. Purvis, G. D.; Bartlett, R. J. A Full Coupled-Cluster Singles and Double Model—The Inclusion of Disconnected Triples. *J. Chem. Phys.* **1982**, *76*, 1910.
18. Urban, M.; Noga, J.; Cole, S. J.; Bartlett, R. J. Towards and Full CCSDT Model for Electron Correlation. *J. Chem. Phys.* **1985**, *83*, 4041.
19. Raghavachri, K.; Trucks, G. W.; Pople, J. A.; Head-Gordon, M. A 5th-order Perturbation Comparison of Electron Correlation Theories. *Chem. Phys. Lett.* **1989**, *157*, 479.
20. Watts, J. D.; Gauss, J.; Bartlett, R. J. Coupled-Cluster Methods with Noniterative Triple Excitations for Restricted Open-Shell Hartree-Fock and Other General Single Determinant Reference Functions. Energies and Analytical Gradients. *J. Chem. Phys.* **1993**, *98*, 8718.
21. Noga, J.; Bartlett, R. J. The Full CCSDT Model for Molecular Electronic-Structure. *J. Chem. Phys.* **1987**, *86*, 7041.
22. Lotrich, V.; Flocke, N.; Ponton, M.; Yau, A.; Perera, A.; Deumens, E.; Bartlett, R. J. Parallel Implementation of Electronic Structure Energy, Gradient, and Hessian Calculations. *J. Chem. Phys.* **2008**, *128*, 194104.
23. Perdew, J. P. Accurate and Simple Density Functional for the Exchange Energy-Generalized Gradient Approximation. *Phys. Rev. B* **1986**, *33*, 8822.
24. Becke, A. D. Density-Functional Exchange-Energy Approximation with Correct Asymptotic-Behavior. *Phys. Rev. B* **1988**, *38*, 3098.
25. Lee, C. T.; Yang, W. T.; Parr, R. G. Development of the Colle-Salvetti Correlation-Energy Formula into a Functional of the Electron-Density. *Phys. Rev. B* **1988**, *37*, 785.
26. Perdew, J. P.; Burke, K.; Ernzerholf, M. Generalized Gradient Approximation Made Simple. *Phys. Rev. Lett.* **1996**, 77, 3865.
27. Kohn, W.; Sham, L. Self-Consistent Equations Including Exchange and Correlation Effects. *J. Phys. Rev.* **1965**, *140*, 1133.
28. Vosko, S. H.; Wilik, L.; Nusair, M. Accurate Spin-dependent Electron Liquid Correlation Energies for Local Spin-Density Calculations—A Critical Analysis. *Can. J. Phys.* **1980**, *58*, 1200.
29. Laviosier, A. *Elements of Chemistry in New Systematic Order, Containing All Modern Discoveries*, 1st ed.; Dover Publications: Edinburgh, 1790; translated from French by Robert Kerr.
30. Bartlett, R. J. Exploding the Mysteries of Nitrogen. *Chem. Indus.* **2000**, *4*, 140–143.
31. Cotton, F. A.; Wilkinson, G. *Advanced Inorganic Chemistry*; 3rd ed.; Interscience: New York, 1972.
32. Lauderdale, W. J.; Myers, M. J.; Bernholdt, D. E.; Stanton, J. F.; Bartlett, R. J. In *Proceedings of the High Energy Density Materials Contractors Conference, February 1990, Long Beach, CA, February*, 1990; pp 25–28.
33. Lauderdale, W. J.; Stanton, J. F.; Bartlett, R. J. Stability and Energetic of Metastable Molecules—Tetraazaterahedrane (N_4), Hexaaazabenzene and Octaazacubane (N_8). *J. Phys. Chem.* **1992**, *96*, 1173.
34. Lee, T. J.; Rice, J. E. Theoretical Characterization of Tetrahedral N_4. *J. Chem. Phys.* **1991**, *94*, 1215.
35. Dunn, K. M.; Morokuma, K. Transition-State for the Dissociation of Tetrahedral N_4. *J. Chem. Phys.* **1995**, *102*, 4904.

36. Zeng, J. P.; Waluk, J.; Spanget-Larson, J.; Blake, D. M.; Radziszewski, J. G. Tetrazete (N_4). Can be Prepared and Observed?*Chem. Phys. Lett.* **2000**, *328*, 227.
37. Perera, S. A.; Bartlett, R. J. Couple-Cluster Calculations of Raman Intensities and Their Application to N_4 and $N_5^{(-)}$. *Chem. Phys. Lett.* **1999**, *314*, 381.
38. Lipschitz, J. Synthese Der Pentazol-Verbindungen. I. *Ber. Dtsch. Chem. Ges.* **1915**, *48*, 410.
39. Curtius, T.; Darapasky, A.; Müller, E. Die Sogenannten Pentazol-Verbindungen von J. Lifschitz. *Ber. Dtsch. Chem. Ges.* **1915**, *48*, 1614.
40. Ferris, K. F.; Bartlett, R. J. Hydrogen Pentazole—Does It Exist.*J. Am. Chem. Soc.* **1992**, *114*, 8302 (Communication).
41. Fau, S.; Wilson, K. J.; Bartlett, R. J. On the Stability of N_5^+-N_5^-.*J. Phys. Chem. A* **2002**, *106*, 4639.
42. (a) Hahma, A.; Holmberg, E.; Hore, N.; Tryman, R.; Wallin, S.; Bergeman, H.; Östmark, H. In International Annual Conference of ICT 33rd (Energetic Materials), Vol. 62, 2002; p 1; (b) Östmark, H.; Wallin, S.; Brinck, T.; Carlqvst, P.; Claridge, R.; Hedlund, E.; Yudina, L. Detection of Pentazole Anion (Cyclo-N_5^-) and Decomposition of Solid *p*-Dimethylaminophenylpentazole. *Chem. Phys. Lett.* **2003**, *379*, 539–546.
43. Vij, A.; Pavlovich, J. G.; Willson, W. W.; Vij, V.; Christie, K. O. Experimental Detection of the Pentaazacyclopentadienide (Pentazolate) Anion, N_5^-. *Angew. Chem. Int. Ed. Engl.* **2002**, *41*, 3051–3054.
44. Butler, R. N.; Stephens, J. C.; Burke, L. A. First Generation of Pentazole (HN_5, Pentazolic acid), the Final Azole, and Zinc Pentazolate Salt in Solution: A New N-Dearylation of 1-(p-Methoxyphenyl) Pyrazoles, a 2-(p-Methoxyphenyl) Tetrazole and Application of the Methodology to 1-(p-Methoxyphenyl) Pentazole. *J. Chem. Commun.* **2003**, *8*, 1016–1017.
45. Schroer, T.; Haiges, R.; Schneider, S.; Christe, K. O. The Race for the First Generation of the Pentazolate Anion Is Far from Over. *Chem. Commun.* **2005**, *12*, 1607–1609.
46. Butler, R. N.; Hanniffy, J. M.; Stephens, J. C.; Burke, L. A. A Ceric Ammonium Nitrate N-Dearylation of N-p-Anisylazoles Applied to Pyrazole, Triazole, Tetrazole and Pentazole Rings, Generation of Unstable Pentazole, HN_5/N_5^- in Solution.*J. Org. Chem.* **2008**, *73*, 1354–1364.
47. Gauss, J. Effects of Electron Correlation in the Calculation of Nuclear Magnetic Chemical Shifts.*J. Chem. Phys.* **1993**, *99*, 3629.
48. Perera, S. A.; Gregušová, A.; Bartlett, R. J. First Calculation of ^{15}N–^{15}N Values and New Calculations of Chemical Shifts for High Nitrogen System: A Comment on the Long Search for HN5 and Its Pentazole Anion.*J. Phys. Chem. A* **2009**, *113*, 3197.
49. Gregušová, A.; Perera, S. A.; Bartlett, R. J. Accuracy of Computed ^{15}N Nuclear Magnetic Resonance Chemical Shifts.*J. Chem. Theory Comput.* **2010**, *6*, 1228.
50. Levy, G. C.; Lichter, R. L. *Nitrogen-15 Nuclear Magnetic Resonance Spectroscopy*; Wiley Inter Science Publications: New York, 1979.
51. Butler, R. N. Private communication.
52. Auer, A. A.; Gauss, J.; Stanton, J. F. Quantitative Prediction of Gas-Phase C-13 Nuclear Magnetic Shielding Constants.*J. Chem. Phys.* **2003**, *118*, 10407–10417.
53. Burke, L. A.; Butler, R. N.; Stephens, J. C. *J. Chem. Soc., Perkin Trans.* **2001**, *2*, 1679–1684.
54. Jameson, C. J.; Jameson, A. K.; Oppusunggu, D.; Wille, S.; Burrell, P. M.; Mason, J. N-15 Nuclear Magnetic Shielding Scale from Gas-Phase Studies. *J. Chem. Phys.* **1981**, *74*, 81–86.
55. Schaëfer, A.; Horn, H.; Ahlrichs, R. Fully Optimized Contracted Gaussian-Basis Sets for Atoms Li to Kr.*J. Chem. Phys.* **1992**, *97*, 2571.

56. Christe, K. O.; Wilson, W. W.; Sheehy, J. A.; Boatz, J. A. $N_5^{(+)}$: A Novel Homoleptic Polynitrogen Ion as a High Energy Density Material. *Angew. Chem. Int. Ed. Engl.* **1999**, *38*, 2004.
57. Fau, S.; Bartlett, R. J. Possible Product of the End-On Addition of $N_3^{(-)}$ and $N_5^{(+)}$. *J. Phys. Chem. A* **2001**, *105*, 4096–4106.
58. A survey of results for neutral and ionic polynitrogen compounds from N_2 to N_{10} and hydrazine and higher derivative of hydrazines and Li substituted hydrazine is available from: http://www.clas.ufl.edu/users/rodbartl/.
59. ACES II is a Product of the Quantum Theory Project, University of Florida, Authors: Stanton, J. F.; Gauss, J.; Perera, S. A.; Yau, A.; Watts, J. D.; Nooijen, M.; Oliphant, N.; Szalay, P. G.; Lauderdale, W. J.; Gwaltney, S. R.; Beck, S.; Balková, A.; Bernholdt, D. E.; Baeck, K.-K.; Rozyczko, P.; Sekino, H.; Huber, C.; Pittner, J.; Bartlett, R. J. Integral packages included are VMOL (Almölf, J.; Taylor, P. R.), VPROPS (Taylor, P. R.), and ABACUS (Helgaker, T.; Jensen, H. J. Aa.; Jørgensen P.; Olsen J.; Taylor, P. R.).
60. Molt, R. W., Jr.; Watson, T., Jr.; Lotrich, V. F.; Bartlett, R. J. RDX Geometries, Excited States, and Revised Energy Ordering of Conformers via MP2 and CCSD(T) Methodologies: Insights into Decomposition Mechanisms. *J. Phys. Chem. A* **2011**, *115*, 884.
61. Molt, R. W., Jr.; Bartlett, R. J.; Watson, T.; Bazanté, A. P. Conformers of CL-20 Explosive and Ab Initio Refinement Using Perturbation Theory: Implications to Detonation Mechanisms. *J. Phys. Chem. A* **2012**, *116*, 12129.
62. Molt, R. W., Jr.; Watson, T.; Bazanté, A. P.; Bartlett, R. J. The Great Diversity of HMX Conformers: Probing the Potential Energy Surface Using CCSD(T). *J. Phys. Chem. A* **2013**, *117*, 3467.
63. Cohen, R. E.; Naumov, I. I.; Hemley, R. J. Electronic Excitations and Metallization of Dense Solid Hydrogen. *Proc. Natl. Acad. Sci. U. S. A.* **2013**, *110*, 13757.
64. Eremets, M. I.; Gavriliuk, A. G.; Trojan, I. A.; Dzivenko, D. A.; Boehler, R. Single Bonded Cubic Form of Nitrogen. *Nat. Mater.* **2004**, *3*, 558.
65. Eremets, M. I.; Gavriliuk, A. G.; Trojan, I. A. Single-Crystalline Polymeric Nitrogen. *Appl. Phys. Lett.* **2007**, *90*, 171904.

CHAPTER FIVE

Quantum-Informed Multiscale M&S for Energetic Materials

DeCarlos E. Taylor, Betsy M. Rice[1]

Weapons and Materials Research Directorate, Army Research Laboratory, Aberdeen Proving Ground, Aberdeen, Maryland, USA

[1]Corresponding author: e-mail address: betsy.rice.civ@mail.mil

Contents

Abstract

A nonempirical, physics-based modeling and simulation (M&S) capability for energetic materials can only be obtained within a multiscale M&S framework that has a quantum mechanical foundation. In this chapter, quantum mechanical methods and applications to investigate these challenging materials will be given, along with descriptions of how these can be used to enable a quantum-informed multiscale M&S capability for energetic materials. Current capability gaps will be presented, along with suggestions for research advances needed to address these deficiencies.

1. INTRODUCTION

As affirmed in a study by the National Research Council of the National Academies,[1] advanced energetic materials (EMs) are a "key component of the nation's defense strategies." Every type of weaponry utilizes

Advances in Quantum Chemistry, Volume 69
ISSN 0065-3276
http://dx.doi.org/10.1016/B978-0-12-800345-9.00005-2

some form of an EM and its superiority in warfare will be strongly dependent on the performance of the EM within it. It is equally important that advanced EMs have reduced sensitivity to support regulatory policies for insensitive munitions[2] and environmental hazard.[3] Within these constraints, the development and fielding of a new EM is fraught with technical challenges to overcome, resulting in a costly developmental cycle that historically requires a minimum of 20 years. Needless to say, advanced modeling and simulation (M&S) can play a significant role in expediting design and development of new EM through virtual screening, virtual optimization, and virtual exploration of novel energy-releasing concepts.

Exploring EM response to insult using M&S is not new; continuum-level M&S of EM response began in the mid-1960s[4]; historical background, methods, refinements, and advances are better described elsewhere (see, e.g., the series of articles in Ref. 5). Although significant advances have been made, current models are still physics-limited: descriptions are overly simplified and often based on assumption, particularly with regard to chemical reaction, deformation and failure mechanisms, and mechanochemical coupling effects. Parameters associated with such models are often defined empirically, rather than being derived from first principles, thus rendering simulation tools that are heavily parameterized using results from very expensive and time-consuming testing. Even worse, the immense energies and pressures that are released over extremely short time scales have precluded direct experimental measurement of the underlying mechanisms that control initiation and consumption of EM. At the other end of the temporal and spatial scales, atomistic simulations using classical molecular dynamics methods have been performed since the 1970s[6–8] in efforts to understand initiation mechanisms to detonation. Like their continuum counterparts, however, the early models used in these explorations were highly idealized with chemical reaction descriptions overly simplified and thus limited in producing the information needed to understand the energy release of an EM subjected to initiation. In recognizing the deficiencies in the atomistic models of EMs, significant efforts have been made to augment their ability to accurately depict chemical reactivity; these developments will be described hereafter. Notably, these models rely upon quantum mechanical (QM) descriptions of physical and chemical properties of the materials and have been shown to reasonably describe shock-initiated chemistry in real EM.[9–13] Unfortunately, advanced model development at the continuum and atomistic scales has, for the most part, been made independently of one another. There have been a few attempts to populate the continuum-level models with atomistic information, mainly by generation of mechanical and physical properties that allow for the development of constitutive models.[14–26] However, simply having accurate models at the extremes of temporal and spatial scales is not sufficient to enable a predictive M&S capability that describes EM response. Instead, modeling of EM response must cover a rich landscape of phenomena over the range of scales. For instance, information about physics, chemistry, and localization mechanisms must be explored at the atomic (i.e., nano) scale. Material science questions about microstructure must be described by mesoscale models and the continuum level requires proper equations to describe the complex phenomena observed at the macroscale, including fracture, fragmentation, and impact. Bridging the scales is also crucial. While the chemistry of initiation of an energetic occurs at the atomistic level, microstructural features such as defects and "hot spots" (i.e., localized high temperatures) in the material are thought to strongly influence the degree of sensitivity to initiation. Therefore, an accurate capability must involve proper description of the chemistry occurring at the atomistic level and its effect on the microstructure as the energy evolves and transfers throughout the material.

The interested reader is directed to Tadmor and Miller's book[27] for an excellent and thorough description of sequential and concurrent multiscale material modeling methods for solids and how they are connected within a multiscale framework to enable a capability that provides a "complete understanding." Many of the concepts and methods described in the book are applicable to multiscale M&S (MM&S) of EM but are limited to descriptions of thermomechanical material response. These descriptions do not include the dependence of material response due to changes from complex chemistry associated with the initiation and consumption of an EM through deflagration and detonation. Thus, there remain significant challenges in establishing an MM&S capability to predict EM response.

Experimental validation of models at the various scales also poses significant technical challenges. At this time, experimental data that provide an unambiguous depiction of the underlying processes controlling EM response at all relevant scales is unachievable due to the immense energies and pressures that are released over extremely short time scales. These factors have precluded direct experimental measurement of many of the physical and chemical processes occurring at initiation and during consumption of an EM. In light of such formidable challenges, a sound strategy for model development is to ensure that all models be derivable purely from the physical and mathematical considerations underlying the problem of interest.

This can be accomplished by establishing a quantum-informed multiscale modeling capability in which all higher-scale models are derived from information obtained at the lower scales, with the smallest scale being described by QM theory. Careful adherence to this strategy will produce a predictive physics-based capability for M&S of advanced EM. With this in mind, we present our perspective for the construction of a quantum-informed MM&S capability; this chapter will first describe appropriate QM methods, followed by applications of these methods for upscaling to higher-level models. We conclude with a discussion of some remaining technical challenges in MM&S of advanced EM and highlight potential paths forward.

2. QM METHODS FOR EM RESEARCH

2.1. *Ab initio* wavefunction QM methods

At the foundation of any hierarchical multiscale modeling framework for the design of advanced EM must lie quantum mechanics which enables a description of the electronic degrees of freedom within a material via solution of the time-independent Schrödinger[28–30] equation

$$\hat{\mathbf{H}}\Psi = E\Psi$$

where $\hat{\mathbf{H}}$ is the Hamiltonian operator and E is the energy eigenvalue corresponding to the wavefunction Ψ. Knowledge of the wavefunction (or equivalently the electron probability density $\Psi^*\Psi$) yields a variety of EM properties which may be computed as expectation values of Ψ with an operator **Op** (e.g., $\langle\Psi \mid \mathbf{Op} \mid \Psi\rangle$) corresponding to the observable of interest. EM properties are also related to derivatives of the QM energy, with the system stress, defined as the first energy derivative with respect to strain,[31] and vibrational frequencies, obtained from second derivatives of the energy with respect to atomic displacements, serving as examples. The power of quantum mechanics in the design of advanced EM lies in the fact that it is a *formally exact* construct containing the necessary physics required for an accurate computational prediction of properties without regard to experiment or empiricism. This is in distinct contrast to classical potentials which are often heavily parameterized with no application to excited-state chemistry which is an absolute necessity when, for example, analysis of the visual signature of an EM is desired. The accuracy of quantum mechanics in EM research is only limited by the fact that the exact solution to the Schrödinger equation in a given basis, the so-called "full configuration interaction,"[32] is only obtainable for comparatively small molecules and purely out of computational necessity, approximate (though still quite accurate!) solutions to the Schrödinger equation must be obtained in actual applications.

Wavefunction[32] and density functional[33]-based QM approaches, of both *ab initio* and semiempirical form, have seen widespread application in EM research. When considering a QM method in the context of a quantum mechanically apprised MM&S framework, a *nonexhaustive* list of desirable features would require that the QM method be:

- accurate in all regions of the potential energy surface
- computationally efficient
- systematically improvable
- amenable to exploration of potential energy surfaces and molecular dynamics simulation using analytic derivatives
- applicable to the condensed phase with some incorporation of periodicity

All of the currently existing QM methods satisfy some, *but not all*, of these criteria. As an example, coupled cluster (CC) theory,[34,35] followed by the closely related analog many body perturbation theory[36] (MBPT), is the systematically improvable "gold-standard" of quantum chemistry. In CC theory, the ground state wavefunction, Ψ_g is given by

$$|\Psi_g\rangle = e^{T}|\phi_o\rangle \tag{5.1}$$

where ϕ_o is a Slater determinant (e.g., Hartree–Fock reference) and T is an excitation operator which creates single, double, up to n-tuply excited configurations from the reference determinant. T can be written as

$$T = T_1 + T_2 + \cdots = \sum_{i,a} t_i^a a^+ i + \frac{1}{2}\sum_{i,j,a,b} t_{ij}^{ab} a^+ i b^+ j + \cdots \tag{5.2}$$

where the standard creation/annihilation operators have been used and summation indices $(i,j,\ldots)$ $(a,b,\ldots)$ represent occupied and virtual molecular orbitals, respectively. A CC calculation reduces to a determination of the "t amplitudes" (see Eq. 5.2) present in each level of excitation and typically the excitations are treated completely through single and double excitation levels[37] (CCSD), and a partial inclusion of triple excitations is done via perturbative techniques which defines the CCSD(T) method.[38] The advantage of CC theory is the use of the exponential ansatz for the wavefunction which yields a more rapidly convergent series of approximations toward the exact solution of the time-independent Schrödinger equation than in the linear configuration interaction[32] approach. Further, it can be shown that the exponential wave operator results in *products* of T operators; therefore, certain classes of excitations enter the CC equations in a nonlinear way and effects from higher excitation manifolds are included in a convenient and compact manner. For example, even when the CC equations are truncated at the single and double excitation level, classes of *triple* excitation effects are included through terms such as $T_1 * T_2$; however, the number of independent amplitudes to be determined remains the same as in the singles and doubles methods. These products of terms also play a role in ensuring size extensivity of the method, and as such, CC is the most robust QM methodology currently available. CC accurately treats molecular configurations near equilibrium, along the repulsive wall, and along the van der Waals tail, the latter of which represents an absolutely critical regime for energetic molecular crystals whose condensed-phase structures are often dictated by long-range weak interactions between molecules.[39] In EM research, CC methods have been used for gas-phase analyses of energy-releasing chemical reaction pathways,[40] identification of low energy conformers,[41] and excitation energies.[42] However, CC, even when truncated at the single and double excitation level, already scales as the sixth power of the molecule's size which renders it far too computationally demanding for condensed-phase applications. This is particularly true in EM research where some systems contain over 100 atoms within a single unit cell as is the case for the energetics cyclotrimethylene trinitramine[43] (RDX) with 168 atoms per unit cell or trinitrotoluene[44] which has 160 atoms. Further, although analytic derivatives are available for CC methods,[45–47] picosecond (or longer) timescale molecular dynamics simulations, which may require several hundred thousand evaluations of the gradient, are absolutely not(!) feasible at the CC/MBPT level of theory and they are best used for accurate treatments of unimolecular/bimolecular processes occurring in the gas phase and to provide benchmark data for parameterizing classical force fields,[48] as discussed below.

2.2. Semiempirical quantum methods

Due to the computational expense of wavefunction-based *ab initio* methods, semiempirical[49–51] quantum chemistry approaches such as tight binding[52–54] (TB) and neglect of diatomic differential overlap[55] (NDDO) methods have also seen application in EM research. Whereas all-electron *ab initio* approaches often employ large, overlapping, Gaussian basis sets with all one, two, three, and four center integrals computed analytically using methods such as the Obara and Saika[56] or McMurchie and Davidson[57] recursion relations, the TB and NDDO semiempirical methods are minimal basis, valence electron only approaches that are limited to one- and two-center interactions which significantly reduces their computational demand relative to *ab initio* techniques. Further, many of the remaining integrals required to construct the TB/NDDO Hamiltonians are replaced by parameters which further reduces the computational cost of the methods. For example, in the AM1[50]/PM3[58] variants of NDDO, which have been used to study EM heats of formation,[59,60] crystal densities,[61] impact sensitivities,[62] decomposition pathways,[63] binding energies with polymers,[64] and hyperpolarizabilities,[65] the one-electron contribution to the Hamiltonian for a charge distribution $|\mu\nu)$ is given compactly by

$$H_{\mu\nu}^{AA} = U_{\mu\nu}^{AA}\delta_{\mu\nu} - \sum_B Z_B\left(\mu^A\nu^A|s^Bs^B\right) \tag{5.3}$$

$$H_{\mu\nu}^{AB} = \frac{1}{2}\left(\beta_\mu + \beta_\nu\right) \times S_{\mu\nu} \tag{5.4}$$

where the superscripts A and B denote the atoms upon which the basis functions μ and ν are centered with Z_B representing the atomic charge of atom B. The one-center atomic orbital energy $U_{\mu\nu}$ (Eq. 5.3), "resonance" parameter β (Eq. 5.4), Slater orbital exponents (used to compute overlap integrals $S_{\mu\nu}$ of Eq. 5.4), and all nonzero one-center two-electron integrals are treated as parameters. The two-center two-electron integrals are computed in terms of the parametric one-center two-electron integrals, using the point charge model of Thiel and Dewar,[66] and the core–core repulsion $V(A, B)$ between two atoms A and B of charge Z_A and Z_B, respectively, is given by

$$\begin{aligned} V(A,B) &= Z_A Z_B(s_A s_A|s_B s_B)\left(1 + e^{-\alpha_A R_{AB}} + e^{-\alpha_B R_{AB}}\right) \\ &+ \frac{Z_A Z_B}{R_{AB}}\left(\sum_k a_{kA}\, e^{-b_{kA}(R_{AB}-c_{kA})^2} + \sum_k a_{kB}\, e^{-b_{kB}(R_{AB}-c_{kB})^2}\right) \end{aligned} \tag{5.5}$$

where α, a, b, and c are also taken as variable parameters. TB also uses analytically compact functional forms to construct the Hamiltonian[67,68] and, in practice, TB matrix elements are pretabulated, stored, and interpolated during calculations which render them marginally faster than NDDO models. The parameters in these empirical model Hamiltonians may be determined by direct fitting to properties obtained from experiment or fitted directly to distant-dependent matrix elements, energies, potential curves, etc. extracted from first-principles calculations. Semiempirical methods satisfy several of the aforementioned criteria required to construct a quantum mechanically founded MM&S framework including computational efficiency and applicability to periodic, condensed-phase systems.[69] However, although they are formulated within a QM framework and, at least in the case of self-consistent charge TB, are derivable from first principles,[54] these methods are still empirical and therefore not systematically improvable due to their inherent dependence on the parameters used to define the models. The accuracy of the semiempirical methods has been the subject of many studies[70–72] and, as with any parametric function, is directly contingent upon the quality of the parameterization. Parameters extracted from first-principles calculations may render a semiempirical Hamiltonian possibly suffering the same maladies as the underlying *ab initio* method upon which it was parameterized. Those obtained using ambient state experimental data may not be transferable to nonequilibrium configurations, such as those which may arise during shock simulations of EM, where molecules may be significantly compressed or undergo chemical reactions along decomposition pathways not considered during fitting of the Hamiltonian.

2.3. Density functional theory

Although wavefunction-based *ab initio* and semiempirical methods have seen numerous applications to energetic compounds, it goes without question that the most often applied QM method in EM research is density functional theory (DFT) which provides a balance between formal rigor (parameters contained within the exchange-correlation potential aside), accuracy, and computational efficiency. DFT has been shown to reasonably describe properties and reactions of isolated energetic molecules,[73–76] and extensive theoretical studies have been performed to understand how processes associated with detonation are influenced by isolated molecule chemistry. There have been numerous condensed-phase DFT studies as well[77–80]; however, in this portion of the chapter, we have chosen not to present a mere summary of the plethora of applications of DFT in EM research, but instead discuss features of the DFT formalism that are critical to accurate first-principles modeling of EM.

DFT reduces solution of the Schrödinger equation from determination of the wavefunction Ψ, which depends on $3N_{el}$ electronic degrees of freedom, to determination of a much simpler quantity, the electron density ρ, which depends on only three spatial coordinates. (The reader seeking a more in-depth derivation of the DFT formalism and Kohn–Sham approach is invited to review the excellent exposition given by Parr and Yang.[33]) The inherent power of DFT is that it is a *formally* exact method which, given knowledge of the exact exchange-correlation and kinetic energy functionals, would yield an exact solution to the many body Schrödinger equation within a computationally efficient one-particle method. This is in contrast to the wavefunction-based Hartree–Fock theory which, similar to DFT, depends on a one-particle operator but does *not* yield an exact solution to the full Schrödinger equation since dynamic electron correlation is not accounted for and must be included using computationally expensive post-Hartree–Fock methods. DFT is an excellent option for the QM regime within an MM&S framework since it is computationally efficient (though still not competitive with classical potentials) and, when implemented in scalable software programs running on massively parallel computers, has been applied to system sizes in excess of 1 million atoms.[81–83] Although DFT methods are generally not systematically improvable, periodicity is available via *k*-point sampling of the Brillioun zone in implementations such as VASP[84] and CRYSTAL[85] or via the supercell approach as in the CP2K[86] software package.

Due to the approximate nature of the applied exchange-correlation functionals in DFT, the method can often be of variable accuracy with functionals sometimes performing very well for some properties and completely failing for others. There exist a large number of methodological surveys in the literature focusing on the accuracy of density functionals for a variety of properties such as structures,[87,88] lattice energies,[89,90] bond energies,[91,92] polarizabilities,[93,94] etc.; however, of particular importance when modeling condensed-phase energetic molecular crystals is the method's utility for the treatment of noncovalent interactions (NCI). The work by Byrd et al.[95] demonstrated the poor performance of conventional density functionals when applied to energetic molecular crystals with density errors on the order of 10–15% compared to experiment. Errors of this magnitude are particularly problematic during computational assessment of EM performance since critical metrics such as the detonation velocity and detonation pressure are directly proportional to the crystal density. The failures of conventional density functionals when applied to molecular crystals have been attributed to the well-known inability of DFT to describe the dispersion contribution to the total energy which is often the primary binding force within energetic molecular crystals.[39] As a concrete example, we have computed the electrostatic, induction, exchange-repulsion, and dispersion energies of four selected configurations corresponding to minima on the triamino-trinitrobenzene[96] (TATB) dimer potential energy surface using symmetry-adapted perturbation theory[97,98] (SAPT). TATB is an exceedingly insensitive explosive used as an initiator in nuclear weapons and, as shown in Table 5.1, dispersion constitutes the largest *binding* contribution by a factor of ≈ 3 or more relative to the electrostatic/induction energies for each dimer. However, it has been shown[99] that conventional B3LYP significantly underestimates, by $\approx 70\%$, the critical dispersion contribution for gas-phase π-stacked TATB configurations and in the condensed phase, the PBE density functional yields a TATB crystal density in error by as much as 15%.[100] These errors can be largely reduced using dispersion-corrected density functionals, as discussed below.

The failures of many conventional DFT approaches for NCIs have spawned intense research activity throughout the international community and strategies for incorporating dispersion into the DFT framework generally fall into the following categories which include:

- Development of new functionals[101,102]
- Addition of empirical corrections to existing functionals[103,104]
- Optimization of pseudopotential parameters[105]
- Application of "post-DFT" methods, for example, inclusion of virtual orbitals within a suitable energy expression[106,107]
- Development of nonlocal correlation functionals[108,109]

All of these are very active areas of research; however, in the spirit of brevity, we will limit the current discussion to two dispersion corrections that have seen specific application in our own EM research: dispersion-corrected atom-centered potentials (DCACP) and DFT-D3.

In the DCACP[105] approach, effective, nonlocal, core potentials are assigned to atoms and fitted to benchmark energies obtained using wavefunction-based methodologies such as CC or configuration interaction.

Table 5.1 SAPT interaction energy components (kcal/mol) for selected minima on the TATB dimer potential energy surface

Dimer	Electrostatic	Exchange	Induction	Dispersion[a]	E_{Total}
1	−4.892	16.616	−5.301	−16.468	−10.046
2	−5.428	24.117	−7.876	−22.465	−11.652
3	0.439	7.161	−2.681	−8.523	−3.603
4	−4.467	11.302	−3.711	−9.215	−6.091

[a]Dispersion is the dominant binding contribution in all cases.

The fitted core potentials are used in conjunction with conventional functionals and since the DCACP contribution to the total energy is included as part of the self-consistent determination of the electron density, these potentials have been demonstrated to yield improved electronic properties (dipole/quadrupole moments, polarizabilities) as well as a more accurate treatment of NCIs.[105] The improved description of the electronic structure is in contrast to non-self-consistent corrections such as DFT-D,[103] where the converged electron density for a given geometric configuration is equivalent to that obtained using the uncorrected functional. Balu et al.[110] presented a study detailing the accuracy of DCACP for crystal structures/densities of a series of energetic molecular crystals including RDX,[43] TATB,[96] 1,3,5,7-tetranitro-1,3,5,7-tetraazacyclooctane,[111] (HMX) and pentaerythritol tetranitrate[112] (PETN). Whereas the conventional PBE functional yielded errors in the crystal density of 5–10%, the inclusion of the DCACP correction reduced the error to ≈2% relative to experiment. Of concern is that DCACP potentials do not, in general, exhibit the correct asymptotic R^{-6} behavior. However, recent work[113] has yielded evidence that R^{-6} asymptotics may be incorporated using a multicenter functional approach.

In the DFT-D3 method, conventional density functionals are supplemented with a pairwise additive dispersion energy term of general form $\sum_n C_n R^{-n}$, which introduces the correct R^{-6} asymptotic behavior for large interatomic separation. DFT-D3 uses the form

$$E_{\text{dispersion}} = -\frac{1}{2}\sum_{A\neq B}\sum_{n=6,8,\ldots} s_n \frac{C_n^{AB}}{R_{AB}^n} f_{\text{damp}}(R_{AB}) \tag{5.6}$$

which includes a damping function

$$f_{\text{damp}}(R_{AB}) = \frac{1}{1+6\left(R_{AB}/\left(sR_0^{AB}\right)\right)^{-\alpha}} \tag{5.7}$$

to avoid singularities at small interatomic separation and double counting of electron correlation in intermediate ranges. R_0^{AB} is a cutoff radius for atom pair AB, s is a radii scaling factor, and α controls the steepness of the function at small interatomic separation. Other damping functional forms are also available. The C_6 coefficients in the series are a function of the coordination numbers obtained from the molecular structure and adapt continuously to changes in hybridization by on-the-fly interpolation of pretabulated reference values. Figure 5.1 presents results we have obtained comparing B3LYP, with and without the D3 dispersion correction, to complete basis set (CBS) extrapolated interaction energies for dimers of the energetic molecules nitromethane (NM),[114] 1,1-diamino-2,2-dinitroethylene[115] (FOX-7), ethylenedinitramine (EDNA), and nitrobenzene. The gross errors relative to the CBS benchmarks using conventional B3LYP are largely remedied when the D3 dispersion correction is included; yet the wall clock time for evaluation of the energy at the B3LYP-D3 level is on the order of minutes compared to several days for the CBS values! It is this balance of accuracy and efficiency that renders *van der Waals corrected* DFT as the most viable QM method to serve as the foundation of an MM&S framework for EM design.

2.4. Linear scaling self-consistent solution of one-particle equations

Be it dispersion-corrected DFT or a semiempirical NDDO/TB QM method that resides at the core of an MM&S framework, the self-consistent solution of the associated one-particle equations quickly becomes a computational bottleneck for large systems. Approaches which enable linear scaling QM treatments of multimillion atom systems include divide and

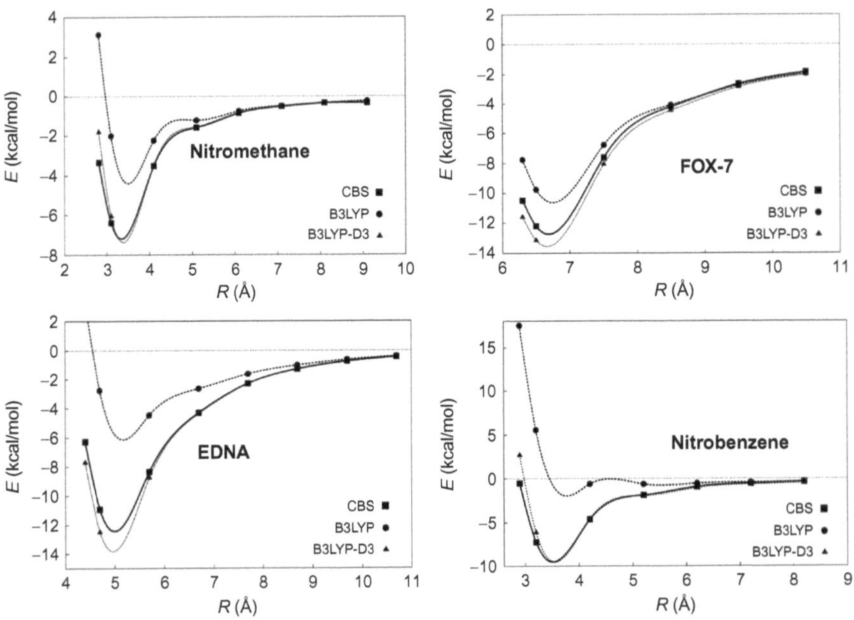

Figure 5.1 Comparison of conventional (B3LYP) and dispersion-corrected (B3LYP-D3) DFT to complete basis set extrapolated interaction energies of four energetic molecule dimers. Curves are guides for the eye.

conquer[116,117] (D&C) and orbital free[83] DFT (OFDFT) methods. In D&C algorithms, the full system is decomposed into a collection of smaller, more computationally tractable subsystems with the density for the full system obtained as a weighted combination of local densities from each subdomain. In OFDFT, given a suitable kinetic energy density functional, materials are modeled solely by the electron density. As such, no one-electron functions, which are necessary to treat the noninteracting kinetic energy in Kohn–Sham DFT, are required which reduces the computational scaling. Other approaches for treating large systems fully quantum mechanically include the fragment molecular orbital method[118] and the emerging approach, exact embedded DFT.[119,120]

Another very promising route toward enabling QM treatments of multimillion atom systems focuses on removal of the matrix diagonalizations that are required in the conventional self-consistent field (SCF) procedure. For small/medium-sized systems, the Kohn–Sham, Hartree–Fock (NDDO methods), or TB QM methods require construction of the associated one-particle Hamiltonian matrix with an initial guess density followed by diagonalization to obtain a new set of molecular orbitals from which a new density is constructed. An updated Hamiltonian based on the new density is diagonalized yielding updated orbitals/density, and this process repeats iteratively until the output density obtained by diagonalization is equivalent to the input density, indicating convergence. Matrix diagonalization scales cubically with system size and for very large systems, such as those which may arise in the QM region of a multiscale simulation, it quickly becomes computationally prohibitive. As an alternative, energy minimization schemes have been developed which involve direct optimization of the total energy with respect to the molecular orbitals/density matrix within an iterative SCF procedure.[121–125] Linear scaling direct energy minimization DFT calculations on systems containing up to 1 million atoms have been reported[81] and methods vary in how orthonormality of the molecular orbitals or N-representability and idempotency of the density matrix are maintained during the optimization. Based on an original algorithm from Li et al.,[124] Millam and Scuseria[122] optimized elements of the density matrix directly using a conjugate gradient algorithm and introduced a Lagrangian constraint on the energy functional to enforce the correct number of electrons at each optimization step. Idempotency of the density matrix was maintained using the McWeeny[126] purification and when a threshold criterion was enforced on elements of the density matrix, linear scaling SCF was achieved at the DFT level using the local density approximation. Weber et al.[121] introduced a direct energy minimization approach whereby the QM energy functional, which is constrained by molecular orbital orthogonality, is mapped to an approximate unconstrained functional. By combining the unconstrained energy functional with a robust conjugate gradient minimization method, they were able to perform a single-point energy + gradient calculation using DFT on a DNA fragment with 12,661 atoms (103,333 basis functions). Approaches based on the matrix sign function[127] and auxiliary density matrices[82] with sparsity enforced using a hard cutoff criterion have been used to generate single-point energies for systems exceeding 1 million atoms with linear scaling up to thousands of cores.

3. APPLICATIONS OF QM FOR UPSCALING

Although MM&S is an active research area in the material science community, and a plethora of methods and models are developed to enable sophisticated, scale-bridging, concurrent schemes, there are only a few approaches utilized within EM research. Typical multiscale approaches used in EM research have been sequential hand-shake methods described herein, with only a few concurrent approaches emerging. Within the range of scales to be interrogated in an MM&S framework, the atomistic and continuum models are the most mature, and it is natural that the simplest multiscale simulations couple the two scales in a hand-shake fashion. This is accomplished by populating the hydrocodes used to model thermal or shock response of EM at the continuum scale with atomistically derived information. Currently, the hydrocodes utilize reactant and product equations of state (EOS), reaction models, and constitutive models. Effects of microstructural material heterogeneity can be modeled by hydrocodes through finer resolution such that individual grains are resolved but models used in the hydrodynamic simulations are often empirically derived and lack calibration data for fitting. Using HMX as an example, Sewell and Menikoff[128] provided a description of some material properties needed for grain-resolved mesoscale simulations, many of which can be obtained from atomistic simulations. These include pressure-dependent chemical reaction rates, phase transitions (including pressure-dependent melting), thermal, mechanical, and transport properties. Although some structural, mechanical, or energetic properties can be calculated directly using large-scale QM calculations for simple systems (e.g., ideal crystals, or crystals having small simple defects), more complex properties and processes (e.g., transport properties and mechanisms, formation, migration, or coalescence of defects) can only be obtained through large-scale classical molecular dynamics simulations using force fields parameterized with either QM or experimental information. Subscale information can be used for more than simply populating assumed constitutive relations in engineering models. They can also be used to develop more physically realistic constitutive relations that more appropriately describe material response. This section will describe direct large-scale QM simulations to predict constituent properties needed for higher-scale models. Additionally, upscaling can also be accomplished through transition of results from classical atomistic and mesoscale simulations using quantum mechanically derived potentials; thus, these models will also be described.

3.1. Full-resolution QM calculations of condensed-phase EM

QM calculations of condensed-phase EMs have been performed since the 1990s, as computational capabilities, theoretical methods, and software had advanced to the point that these formerly prohibitive calculations could be attempted. The earliest of studies were ambitious attempts to explore possible initiation mechanisms involving both vibrational and electronic excitations[129–131] from condensed-phase QM calculations rather than providing information that could be utilized in higher-scale models. However, system size limitations and functionality of the software precluded adequate depiction of microstructural features on the atomistic properties and processes being explored, rendering these results as representing extremely idealized and simplified systems. Additionally, the early studies used low levels of QM theory (both wavefunction and DFT-based); therefore the results had limited accuracy due to inadequate treatment of electron correlation and/or small basis sets. Although these early calculations had limited utility, they highlighted the potential of condensed-phase QM studies to expand the fundamental understanding of the complex processes associated with initiation to detonation of an EM. Even now, more sophisticated QM investigations of complex condensed-phase chemical processes still face limitations of system size and complexity. For example, as discussed in a thoughtful paper by Sharia and Kuklja,[132] the technical challenges in modeling condensed-phase chemistry of EMs using solid state methods is formidable even though DFT methods have been shown to reasonably depict gas-phase chemistry. In this study, rates of gas-phase and condensed-phase decomposition reactions of HMX were calculated using DFT and transition state theory. While the work demonstrates differences in reaction rates and mechanisms between the two phases, it neglects certain material science aspects (e.g., defects and dislocations) that might dramatically influence condensed-phase reaction. This work also demonstrated that gas-phase concepts for chemical reaction of an EM do not necessarily transition to the condensed phase. Studies such as these highlight the need for significant developmental investments for accurate, full-resolution treatments that can routinely handle larger systems beyond the current state of the art (~10,000 atoms in quantum molecular dynamics (QMD) simulations), and in multiresolution methods to mimic even larger system sizes. Only then will such QM capabilities enable direct first-principles explorations into the role of material science influences on EM properties and behavior. Until then, however, QM studies can provide very important information for development or population of relevant constitutive models used in the continuum and grain-scale simulations. We will limit our discussion herein to QM studies for these applications.

A very large number of DFT studies have been performed to calculate isothermal EOS information for several well-known EMs [110,133–146] such as RDX, HMX, PETN, TATB, FOX-7, and CL-20, often to pressures higher than have been measured experimentally.[141,147–156] Until the last couple of years, the majority of the studies used conventional DFT (i.e., those that do not include van der Waals corrections), with most reporting unacceptably large errors in densities for low-pressure structures.[95] It was shown in one study,[100] however, that error in the conventional DFT calculations is significantly reduced for EMs at high pressures, indicating that van der Waals interactions are not dominant at large compression. The more recent studies are now using van der Waals DFT methods, with the results having good agreement with experiment at all degrees of isotropic compression. Structural and physical properties can be obtained from these calculations (e.g., bulk elastic properties). A few studies have explored anisotropic response of a material by performing DFT calculations of the material under uniaxial compression,[20,22,138,146,157] thus providing crucial information for developing models that properly describe material deformation. Although most of the EOS simulations have been performed on well-known systems to validate the methods, the quality of the more recent results indicates that current DFT methods are available to provide reliable EOS information for systems for which no experimental data exist. In such a case, *a priori* knowledge of the crystal structure is required to generate the EOS. In lieu of experimentally derived information, computational approaches to predict crystal structures can be invoked. While computationally formidable, several methods have been developed that can, in principle, predict crystal structures of molecular crystals using structural information of an isolated molecule. Development of crystal structure prediction (CSP) methods has been progressing since at least the 1990s, with progress monitored through a series of blind tests designed to measure the ability of the methods to reliably predict crystal structures. CSP follows essentially the same prescription: A series of trial crystal structures are generated (in a variety of ways) using information about a single molecule, followed by an energy minimization. The optimized candidates are then ranked, usually in order of lattice energy (thus neglecting important influences in crystal growth, such as kinetic and solvent effects). The majority of methods rely upon empirically derived force fields to describe interatomic interactions used in the energy minimization; these often produce erroneous results, as detailed in the blind tests.[158] As reliable van der Waals DFT methods are emerging, these are now beginning to be used in CSP methods to overcome inaccuracies of the approach due to incorrect interatomic interactions, with results suggesting that further advances in candidate structure generation combined with more rapid evaluation of structures will make the ability to predict EOS of new EM without reliance on experimental information a reality.[159–162]

3.2. QM-based force fields for pure EM

Although the crucial EOS information can be reliably generated using DFT, several other important properties and behaviors can be obtained only through large-scale classical MD simulations at this time, due to previously described computational limitations. As is well known, the reliability of the results generated from classical MD simulations is largely dependent on the quality of the model of the interatomic interactions (i.e., the force fields) describing the system. Unfortunately, the complexity of the crystal structures of most EM makes development of realistic models quite difficult; thus, the earliest MD simulations of EM used models of condensed-phase explosives that had overly simplified structures and chemistry. While useful in developing simulation protocols and providing insights into phenomena associated with initiation of an EM, these early models were unable to properly depict material features that would strongly influence material response (including complex chemistry).

Development of more realistic models of EM (described elsewhere[163]) progressed with the earliest being developed for rigid-molecule condensed-phase MD simulations that were parameterized mainly using empirical information. Molecular flexibility has also been incorporated into the models, using gas-phase QM calculations of monomers or dimers of the energetic molecule and/or empirical information. Two such models are mentioned that have been used to develop coarse-grained (CG) models for use in particle-based mesoscale dynamics simulations, described hereafter. The first model[164] (denoted SRT hereafter) was parameterized using a combination of empirical low-temperature crystalline structural information and quantum mechanically determined energy and energy derivatives for the isolated molecule at equilibrium. The resulting model predicts NM over a wide range of temperature and pressures in both liquid and solid phases, and predicts the experimental shock Hugoniot. The second, developed for HMX and denoted hereafter as the SB model,[165] is composed of empirically derived nonbonded interactions coupled with intramolecular terms parameterized using QM calculations of energies of various conformations of nitramine molecules. In addition to accurately predicting a variety of nonreactive processes of HMX in both the solid and liquid states, this model was found to be transferable to describe polymorphs of crystalline RDX for which an extensive series of MD simulations have been done to explore mechanical response at a variety of conditions.[17]

The general COMPASS force field[166] is another flexible force field parameterized using a combination of *ab initio* and experimental information for a diverse set of chemical systems, including nitrate esters,[167] a class of EMs. It has been used extensively to calculate structural and mechanical properties of pure energetic crystals and plastic-bonded explosives (PBXs); however, the degree of success in predicting properties is mixed.[153,168–220] A recent report on the performance of the COMPASS force field in predicting the lattice parameters, EOS, and mechanical properties for polymorphs of CL-20 at high pressure and temperature[189] gave reasonable agreement with experiment. Likewise, lattice parameters simulated by the COMPASS force field for HMX polymorphs[196] were in extremely good agreement with experiment. However, application of the COMPASS force field in MD simulations of bicyclo-HMX produced error in the three lattice vectors of −7.5%, 5.8%, and −2.4%, yet offsets in the signs of the lattice vector errors rendered a density error of only 1.5%. Additionally, the COMPASS force field was shown to poorly describe TATB in MD simulations.[207,220]

The classical models discussed above rely in part on experimental data for calibration of parameters. However, the ultimate goal for computational development of EM is a *predictive* capability which is critical for notional EMs where no experimental data is available, or for properties and behavior of EM at extreme temperatures/pressures where experimentation may be difficult. A successful strategy to circumvent this limitation has been parameterization of classical force fields based solely on quantum mechanically derived reference data with no experimental information included. The development of classical force fields for energetic molecular crystals, with no regard to empiricism, is particularly challenging due to the variety of intermolecular forces within the crystal structures which include hydrogen bond and weak van der Waals interactions. Further, there is often significant anisotropy with strong electrostatic interactions prominent along one crystallographic axis and dispersion dominating along others. Due to the complexity of the intermolecular forces in energetic molecular crystals, highly accurate QM methods are required to properly characterize these critical interactions and a combined DFT and SAPT approach known as SAPT(DFT)[98] has proven to be an invaluable tool for first-principles force field development. SAPT[97] is a method where the interaction energy between two molecules is computed perturbatively, thereby avoiding the basis set superposition error inherent in the supermolecular approach. In SAPT(DFT), given the unperturbed monomer DFT densities, the intermolecular interaction can be evaluated at increasing orders of the intermolecular perturbation and the total interaction energy is given as a sum of components including electrostatic, exchange, induction, and dispersion contributions. This energy decomposition yields more physical insight into the nature of the intermolecular interaction than does the single number that results from the supermolecular approach and this can be exploited in EM force field fitting since individual components of the potential energy function can be fitted to distinct contributions. Specifically, very accurate atomic charges can be determined by fitting to large grids of SAPT(DFT) electrostatic energies and a similar approach can be applied to obtain the critical dispersion coefficients as well. SAPT(DFT) has been used to develop classical intermolecular potentials for several EMs including RDX,[221] FOX-7,[222] and TATB.[223] FOX-7 and TATB are particularly challenging due to the complexity of the crystal structures. As shown in Fig. 5.2, the FOX-7 crystal is characterized by "zig-zag" layers within molecular sheets with extensive hydrogen bonding between monomers. On the contrary, the TATB crystal (Fig. 5.2) not only consists of essentially planar graphitic-like sheets but also contains extensive hydrogen bonding between monomers. In both materials, interlayer interactions are dominated by weak van der Waals interactions and potentials were developed empirically by Sorescu et al.[224] for FOX-7 and Bedrov et al.[225] for TATB with some parameters fit directly to experimental data. Gee et al.[220] developed a nonempirical force field for TATB using dimer interaction energies obtained at the MP2 level of theory; however, the resulting potential gave large errors for the sublimation enthalpy compared to experiment. Completely predictive models[222,223] for both of these systems were formulated by computing grids (1008 and 880 points for FOX-7 and TATB, respectively) of dimer interaction energies using SAPT(DFT) and fitting to an exponential-6 functional form. The resulting potentials, when used in molecular dynamics simulations, yielded excellent agreement with the ambient state experimental crystal structures with errors less than 3%. Further, the thermal and pressure response using the SAPT(DFT) fitted potentials were also in agreement with experiment and properly reproduced the bulk modulus, thermal expansion coefficients, and sublimation enthalpy, all without empiricism. Although accurate at the ambient state and under moderate ranges of compression, a very severe limitation of models derived using SAPT(DFT) is that they are, by necessity, nonreactive and nonflexible. Specifically, SAPT(DFT) interaction energies, and the classical potentials derived from them, assume a specific intramonomer geometry; if the internal orientation changes as a result of compression or shock induced chemical reaction, the atomic charges and potential parameters derived using SAPT(DFT) may not be applicable to the new configurations.

Geometric and chemical changes resulting from compression or shock loading require flexible and reactive force fields for appropriate depiction of material response. The Reax force field[226] (ReaxFF) is the most well-developed general reactive force field for use in classical MD simulations of condensed-phase materials and was originally parameterized using QM reference data for a large set of hydrocarbon compounds. ReaxFF is based on a bond length/bond order relationship where all connectivity-dependent

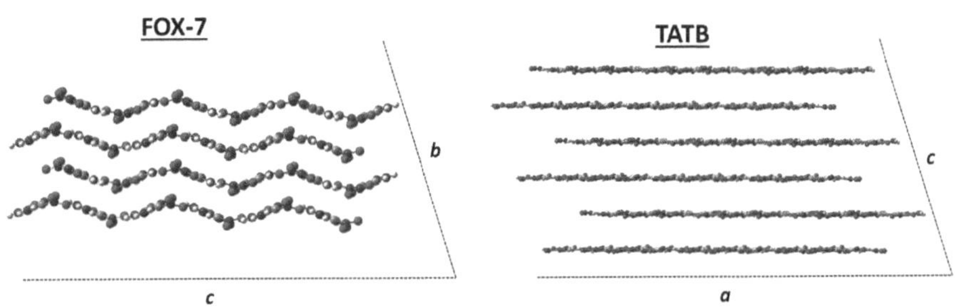

Figure 5.2 FOX-7 and TATB crystal structures. (For color version of this figure, the reader is referred to the online version of this chapter.)

interactions, such as valence and torsion angles, are bond order dependent which ensures that they do not contribute at dissociation. The Reax energy:

$$\begin{aligned} E_{\text{Reax}} = & E_{\text{bond}} + E_{\text{lp}} + E_{\text{over}} + E_{\text{under}} + E_{\text{val}} + E_{\text{pen}} + E_{\text{coa}} \\ & + E_{\text{tors}} + E_{\text{conj}} + E_{\text{Hbond}} + E_{\text{vdW}} + E_{\text{Coul}} \end{aligned} \tag{5.8}$$

consists of bond energy, lone pair energy, over coordination penalty, under coordination penalty, valence angle energy (including a penalty function E_{pen}), angle conjugation energy, torsion angle energy, conjugation energy, hydrogen bond energy, van der Waals energy, and Coulomb energy terms, respectively, and has been used in shock simulations of a variety of energetics including RDX,[11] PETN,[13] HMX,[227] TATB,[227] NM,[228] and tri-acetonetriperoxide.[229] As discussed previously, a proper accounting of the dispersion contribution in energetic molecular crystals is critical. However, the original Reax parameterization was based on QM reference data obtained with conventional density functionals (lacking a van der Waals correction); simulations using this model yielded crystal densities for energetics that were significantly smaller than the experimental values[230] (see Table 5.2). However, a London dispersion ("low-gradient") correction term[230] was subsequently added; simulations using this form produced results having significant improvements for RDX and PETN in particular.

3.3. QM-based force fields for composite EMs

The energetic fill within in-service munitions consist of multicomponent "formulations" containing EMs mixed with polymeric binders, plasticizers, or stabilizers. Depending on the application, the formulation can be used for gun or rocket propulsion or within explosive warheads. However, a significant portion of the energetic formulation design process follows an Edisonian "trial and error" approach with little fundamental understanding,

Table 5.2 Percent error in computed crystal density, relative to experiment, for ReaxFF and van der Waals corrected ReaxFF-lg

Energetic	Density error (%)	
	ReaxFF	ReaxFF-lg
Nitromethane	−8.04	6.69
RDX	−28.45	−0.82
PETN	−16.88	−0.89
TATB	−6.30	4.09

a priori, of the interactions that determine the miscibility of the component ingredients. As for the pure energetic component, the experimental development, testing, and disposal of poor-performing candidate formulations is a hazardous, expensive, and time-consuming process. Computational modeling can expedite the procedure and readily computable quantities such as the elastic constant tensor and heat of formation have been used to assess the mechanical response and detonation velocity, respectively, of candidate formulations.[202,216]

Of critical importance for PBX design is the behavior of the composite material at the energetic-polymer interface. At the atomistic level, binding energies[202,216] between the formulation components have been used to quantify the affinity (or lack thereof) of the ingredients which is clearly a necessary feature for a deployable formulation. Further, the effects of polymer functionalization and crystal surface morphology on miscibility can be readily studied atomistically. However, a variety of heterogeneities are found in EM composites including variable grain size distributions, intra- and intergranular voids, and intragranular cracks, all of which are beyond currently accessible atomistic capabilities and require mesoscale and continuum-level models for analysis. At the continuum level, traction–displacement relationships along the energetic-polymer interface may be represented by cohesive zone models[231] (CZMs) which are used in conjunction with the finite element method to study fracture in materials. From the CZM viewpoint, fracture is a gradually occurring process where the separation occurs across an extended crack tip (or "cohesive zone") whose growth is resisted by cohesive forces between the separating components. As the displacement increases, eventually resulting in complete separation at the interface, the traction increases up to a maximum and decreases to zero with the area under the curve corresponding to the work of separation. Multiscale approaches coupling the atomistic regime with the continuum for study of interfacial crack propagation and delamination require calculation of the traction–displacement relationship using atomistic MD simulation which is then used to populate continuum-level CZMs for macroscale finite element analysis of dynamic material response. With regard to the aforementioned quantum mechanically informed MM&S strategy outlined in this chapter, such an approach first requires development of atomistic potentials using first-principles quantum mechanics followed by large-scale MD simulations, to obtain the requisite traction–displacement curves. In the following, we present a demonstration of this multiscale atomistic to continuum approach for a real energetic formulation, Composition A3 Type II which is nominally a two-component mixture of 91% RDX and 9% polyethylene (PE) emulsion.

Atomistic modeling of Composition A3 Type II requires not only a proper description of the intracomponent (RDX–RDX and PE–PE) interactions but also the intercomponent cross terms (RDX–PE) present at the interface. For the current demonstration, the previously discussed SB[165] and the COMPASS force fields[166] have been applied to model atoms in the RDX and PE layers, respectively. These potentials were selected based on their accuracy for properties of the individual components and because they were developed, in part, using QM calculations. Therefore, they are concomitant with the quantum mechanically informed MM&S philosophy we adhere to. Completion of the model for this simple formulation requires development of a quantum mechanically derived RDX–PE interaction potential.

Based on our success in the development of intermolecular potentials for the pure FOX-7[222] and TATB[223] energetics, the exponential-6 functional form (Eq. 5.9) was chosen for the RDX–PE cross terms, where the "*i*" index labels atoms of the RDX layer and "*j*" corresponds to those of PE:

$$V(r)_{\mathrm{RDX-PE}} = \sum_{i \in \mathrm{RDX}, j \in \mathrm{PE}} \frac{q_i q_j}{r_{ij}} + A_{ij} \exp\left(-\frac{r_{ij}}{B_{ij}}\right) - \frac{C_6^{ij}}{r_{ij}^6} \tag{5.9}$$

The set of atomic charges $\{q\}$ for all RDX and PE atoms were fixed at the values defined in the individual SB and COMPASS potentials, respectively, with the remaining parameters fitted to 250 DFT interaction energies of a randomly configured, 87-oligomer PE chain interacting with a 4200-atom RDX slab (see Fig. 5.3). The (A_{ij}, B_{ij}, C_{ij}) parameters were fitted with a genetic algorithm using a population of 100 individuals that evolved for 500 generations with fitness scoring for each individual determined by the magnitude of the root-mean-square deviation from the reference DFT interaction energies. A comparison of the fitted and reference DFT energies is given in Fig. 5.4, and as shown, there is very good agreement between the two data sets. We have also verified the transferability of our potential by computing interaction energies of configurations containing multiple PE chains on the RDX surface and the potential remains accurate for those configurations as well.

The required traction–displacement curve for construction of the continuum CZM is obtained via uniaxial tension molecular dynamics simulations

Figure 5.3 Sample configuration of PE chain on RDX surface used for fitting of RDX–PE cross terms. (For color version of this figure, the reader is referred to the online version of this chapter.)

of a 116,832 atom sample of A3 Type II (Fig. 5.5). For this calculation, a constant strain acting normal to the interface is applied to all RDX and PE atoms in regions 1 and 3, respectively. At each timestep, the normal stress on all atoms residing within an ≈ 60 Å layer (region 2) above/below the interface is tabulated with the strain given by $(L - L_o)/L_o$ with L_o being the original gauge length of region 2. This was done for three different interface configurations; the resulting traction–displacement curve for each cell is given in Fig. 5.6. Thermal oscillations were removed by averaging which yielded the final traction–displacement curve (solid curve in Fig. 5.6) used to populate the CZM.

The CZM derived from the atomistic simulation described above was used in polycrystalline finite element simulations of A3 Type II under tension in conjunction with a previously developed RDX plasticity model.[232] In Fig. 5.7, we present the fractured polycrystalline sample resulting from the FE simulation. The images are colored by temperature (left) and stress (right) and regions of localized heating and stress concentration are evident near the fracture surfaces.

The use of this atomistically derived CZM using a quantum-based interaction potential in a grain-scale simulation is a small step toward enabling a robust, physics-based multiscale simulation of EM response. However, the material model used in the simulations does not contain critical information about other important microstructural properties (e.g., defect sizes and densities, grain and particle sizes, grain boundaries,

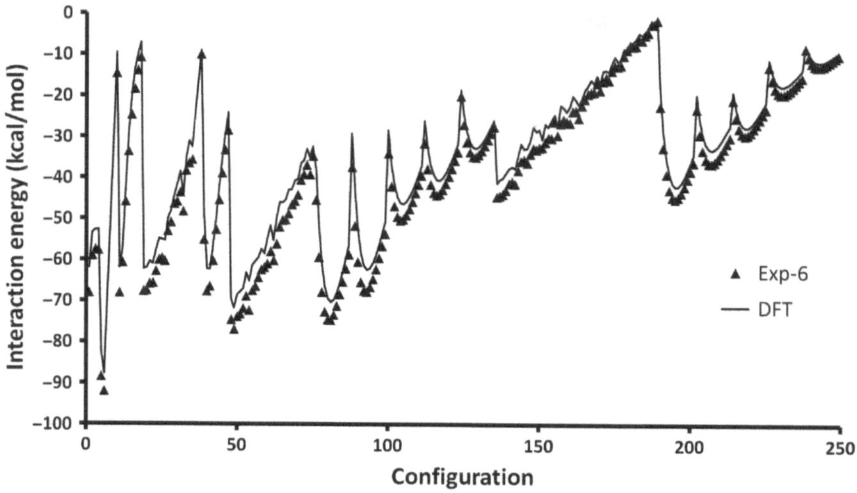

Figure 5.4 Comparison of forcefield energies to DFT values.

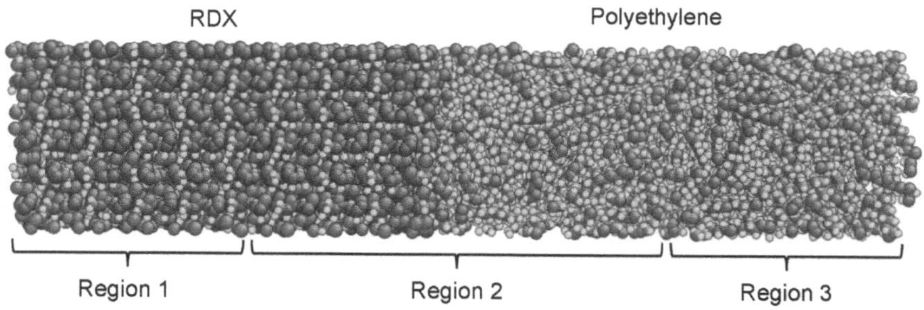

Figure 5.5 A3 Type II simulation cell. Strain is applied to all atoms in regions 1 and 3. Stress is tabulated on all atoms in region 2 to determine cohesive zone model. (For color version of this figure, the reader is referred to the online version of this chapter.)

compositional heterogeneities) that strongly influence material response. Realistic modeling of the complex microstructure of an EM cannot be done with molecular dynamics simulations using all-atom force fields (e.g., models in which interactions for all atoms are explicitly described). Such simulations are limited to the nanoscale due to computational costs, and thus cannot be used to explore the critical micron-scale regime associated with "hot spots."

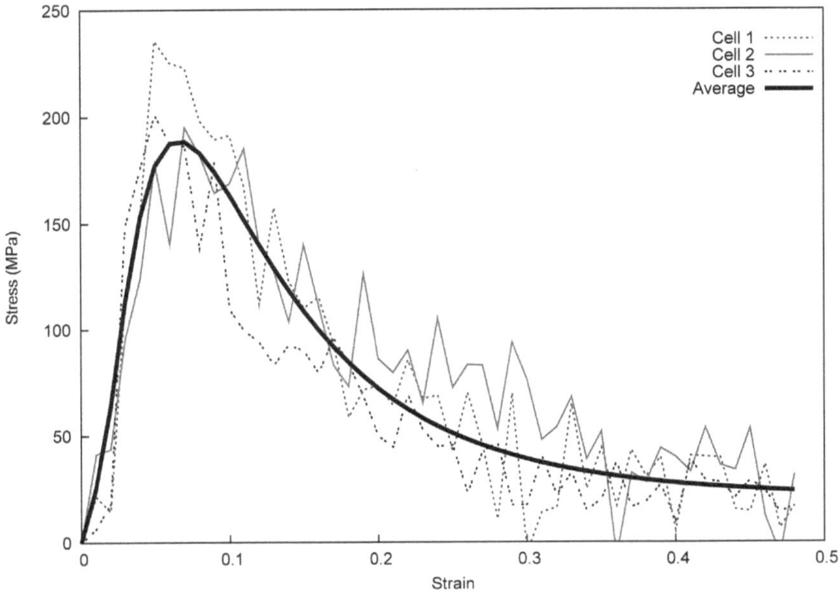

Figure 5.6 Traction–displacement curves for three simulation cells of A3 Type II. Solid black curve is the average of the three curves. (For color version of this figure, the reader is referred to the online version of this chapter.)

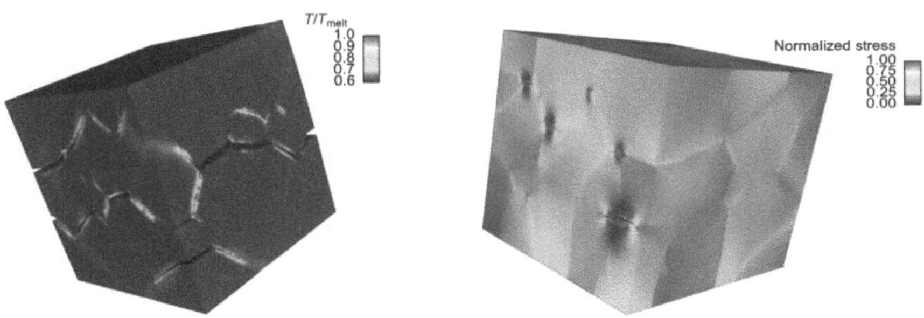

Figure 5.7 Fractured sample resulting from finite element simulations of Composition A3 Type II under tension. (For color version of this figure, the reader is referred to the online version of this chapter.)

3.4. Coarse-graining quantum-based atomistic models for meso-level simulations

In order to go beyond the time and spatial scales that can be accessed with atomistic simulations, lower-resolution CG descriptions of the forces must be invoked to reduce the computational cost associated with all-atom models.[233] Some of the simplest forms of coarse-graining when applied to

EM are those using united atom potentials, those in which a few atoms are combined to form a single "united atom" (UA) interaction site. Simulations using UA models maintain accuracy relative to all-atom simulations, provided the results are not strongly dependent on the degrees of freedom that were coarsened. A recent application of this for EMs is seen in Ref. 234 for the calculation of physical properties used to predict environmental fate of nitroaromatic energetics. In this model, each carbon atom and its bonded hydrogen atom were grouped into a single UA. Similarly, in studies using atomistic simulation to explore shock response and detonation of liquid NM,[235–237] the methyl group is CG into a single interaction site. A slightly more aggressive coarsening of an explosive molecule for the purpose of exploring microstructure was performed for the nitrate ester explosive PETN.[238] This molecule was coarsened into a moiety composed of one A and four B pseudoatoms, where each B pseudoatom represented a single NO_2 group and the single A pseudoatom was composed of the remaining $C(CH_2O)$ portion of the molecule. The model was first used to predict crystallographic parameters, bulk modulus, and sublimation energy for comparison with all-atom MD simulations using the COMPASS force field.[166] These authors showed that calculations using the CG model produced results that were in good agreement with those using all-atom models, but had a computational time that was smaller by more than two orders of magnitude. The results indicated that the important interactions were appropriately described in the CG model and could be used to understand the relation between performance and the morphology of the crystal. Since this model was developed, coarse-grain MD (CG-MD) simulations, in which the classical equations of motion are integrated using pseudoatoms, have been used to explore proposed mechanisms for morphology changes[239] and to simulate growth, morphology, and particle coarsening of PETN powders.[240] This model has demonstrated impressive performance, considering that it was fitted to only two experimental data points, and has given significant credence to such particle-based coarse-grain modeling for exploring microstructural detail.

While the CG-PETN model is an example of "top-down" coarse-graining, that is, the description of CG interactions was fitted to experimental observables, "bottom-up" approaches for coarse-graining exist, in which CG models are developed by matching atomistic and CG system structures (structure matching) or forces (force matching). Izvekov and coworkers have utilized a powerful bottom-up coarse-graining approach called the multiscale coarse-graining (MS-CG) method[241] to develop CG models of two explosives directly from all-atom MD simulations through force matching. The method, initially applied to soft materials, was applied to the explosives NM[242] and RDX[243] using the quantum-based atomistic models described previously.[164,165] For NM, one- and two-site CG models were generated, with the centers of mass of the molecule being used for the one-site CG model, and the centers of mass of the CH_3 and NO_2 groups being used for the two-site CG models. The CG models were force-matched to liquid phase simulations over a range of temperatures and pressures and included a particle-density dependence in order to improve the transferability of the potentials in describing a variety of thermodynamic states. When compared to the atomistic simulation results, the CG models reproduced a variety of properties, including the melting point and liquid structures over a range of pressures. In a notable demonstration of transferability, the two-site model reasonably predicted the low-temperature crystal structure of NM, although all-atom crystal simulations were not used in the force matching.

Development of an MS-CG model for RDX was more challenging, since the ambient state of RDX is crystalline, whereas the ambient state of NM is liquid. For RDX, one- and four-site density-dependent models were force-matched to crystal and molten phase all-atom simulations; the one-site model reproduced crystal structural information over a range of pressures, the melting point, and elastic and vibrational properties. Also, the two-site density-dependent NM and RDX models both reproduced the unreacted shock Hugoniot in good agreement with MD results using the reference atomistic models; all were in good agreement with experiment. Demonstrations using these models indicate success in retaining salient features upon coarsening of the atomistic information. Although the all-atom MD simulations used quantum-based potentials for developing the MS-CG models through force matching, MS-CG models can be force-matched to direct QMD simulations; the only limitations at this time are computational resources to perform sufficiently long all-atom QMD simulations of a condensed-phase system needed to converge the MS-CG models.

CG potentials used in molecular dynamics simulations of the "meso" particles can successfully predict various static properties of a system, as demonstrated above using the NM and RDX CG potentials. However, the loss of information resulting from coarse-graining results in the inability of CG-MD simulations to properly describe dynamic properties and response.[244–247] The limitation in this methodology precludes it from being used to predict microstructure evolution, emphasized in Section 1 as being one of the key material features controlling sensitivity to initiation. There is, however, a particle-based mesoscale simulation method that can properly depict the dynamics of a CG system, and which has recently been used to simulate EM response to thermal and shock loading.[248]

The dissipative particle dynamics (DPD) method,[249] initially developed to study hydrodynamic phenomena of complex fluids, is similar in spirit to conventional molecular dynamics in that it involves integration of equations of motion for particles that interact through conservative forces. However, the DPD method differs from conventional MD in that the mesoparticles also interact via frictional and random forces, to recover information lost upon coarse-graining, thus resulting in proper depictions of dynamic properties. The original form of DPD is not suitable to explore EM response to shock or thermal loading as it does not allow heat exchange between particles or chemical reaction. Inability to properly depict thermal response led to an extension of the method, known as the constant-energy DPD-E method,[250,251] in which each particle is assigned an additional variable u_i that corresponds to its internal energy. Integration of the equation of motion for u_i allows a dynamic description of the "internal state" (as defined by a mesoparticle equation of state) in a simulation, allowing a means for exchange of momentum and heat between particles. There are also ongoing efforts by Brennan and coworkers[252] to extend the method further to include chemical reactivity in a manner that is similar in spirit to the DPD-E variant (denoted as DPD-Rx). In this case, however, each mesoparticle is assigned a new reaction progress variable that renders it as a microreactor; integration of the equation of motion that includes the reaction progress variable allows a dynamic description of chemical reactions of the system. For the DPD method and its extensions, QM or classical atomistic information can be used to parameterize the conservative forces (as done using the MS-CG method), as well as the mesosparticle equation of state[253] or the chemical reactions used in the DPD-Rx scheme.

In a demonstration of the method, both CG-MD and the DPD-E method using the MS-CG one-site model of RDX have been used to explore the nonreactive shock response of the crystal, with results compared to atomistic simulations.[254] In the simulations, the CG-MD model overestimates both temperature (>4000 K) and density ($\rho/\rho_0 \sim 1.3$) relative to the atomistic values (1500 and 1.22 K, respectively). The DPD-E simulations, on the other hand, are in quantitatively good agreement with the atomistic values at the shock front as shown in Fig. 5.8.[255]

As noted in a report on a 2008 workshop on the DPD method,[256] it is hoped that the DPD methodology will provide a critical link between the atomistic and continuum scales. The mesoscale information can be used, of course, to parameterize constitutive models; however, it can also be used in a concurrent multiscale scheme. A demonstration of such a directly coupled DPD-continuum simulation was accomplished,[257] where thermodynamic states of localized regions, obtained from DPD-E simulations of a model HMX system subjected to shear at various rates, were directly used in continuum-level modeling of shear band nucleation and propagation using the CTH hydrocode.[258] The deleterious effects resulting from the approximations contained in constitutive material models are avoided in such an approach since material properties on a local scale are directly described by DPD simulations and included in the continuum calculations. The accuracy of this approach, obviously, depends on the mesoscale model used in the DPD-E simulations. The model used in this demonstration example was a CG representation of HMX, where one mesoparticle represented a single HMX molecule. This simple model was parameterized in a top-down

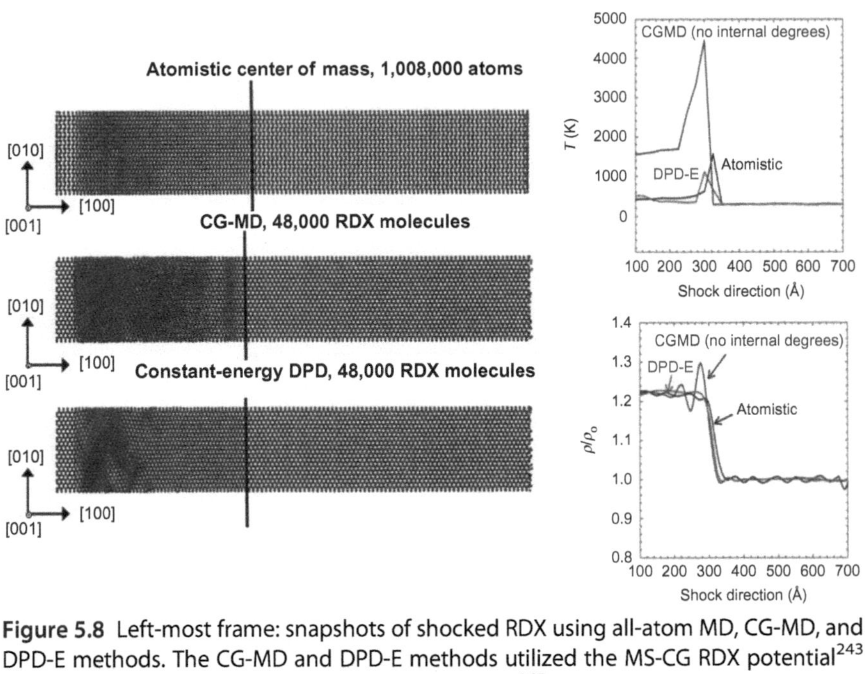

Figure 5.8 Left-most frame: snapshots of shocked RDX using all-atom MD, CG-MD, and DPD-E methods. The CG-MD and DPD-E methods utilized the MS-CG RDX potential[243] and the all-atom MD simulations used the SB model.[165] Right frames show temperature and density profiles corresponding to the snapshot. (For color version of this figure, the reader is referred to the online version of this chapter.)

approach using a few experimental observables, thus limiting its predictive capability. However, the development of models based on first-principles quantum mechanics through bottom-up approaches, such as those developed by Izvekov and coworkers,[241] will undoubtedly better capture the fundamental physical and chemical features at the coarse-grain scale needed to produce high-fidelity results in a continuum-scale simulation.

3.5. Multiresolution methods

Although much progress has been made in QM modeling of condensed-phase EMs, there still exist a variety of technical challenges which must be addressed before a truly quantum mechanically based MM&S paradigm can be realized. Perhaps the biggest technical challenge is the seamless coupling of models representing the disparate length scales and how the physics/chemistry occurring at one length scale can be incorporated dynamically into other regions. Regarding the QM region, which is typically embedded within a larger cluster of atoms treated using classical potentials, a proper accounting of the Coulomb and exchange interactions, which have been artificially truncated at the quantum–classical interface, is required. There have been many approaches for coupling quantum and classical regions ranging from a simple inclusion of hydrogen link atoms[259] to schemes that include localized orbitals and effective group potentials.[260] Link atoms are applied most often, due to the simplicity of the method; however, link atoms may introduce artificial interactions at the boundary which may perturb the dynamics of the QM region.[261] An alternative technique that has proven beneficial for quantum–classical coupling is the consistent embedding approach of Mallik et al.[262,263] where the quantum domain is terminated by effective atoms to which an effective core potential (ECP) is assigned with a multipole representation of the extended medium. The coefficients and exponents which parameterize the ECP are optimized so that the atomic forces and total charge density within the truncated quantum domain match those obtained for QM treatment of the full system. This promising approach was used with success in multiscale simulations of a silica nanorod subjected to small strain; however, application of the approach to molecular crystals has not been attempted.

Another emerging method for QM embedding is "exact embedded density functional theory" (EE-DFT).[119,120] EE-DFT is a formally exact orbital-free approach for partitioning a large system into smaller, more computationally tractable units which are then combined using a nonadditive kinetic potential (NAKP) term. The NAKP term enables an accurate, scalable protocol for partitioning a system into subsystems such that it is possible to get a *formally exact solution* using only the information from the combined individual subsystems, which is a particularly attractive feature of the formalism. EE-DFT allows for a true MM&S approach, as the various subsystems (which could, in principle, range from the atomic to the continuum scale) only interact through their relative electron densities. This necessitates an appropriate description of the electronic densities at all length scales and for the macro and micron scales; this representation should be fairly well represented by standard field models. The densities at the atomic level can be evaluated readily from quantum mechanics; however, electronic density descriptors at the subgrain, particle-based meso-level have not been developed at present.

The multiscale shock technique (MSST) is a very powerful multiresolution method coupling the molecular dynamics method with the Navier–Stokes equations for compressible flow to provide a computationally efficient means to study response of a shocked material at the atomistic level.[264] MSST, using the self-consistent charge density-functional tight binding method (SCC-DFTB)[54] to describe the forces, has been applied for studies of shock response of NM,[265] TATB,[266] hydrazoic acid,[267] and HMX.[268,269] The technique has the advantage that it can simulate the evolution of a shocked system using considerably fewer atoms than what would be required for a direct large-scale molecular dynamics shock simulation. For example, large-scale MD simulations of shocked PETN using ReaxFF required 352,640 atoms,[13] whereas MSST simulations using ReaxFF required 3712 atoms, and produced pressure profiles in good agreement with the direct MD simulations.[270] The substantial computational savings, therefore, allow for longer-time trajectories and computationally intensive descriptions of interatomic forces (such as DFT) in the integration of the equations of motion. For example, MSST using SCC-DFTB[54] was used to simulate shocked TATB for 0.43 ns in order to monitor the chemistry resulting from the shock. In addition to providing a computationally efficient means to study shocked materials using QM descriptions of forces, an extended version of MSST that includes electron–ion coupling enabled computational exploration of the role of electronic excitations in shock compression. The results showed that chemical reactivity of shocked hydrazoic acid was enhanced by 30% due to electronic excitations, indicating that electronic excitation must be considered when characterizing phenomena behind a detonation front.[271]

4. OTHER CHALLENGES AND PATHS FORWARD

Perhaps the most difficult property to predict for an EM is its sensitivity to initiation, which is a critical metric in the performance assessment of a candidate explosive. There have been numerous efforts to establish correlations to impact sensitivity with QM descriptors such as partial atomic charges,[272,273] activation energies,[274] bond orders,[275] heats of reaction,[276] electrostatic potentials,[277] and features of the electron density including bond critical points and bond bundles.[278,279] Much attention has been paid to intramolecular chemistry with particular emphasis on the strength of $X—NO_2$ bonds (X = C, N) which, upon rupture, may be an initial trigger for initiation to detonation. Further, reports suggesting the importance of *cooperative* effects and the strength of intermolecular interactions[280] on impact sensitivity have been presented. Correlations based on the rate of energy transfer between phonons and vibrons[281] have also appeared (though not at the QM level). Part of the difficulty in computationally predicting EM sensitivity to impact at the QM level is accounting for energy localization (i.e., "hot spots") due to microstructural defects (e.g., dislocations, voids, and cracks) which occur on longer length scales than are quantum mechanically accessible. Mechanical energy imposed upon such a localized region through shock may couple with intramolecular vibrations resulting in molecular decomposition and energy release, thereby destabilizing the energetic. Suggested features associated with hot-spot formation include intra/intercrystalline voids, shear banding, and dislocation pile-up, all of which are well beyond current QM level treatments and remain quite challenging at the MD level. Orbital-free and quasicontinuum DFT approaches provide a potential route toward QM descriptions of physically realistic defect structures corresponding to these microstructural features; further development of these methods could allow for coupling of currently established QM sensitivity descriptors with the microstructural influences that play a critical role in determining EM response to insult.

Other challenges in QM modeling of condensed-phase EM lie in analysis of the results. For example, there is a need to effectively interrogate the electron density in a highly compressed material (such as that which is behind a shock front) in order to identify covalent bonds. It is within this region that chemistry of a condensed-phase EM subjected to shock begins, and under correct conditions, leads to self-sustained detonation. Before initiation mechanisms can be understood, chemical species must be identified. Similar issues arise in reactive modeling of condensed-phase EM using the ReaxFF force field;[226] however, a covalent bond between pairs of atoms is defined by its bond order, a calculable feature in ReaxFF that is based on local atomic environment. Thus, molecular species can be identified and monitored in a ReaxFF MD simulation. For QM systems, electron density partitioning schemes such as that given in Ref. 279 might allow for such analyses required for mechanistic determination.

Solving these difficult technical challenges will advance the state of the art and will enable a quantum-informed MM&S computational capability to characterize EM properties and response. MM&S will provide the understanding of the role of microstructural features coupled with the chemistry of initiation on overall performance; thus, providing the EM formulator with critical information needed for processing EM with specific performance properties. This modeling capability offers unsurpassed design and screening technologies for faster, safer, and less costly implementation of advanced insensitive EM. It is crucial that further research investments are made, which will lead to breakthrough capabilities in the development of new, optimally performing EM.

REFERENCES

1. National Research Council. *Advanced Energetic Materials;* The National Academies Press: Washington, DC, 2004.
2. *MIL-STD-2105C, Department of Defense Test Method Standard: Hazard Assessment Tests For Non-Nuclear Munitions*; United States Department of Defense, 14 July 2003.
3. *MIL-STD-810G, Department of Defense Test Method Standard for Environmental Engineering Considerations and Laboratory Tests*; United States Department of Defense, 31 Oct 2008.
4. Mader, C. L. *Numerical Modeling of Explosives and Propellants*, 3rd ed.; Boca Raton: CRC Press, 2007.
5. Horie, Y. Ed.; *Shock Wave Science and Technology Reference Library, Vol. 2: Solids I and Vol. 3: Solids II*; Springer: Berlin, Heidelberg, 2007.
6. Karo, A. M.; Hardy, J. R. Molecular Dynamics of Shock-Initiated Detonations. *Int. J. Quantum Chem.* **1978**, *12*(Suppl. 1), 333.
7. Karo, A. M.; Hardy, J. R.; Walker, F. E. *Theoretical Studies of Shock-Initiated Detonations: Technical Report*; Lawrence Livermore National Laboratory: Livermore, CA, 1977.
8. Karo, A. M.; Hardy, J. R. *Molecular Dynamics of Shock-Initiated Detonations:Technical Report*; Lawrence Livermore National Laboratory: Livermore, CA, 1977.
9. Zhou, T. T.; Zybin, S. V.; Liu, Y.; Huang, F. L.; Goddard, W. A., III. Anisotropic Shock Sensitivity for Beta-Octahydro-1,3,5,7-Tetranitro-1,3,5,7-Tetrazocine Energetic Material under Compressive-Shear Loading from Reaxff-Lg Reactive Dynamics Simulations. *J. Appl. Phys.* **2012**, *111*, 124904.
10. Zhang, L.; Zybin, S. V.; Van Duin, A. C. T.; Goddard, W. A., III. Modeling High Rate Impact Sensitivity of Perfect RDX and HMX Crystals by Reaxff Reactive Dynamics. *J. Energ. Mater.* **2010**, *28*(Suppl. 1), 92.

11. Nomura, K.; Kalia, R. K.; Nakano, A.; Vashishta, P.; van Duin, A. C. T.; Goddard, W. A., III. Dynamic Transition in the Structure of an Energetic Crystal During Chemical Reactions at Shock Front Prior to Detonation. *Phys. Rev. Lett.* **2007**, *99*, 148303.
12. Strachan, A.; van Duin, A. C. T.; Chakraborty, D.; Dasgupta, S.; Goddard, W. A., III. Shock Waves in High-Energy Materials: The Initial Chemical Events in Nitramine RDX. *Phys. Rev. Lett.* **2003**, *91*, 098301.
13. Budzien, J.; Thompson, A. P.; Zybin, S. V. Reactive Molecular Dynamics Simulations of Shock Through a Single Crystal Of Pentaerythritol Tetranitrate. *J. Phys. Chem. B* **2009**, *113*, 13142.
14. See, for example, Sewell, T. D.; Menikoff, R.; Bedrov, D.; Smith, G. D. A Molecular Dynamics Simulation Study of Elastic Properties of HMX. *J. Chem. Phys.* **2003**, *119*, 7417.
15. See, for example, Munday, L. B.; Solares, S. D.; Chung, P. W. Generalized Stacking Fault Energy Surfaces in the Molecular Crystal Alpha RDX. *Philos. Mag.* **2012**, *92*, 3036.
16. Izvekov, S.; Chung, P. W.; Rice, B. M. Non-Equilibrium Molecular Dynamics Simulation Study of Heat Transport in Hexahydro-1,3,5-Trinitro-*S*-Triazine (RDX). *Int. J. Heat Mass Transf.* **2011**, *54*, 5623.
17. Mathew, N.; Picu, C. R.; Chung, P. W. Peierls Stress of Dislocations in Molecular Crystal Cyclotrimethylenetrinitramine. *J. Phys. Chem. A* **2013**, *117*, 5326.
18. Long, Y.; Chen, J.; Liu, Y. G.; Nie, F. D.; Sun, J. S. A Direct Method to Calculate Thermal Conductivity and Its Application in Solid HMX. *J. Phys. Condens. Matter* **2010**, *22*, 185404.
19. Long, Y.; Liu, Y. G.; Nie, F. D.; Chen, J. A Method to Calculate the Thermal Conductivity of HMX Under High Pressure. *Philos. Mag.* **2012**, *92*, 1023.
20. Conroy, M. W.; Oleynik, I. I.; Zybin, S. V.; White, C. T. First-Principles Anisotropic Constitutive Relationships in Beta-Cyclotetramethylene Tetranitramine (Beta-HMX). *J. Appl. Phys.* **2008**, *104*, 053506.
21. Conroy, M. W.; Oleynik, I. I.; Zybin, S. V.; White, C. T. Density Functional Theory Calculations of Anisotropic Constitutive Relationships in Alpha-Cyclotrimethylenetrinitramine. *J. Appl. Phys.* **2008**, *104*, 113501.
22. Conroy, M. W.; Oleynik, I. I.; Zybin, S. V.; White, C. T. First-Principles Investigation of Anisotropic Constitutive Relationships in Pentaerythritol Tetranitrate. *Phys. Rev. B* **2008**, 77, 094107.
23. Oleynik, I. I.; Conroy, M.; White, C. T. Anisotropic Constitutive Relationships in Energetic Materials: Nitromethane and RDX. *AIP Conf. Proc.* **2007**, *955*, 401.
24. Zerilli, F. J.; Kuklja, M. M. Equation of State of 1,1-Diamino-2,2-Dinitroethylene from First Principles. *AIP Conf. Proc.* **2006**, *845*, 183.
25. Conroy, M.; Oleynik, I. I.; Zybin, S. V.; White, C. T. Anisotropic Constitutive Relationships in Energetic Materials: PETN and HMX. *AIP Conf. Proc.* **2007**, *955*, 361.
26. Klimenko, V. Y.; Kozyreva, I. Y. Numerical Model for Dislocation Mechanism of Detonation. *J. Energ. Mater.* **2010**, *28*(Suppl. 1), 249.
27. Tadmor, E. B.; Miller, R. E. *Modeling Materials—Continuum, Atomistic and Multiscale Techniques*; Cambridge University Press: Cambridge, 2012.
28. Schrödinger, E. An Undulatory Theory of the Mechanics of Atoms and Molecules. *Phys. Rev.* **1926**, *28*, 1049.
29. Cohen-Tannoudji, C.; Diu, B.; Laloe, F. *Quantum Mechanics*; John Wiley & Sons: New York, 1977; Vol. 1: pp 9–40.
30. Levine, I. *Quantum Chemistry*; New Jersey: Prentice Hall, 1991; pp 1–18.
31. Nye, J. F. *Physical Properties of Crystals*; Clarendon Press: Oxford, 1957; pp 82–105.

32. Szabo, A.; Ostlund, N. S. *Modern Quantum Chemistry*; Dover: New York, 1957; pp 231–270.
33. Parr, R. G.; Yang, W. *Density-Functional Theory of Atoms and Molecules*; Oxford University Press: New York, 1989; pp 47–69.
34. Bartlett, R. J. Coupled Cluster Theory: An Overview of Recent Developments. In: *Modern Electronic Structure Theory Part II*; Yarkony, D. R. Ed.; Vol. 2, World Scientific: Singapore, 1995; pp 1047–1131.
35. Crawford, T. D.; Schaefer, H. F. An Introduction to Coupled Cluster Theory for Computational Chemists. In: *Reviews in Computational Chemistry*; Lipkowitz, K. B.; Boyd, D. B. Eds.; Vol. 14, VCH: New York, 2000; pp 33–115.
36. Bartlett, R. J. Many-Body Perturbation Theory and Coupled Cluster Theory for Electron Correlation in Molecules. *Ann. Rev. Phys. Chem.* **1981**, *32*, 359–401.
37. Purvis, G. D.; Bartlett, R. J. A Full Coupled-Cluster Singles and Doubles Model: The Inclusion of Disconnected Triples. *J. Chem. Phys.* **1982**, *76*, 1910–1918.
38. Raghavachari, K.; Truck, G. W.; Pople, J. A.; Head-Gordon, M. A Fifth-Order Perturbation Comparison of Electron Correlation Theories. *Chem. Phys. Lett.* **1989**, *157*, 479–483.
39. Eckhardt, C. J.; Gavezzotti, A. Computer Simulations and Analysis of Structural and Energetic Features of Some Crystalline Energetic Materials. *J. Phys. Chem. B* **2007**, *111*, 3430–3437.
40. Molt, R. W.; Watson, T., Jr.; Lotrich, V. F.; Bartlett, R. J. RDX Geometries, Excited States, and Revised Energy Ordering of Conformers via MP2 and CCSD(T) Methodologies: Insights into Decomposition Mechanism. *J. Phys. Chem. A* **2011**, *115*, 884–890.
41. Molt, R. W.; Watson, T., Jr.; Bazante, A. P.; Bartlett, R. J. The Great Diversity of HMX Conformers: Probing the Potential Energy Surface Using CCSD(T). *J. Phys. Chem. A* **2013**, *117*, 3467–3474.
42. Borges, I.; Aquino, A.; Barbatti, M.; Lischka, H. The Electronically Excited States of RDX (Hexahydro-1,3,5-Trinitro-1,3,5-Triazine): Vertical Excitations. *Int. J. Quantum Chem.* **2009**, *109*, 2348–2355.
43. Choi, C. S.; Prince, E. Crystal-Structure of Cyclotrimethylenetrinitramine. *Acta Crystallogr. B* **1972**, *28*, 2857–2862.
44. Vrcelj, R. M.; Sherwood, J. N.; Kennedy, A. R.; Gallagher, H. G.; Gelbrich, T. Polymorphism in 2-4-6-Trinitrotoluene. *Cryst. Growth Des.* **2003**, *3*, 1027–1032.
45. Fitzgerald, G.; Harrison, R. J.; Bartlett, R. J. Analytic Energy Gradients for General Coupled-Cluster Methods and 4th-Order Many-Body Perturbation Theory. *J. Chem. Phys.* **1986**, *85*, 5143–5150.
46. Scuseria, G. Analytic Evaluation of Energy Gradients for the Singles and Doubles Coupled Cluster Method Including Perturbative Triple Excitations—Theory and Applications to FOOF and CR-2. *J. Chem. Phys.* **1991**, *94*, 442–447.
47. Gauss, J.; Stanton, J. Analytic Gradients for the Coupled-Cluster Singles, Doubles, and Triples (CCSDT) Model. *J. Chem. Phys.* **2002**, *116*, 1773–1782.
48. Taylor, D. E. Potential Energy Surface Mapping of Energetic Materials Using Coupled Cluster Theory. In *Proceedings of the 2010 DoD High Performance Computing Modernization Program Users Group Conference, Shaumburg, IL, June 14–17, 2010*; IEEE Computer Society: Los Alamitos, CA, 2011.
49. Pople, J. A.; Beveridge, D. L. *Approximate Molecular Orbital Theory*; McGraw-Hill: New York, 1970; pp 1–100.
50. Dewar, M. J. S.; Zoebisch, E. G.; Healy, E. F.; Stewart, J. J. P. The Development and Use of Quantum Mechanical Molecular Models.76. AM1—A New General Purpose Quantum Mechanical Molecular Model. *J. Am. Chem. Soc.* **1985**, *107*, 3902–3909.

51. Thiel, W. Semiempirical Methods—Current Status and Perspectives. *Tetrahedron* **1988**, *44*, 7393–7408.
52. Seifert, G.; Eschrig, H.; Bieger, W. An Approximation Variant of LCAO-X-Alpha Methods. *Z. Phys. Chem. (Leipzig)* **1986**, *267*, 529–539.
53. Seifert, G.; Porezag, D.; Frauenheim, T. Calculations of Molecules, Clusters, and Solids with a Simplified LCAO-DFT-LDA Scheme. *Int. J. Quantum Chem.* **1996**, *58*, 185–192.
54. Elstner, M.; Porezag, D.; Jungnickel, G.; Elsner, J.; Haugk, M.; Frauenheim, T.; Suhai, S.; Seifert, G. Self-Consistent-Charge Density-Functional Tight-Binding Method for Simulations of Complex Materials Properties. *Phys. Rev. B* **1998**, *58*, 7260–7268.
55. Jensen, F. *Introduction to Computational Chemistry*; John Wiley and Sons: Chichester, 1999; pp 81–97.
56. Obara, S.; Saika, A. Efficient Recursive Computation of Molecular Integrals over Cartesian Gaussian Functions. *J. Chem. Phys.* **1986**, *84*, 3963–3974.
57. McMurchie, L. E.; Davidson, E. R. One-Electron and 2-Electron Integrals over Cartesian Gaussian Functions. *J. Comput. Phys.* **1978**, *26*, 218.
58. Stewart, J. J. P. Optimization of Parameters for Semiempirical Methods.1. Method. *J. Comput. Chem.* **1989**, *10*, 209–220.
59. Xiao, H. M.; Zhang, J. Theoretical Prediction on Heats of Formation for Polyisocyanocubanes—Looking for Typical High Energetic Density Material (HEDM). *Sci. China B* **2002**, *45*, 21–29.
60. Zhang, J.; Xiao, H. M.; Gong, X. D. Theoretical Studies on Heats of Formation for Polynitrocubanes Using the Density Functional Theory B3LYP Method and Semiempirical MO Methods. *J. Phys. Org. Chem.* **2001**, *14*, 583–588.
61. Klapotke, T. M.; Ang, H. Estimation of the Crystalline Density of Nitramine (N-NO_2 based) High Energy Density Materials (HEDM). *Propel. Explos. Pyrotech.* **2001**, *26*, 221–224.
62. Morrill, J. A.; Byrd, E. F. C. Development of Quantitative Structure–Property Relationships for Predictive Modeling and Design of Energetic Materials. *J. Mol. Graph. Model.* **2008**, *27*, 349–355.
63. McClellan, J. J.; Hughes, T. F.; Bartlett, R. J. Application of the Transfer Hamiltonian Formalism to High-Energy Model Systems. *Int. J. Quantum Chem.* **2005**, *105*, 914–920.
64. Xiao, J. J.; Fang, G. Y.; Li, G. F.; Xiao, H. M. Simulation Investigations in the Binding Energy and Mechanical Properties of HMX-Based Polymer-Bonded Explosives. *Chin. Sci. Bull.* **2005**, *50*, 21–26.
65. Nalwa, H. S.; Watanabe, T.; Miyata, S. A Comparative Study of 4-Nitroaniline, 1,5-Diamino-2,4- Dinitrobenzene and 1,3,5-Triamino-2,4,6-Trinitrobenzene and Their Molecular Engineering for Second-Order Nonlinear Optics. *Opt. Mater.* **1993**, *2*, 73–81.
66. Dewar, M. J. S.; Thiel, W. Semiempirical Model for 2-Center Repulsion Integrals in NDDO Approximation. *Theor. Chim. Acta* **1977**, *46*, 89–104.
67. Koskinen, P.; Makinen, V. Density-Functional Tight-Binding for Beginners. *Comput. Mater. Sci.* **2009**, *47*, 237–253.
68. Porezag, D.; Frauenheim, T.; Kohler, T.; Seifert, G.; Kaschner, R. Construction of Tight-Binding-Like Potentials on the Basis of Density-Functional Theory—Application to Carbon. *Phys. Rev. B* **1995**, *51*, 12947–12957.
69. Stewart, J. J. P. Application of the PM6 Method to Modeling the Solid State. *J. Mol. Model.* **2008**, *14*, 499–535.
70. Korth, M.; Thiel, W. Benchmarking Semiempirical Methods for Thermochemistry, Kinetics, and Noncovalent Interactions: OMx Methods are Almost as Accurate and

Robust as DFT-GGA Methods for Organic Molecules. *J. Chem. Theor. Comput.* **2011**, 7, 2929–2936.
71. Stewart, J. J. P. Comparison of the Accuracy of Semiempirical and Some DFT Methods for Predicting Heats of Formation. *J. Mol. Model.* **2004**, *10*, 6–12.
72. Otte, N.; Scholten, M.; Thiel, W. Looking at Self-Consistent-Charge Density Functional Tight Binding from a Semiempirical Perspective. *J. Phys. Chem. A* **2007**, *111*, 5751–5755.
73. Manaa, M. R.; Gee, R. H.; Fried, L. E. Internal Rotation of Amino and Nitro Groups in TATB: MP2 versus DFT (B3LYP). *J. Phys. Chem. A* **2002**, *106*, 8806–8810.
74. Manaa, M. R.; Fried, L. E. Internal Rotation in Energetic Systems: TATB. *J. Phys. Chem. A* **2001**, *105*, 6765–6768.
75. Chakraborty, D.; Muller, R. P.; Dasgupta, S.; Goddard, W. A. A Detailed Model for the Decomposition of Nitramines: RDX and HMX. *J. Comput. Aided Mater. Des.* **2002**, *8*, 203–212.
76. Okovytyy, S.; Kholod, Y.; Qasim, M.; Fredrickson, H.; Leszcynski, J. The Mechanism of Unimolecular Decomposition of 2,4,6,8,10,12-Hexanitro-2,4,6,8,10,12-Hexaazaisowurtzitane. A Computational DFT Study. *J. Phys. Chem. A* **2005**, *109*, 2964–2970.
77. Zhao, J.; Liu, H. High-Pressure Behavior of Crystalline FOX-7 by Density Functional Theory Calculations. *Comput. Mater. Sci.* **2008**, *42*, 698–703.
78. Valenzano, L.; Slough, W. J.; Perger, W. F. Accurate Prediction of Second-Order Elastic Constants from First Principles: PETN and TATB. *AIP Conf. Proc.* **2012**, *1426*, 1191.
79. Manaa, M. R.; Fried, L. E. Nearly Equivalent Inter- and Intramolecular Hydrogen Bonding in 1,3,5-Triamino-2,4,6-Trinitrobenzene at High Pressure. *J. Phys. Chem. C* **2011**, *116*, 2116–2122.
80. Ojeda, O.; Cagin, T. Hydrogen Bonding and Molecular Rearrangement in 1,3,5-Triamino-2,4,6-Trinitrobenzene Under Compression. *J. Phys. Chem. B* **2011**, *115*, 12085–12093.
81. VandeVondele, J.; Borstnik, U.; Hutter, J. Linear Scaling Self-Consistent Field Calculations with Millions of Atoms in the Condensed Phase. *J. Chem. Theor. Comput.* **2012**, *8*, 3565–3573.
82. Bowler, D. R.; Miyazaki, T. Calculations for Millions of Atoms with Density Functional Theory: Linear Scaling Shows Its Potential. *J. Phys. Condens. Matter* **2010**, *22*, 074207.
83. Hung, L.; Carter, E. A. Accurate Simulations of Metals at the Mesoscale: Explicit Treatment of 1 Million Atoms with Quantum Mechanics. *Chem. Phys. Lett.* **2009**, *475*, 163–170.
84. Kresse, G.; Furthmuller, J. Efficiency of Ab-Initio Total Energy Calculations for Metals and Semiconductors Using a Plane-Wave Basis Set. *Comput. Mater. Sci.* **1996**, *6*, 15–50.
85. Dovesi, R.; Orlando, R.; Civalleri, B.; Roetti, C.; Saunders, V. R.; Zicovich-Wilson, C. M. CRYSTAL: A Computational Tool for the Ab Initio Study of the Electronic Properties of Crystals. *Z. Kristallogr.* **2005**, *220*, 571–573.
86. CP2K is freely available from: www.cp2k.org.
87. Zhang, Y.; Ma, N.; Wang, W. Z. Assessment of the Performance of the M05-Class and M06-Class Functionals for the Structure and Geometry of the Hydrogen-Bonded and Halogen-Bonded Complexes. *J. Theor. Comput. Chem.* **2012**, *11*, 1165–1173.
88. Dahlke, E. E.; Olson, R. M.; Leverentz, H. R.; Truhlar, D. G. Assessment of the Accuracy of Density Functionals for Prediction of Relative Energies and Geometries of Low-Lying Isomers of Water Hexamers. *J. Phys. Chem. A* **2008**, *112*, 3976–3984.

89. Zheng, Z. Y.; Zhao, J. J. Lattice Energies and Elastic Properties of Solid Methane: Assessment of Different Density Functionals. *Acta Phys. Chim. Sin.* **2012**, *28*, 1809–1814.
90. Zheng, Z.; Zhao, J.; Yiyang, S.; Zhang, S. Structures and Lattice Energies of Molecular Crystals Using Density Functional Theory: Assessment of a Local Atomic Potential Approach. *Chem. Phys. Lett.* **2012**, *550*, 94–98.
91. Qi, C.; Lin, Q. H.; Li, Y. Y.; Pang, S. P.; Zhang, R. B. C–N Bond Dissociation Energies: An Assessment of Contemporary DFT Methodologies. *J. Mol. Struct.* **2010**, *961*, 97–100.
92. Zhang, J.; Zhou, W.; Peng, B.; Zhang, S.; Gao, H.; Zhou, Z. Density Functional Theory for N–O Bond Dissociation Enthalpies of Quinoxaline-1,4-Dioxide Derivatives: Theoretical Method Assessment and Prediction. *J. Mol. Struct.* **2010**, *957*, 36–40.
93. Zawada, A.; Kaczmarek-Kedziera, A.; Bartkowiak, W. Assessment of DFT Functionals for the Calculation of Interaction-Induced Electric Properties of Molecular Complexes. *Chem. Phys. Lett.* **2011**, *503*, 39–44.
94. Champagne, B.; Perpete, E. A.; van Gisbergen, S. J. A.; Baerends, E. J.; Snijders, J. G.; Soubra-Ghaoui, C.; Robins, K. A.; Kirtman, B. Assessment of Conventional Density Functional Schemes for Computing the Polarizabilities and Hyperpolarizabilities of Conjugated Oligomers: An Ab Initio Investigation of Polyacetylene Chains. *J. Chem. Phys.* **1998**, *109*, 10489–10498.
95. Byrd, E. F. C.; Scuseria, G. E.; Chabalowski, C. F. An Ab Initio Study of Solid Nitromethane, HMX, RDX, and CL20: Successes and Failures of DFT. *J. Phys. Chem. B* **2004**, 77, 13100–13106.
96. Cady, H. H.; Larson, A. C. Crystal Structure of 1,3,5-Triamino-2,4,6-Trinitrobenzene. *Acta Crystallogr.* **1965**, *18*, 485–496.
97. Jeziorski, B.; Moszynski, R.; Szalewicz, K. Perturbation Theory Approach to Intermolecular Potential Energy Surfaces of Van der Waals Complexes. *Chem. Rev.* **1994**, *94*, 1887–1930.
98. Williams, H. L.; Chabalowski, C. F. Using Kohn–Sham Orbitals in Symmetry-Adapted Perturbation Theory to Investigate Intermolecular Interactions. *J. Phys. Chem. A* **2001**, *105*, 646–659.
99. Roszak, S.; Gee, R. H.; Balasubramanian, K.; Fried, L. E. Molecular Interactions of TATB Clusters. *Chem. Phys. Lett.* **2003**, *374*, 286–296.
100. Byrd, E. F. C.; Rice, B. M. Ab Initio Study of Compressed 1,3,5,7-Tetranitro-1,3,5,7-Tetraazacyclooctane (HMX), Cyclotrimethylenetrinitramine (RDX), 2,4,6,8,10,12-Hexanitrohexaazaisowurzitane (CL-20), 2,4,6-Trinitro-1,3,5-Benzenetriamine (TATB), and Pentaerythritol Tetranitrate (PETN). *J. Phys. Chem. C* **2007**, *111*, 2787–2796.
101. Zhao, Y.; Truhlar, D. G. A New Local Density Functional for Main-Group Thermochemistry, Transition Metal Bonding, Thermochemical Kinetics, and Noncovalent Interactions. *J. Chem. Phys.* **2006**, *125*, 194101.
102. Peverati, R.; Truhlar, D. G. Improving the Accuracy of Hybrid Meta-GGA Density Functionals by Range Separation. *J. Phys. Chem. Lett.* **2011**, *2*, 2810–2817.
103. Grimme, S. Accurate Description of Van der Waals Complexes by Density Functional Theory Including Empirical Corrections. *J. Comput. Chem.* **2004**, *25*, 1463–1473.
104. Pernal, K.; Podeszwa, R.; Patkowski, K.; Szalewicz, K. Dispersionless Density Functional Theory. *Phys. Rev. Lett.* **2009**, *103*, 263201.
105. von Lilienfeld, O.; Tavernelli, I.; Rothlisberger, U.; Sebastiani, D. Optimization of Effective Atom Centered Potentials for London Dispersion Forces in Density Functional Theory. *Phys. Rev. Lett.* **2004**, *93*, 153004.

106. Zhu, W. M.; Toulouse, J.; Savin, A.; Angyan, J. G. Range-Separated Density-Functional Theory with Random Phase Approximation Applied to Noncovalent Intermolecular Interactions. *J. Chem. Phys.* **2010**, *132*, 244108.
107. Angyan, J. G.; Gerber, I. C.; Savin, A.; Toulouse, J. Van der Waals Forces in Density Functional Theory: Perturbational Long-Range Electron-Interaction Corrections. *Phys. Rev. A* **2005**, *72*, 012510.
108. Lee, K.; Murray, E.; Kong, L.; Lundqvist, B.; Langreth, D. Higher-Accuracy Van der Waals Density Functional. *Phys. Rev. B* **2010**, *82*, 081101.
109. Dion, M.; Rydberg, H.; Schroder, E.; Langreth, D.; Lundqvist, B. Van der Waals Density Functional for General Geometries. *Phys. Rev. Lett.* **2004**, *92*, 246401.
110. Balu, R.; Byrd, E. F. C.; Rice, B. M. Assessment of Dispersion Corrected Atom Centered Pseudopotentials: Application to Energetic Molecular Crystals. *J. Phys. Chem. B* **2011**, *115*, 803–810.
111. Choi, C. S.; Boutin, H. P. A Study of Crystal Structure of Beta-Cyclotetramethylene Tetranitramine by Neutron Diffraction. *Acta Crystallogr. B* **1970**, *26*, 1235.
112. Zhurova, E. A.; Stash, A. I.; Tsirelson, V. G.; Zhurov, V. V.; Bartashevich, E. V.; Potemkin, V. A.; Pinkerton, A. A. Atoms-in-Molecules Study of Intra- and Intermolecular Bonding in the Pentaerythritol Tetranitrate Crystal. *J. Am. Chem. Soc.* **2006**, *128*, 14728.
113. Tavernelli, I.; Lin, I.; Rothlisberger, U. Multicenter-Type Corrections to Standard DFT Exchange and Correlation Functional. *Phys. Rev. B* **2009**, *79*, 045106.
114. Trevino, S. F.; Prince, E.; Hubbard, C. R. Refinement of the Structure of Solid Nitromethane. *J. Chem. Phys.* **1980**, *73*, 2996.
115. Evers, J.; Klapotke, T. M.; Mayer, P.; Oehlinger, G.; Welch, J. Alpha- and Beta-FOX-7, Polymorphs of a High Energy Density Material, Studied By X-Ray Single Crystal and Powder Investigations in the Temperature Range from 200 to 423 K. *Inorg. Chem.* **2006**, *45*, 4996–5007.
116. Vashishta, P.; Kalia, R.; Nakano, A.; Homan, B.; McNesby, K. Multimillion Atom Reactive Simulations of Nanostructured Energetic Materials. *J. Propul. Power* **2007**, *23*, 688–692.
117. Shimojo, F.; Kalia, R.; Nakano, A.; Vashishta, P. Divide-and-Conquer Density Functional Theory on Hierarchical Real-Space Grids: Parallel Implementation and Applications. *Phys. Rev. B* **2008**, 77, 085103.
118. Fedorov, D. G.; Nagata, T.; Kitaura, K. Exploring Chemistry with the Fragment Molecular Orbital Method. *Phys. Chem. Chem. Phys.* **2012**, *14*, 7562–7577.
119. Goodpaster, J.; Ananth, N.; Manby, F.; Miller, T. Exact Nonadditive Kinetic Potentials for Embedded Density Functional Theory. *J. Chem. Phys.* **2010**, *133*, 084103.
120. Manby, F. R.; Stella, M.; Goodpaster, J. D.; Miller, T. F., III A Simple Exact Density-Functional-Theory Embedding Scheme. *J. Chem. Theor. Comput.* **2012**, *8*, 2564.
121. Weber, V.; VandeVondele, J.; Hutter, J.; Niklasson, A. Direct Energy Functional Minimization under Orthogonality Constraints. *J. Chem. Phys.* **2008**, *128*, 084113.
122. Millam, J.; Scuseria, G. Linear Scaling Conjugate Gradient Density Matrix Search as an Alternative to Diagonalization for First Principles Electronic Structure Calculations. *J. Chem. Phys.* **1997**, *106*, 5569–5577.
123. Van Voorhis, T.; Head-Gordon, M. A Geometric Approach to Direct Minimization. *Mol. Phys.* **2002**, *100*, 1713–1721.
124. Li, X.; Nunes, R.; Vanderbilt, D. Density-Matrix Electronic-Structure Method with Linear System-Size Scaling. *Phys. Rev. B* **1993**, *47*, 10891–10894.
125. Shao, Y.; Saravanan, C.; Head-Gordon, M.; White, C. Curvy Steps for Density Matrix-Based Energy Minimization: Application to Large-Scale Self-Consistent-Field Calculations. *J. Chem. Phys.* **2003**, *118*, 6144.

126. McWeeny, R. Some Recent Advances in Density Matrix Theory. *Rev. Mod. Phys.* **1960**, *32*, 335–369.
127. Nemeth, K.; Scuseria, G. Linear Scaling Density Matrix Search Based on Sign Matrices. *J. Chem. Phys.* **2000**, *113*, 6035–6041.
128. Sewell, T. D.; Menikoff, R. Constituent Properties of HMX Needed For Mesoscale Simulations. *Combust. Theor. Model.* **2002**, *6*, 103.
129. Kuklja, M. M.; Stefanovich, E. V.; Kunz, A. B. An Excitonic Mechanism of Detonation Initiation in Explosives. *J. Chem. Phys.* **2000**, *112*, 3417, and references therein.
130. Kuklja, M. M.; Kunz, A. Barry Compression-Induced Effect on the Electronic Structure of Cyclotrimethylenetrinitramine Containing an Edge Dislocation. *J. Appl. Phys.* **2000**, *87*, 2215.
131. Kuklja, M. M.; Kunz, A. Barry Electronic Structure of Molecular Crystals Containing Edge Dislocations. *J. Appl. Phys.* **2001**, *89*, 4962.
132. Sharia, O.; Kuklja, M. M. Modeling Thermal Decomposition Mechanisms in Gaseous and Crystalline Molecular Materials: Application to Beta-HMX. *J. Phys. Chem. B* **2011**, *115*, 12677.
133. Xu, X.-J.; Zhu, W.-H.; Xiao, H.-M. DFT Studies on the Four Polymorphs of Crystalline CL-20 and the Influences of Hydrostatic Pressure on Epsilon-CL-20 Crystal. *J. Phys. Chem. B* **2007**, *111*, 2090.
134. Hunter, S.; Sutinen, T.; Parker, S. F.; Morrison, C. A.; Williamson, D. M.; Thompson, S.; Gould, P. J.; Pulham, C. R. Experimental and DFT-D Studies of the Molecular Organic Energetic Material RDX. *J. Phys. Chem. C* **2013**, *117*, 8062.
135. Landerville, A. C.; Conroy, M. W.; Lin, Y.; Budzevich, M. M.; White, C. T.; Oleynik, I. I. First-Principles Thermodynamics of Energetic Materials. *AIP Conf. Proc.* **2012**, *1426*, 1199.
136. Flurchick, K. M.; Perger, W. F.; Slough, W. J.; Valenzano, L. The Effect of a Simulated Volumetric Expansion: Calculated Vibrational Properties and Elastic Constants of Pentaerythritol. *AIP Conf. Proc.* **2012**, *1426*, 567.
137. Landerville, A. C.; Conroy, M. W.; Budzevich, M. M.; Lin, Y.; White, C. T.; Oleynik, I. I. Equations of State for Energetic Materials from Density Functional Theory with Van der Waals, Thermal, and Zero-Point Energy Corrections. *Appl. Phys. Lett.* **2010**, *97*, 251908.
138. Conroy, M. W.; Budzevich, M. M.; Lin, Y.; Oleynik, I. I.; White, C. T. Application of Van der Waals Density Functional Theory to Study Physical Properties of Energetic Materials. *AIP Conf. Proc.* **2009**, *1195*, 805.
139. Budzevich, M.; Conroy, M.; Landerville, A.; Lin, Y.; Oleynik, I.; White, C. T. Hydrostatic Equation of State and Anisotropic Constitutive Relationships in 1,3,5-Triamino-2,4,6-Trinitrobenzene (TATB). *AIP Conf. Proc.* **2009**, *1195*, 545.
140. Sorescu, D. C.; Rice, B. M. Theoretical Predictions of Energetic Molecular Crystals at Ambient and Hydrostatic Compression Conditions Using Dispersion Corrections to Conventional Density Functionals (DFT-D). *J. Phys. Chem. C* **2010**, *114*, 6734.
141. Cui, H.-L.; Ji, G.-F.; Zhao, J.-J.; Zhao, F.; Chen, X.-R.; Zhang, Q.-M.; Wei, D.-Q. Ab Initio and Molecular Dynamics Studies of Solid-HMX: Effects of Hydrostatic Pressure and High Temperature. *Mol. Simul.* **2010**, *36*, 670.
142. Gan, C. K.; Sewell, T. D.; Challacombe, M. All-Electron Density-Functional Studies of Hydrostatic Compression of Pentaerythritol Tetranitrate $C(CH_2ONO_2)_4$. *Phys. Rev. B* **2004**, *69*, 035116.
143. Liu, H.; Zhao, J.; Wei, D.; Gong, Z. Structural and Vibrational Properties of Solid Nitromethane Under High Pressure by Density Functional Theory. *J. Chem. Phys.* **2006**, *124*, 124501.

144. Perger, W. F.; Zhao, J.; Winey, J. M.; Gupta, Y. M. First-Principles Study of Pentaerythritol Tetranitrate Single Crystals under High Pressure: Vibrational Properties. *Chem. Phys. Lett.* **2006**, *428*, 394.
145. Lian, D.; Lu, L.-Y.; Wei, D.-Q.; Zhang, Q.-M.; Gong, Z.-Z.; Guo, Y.-X. High-Pressure Behaviour of Beta-HMX Crystal Studied by DFT-LDA. *Chin. Phys. Lett.* **2008**, *25*, 899.
146. Conroy, M. W.; Oleynik, I. I.; Zybin, S. V.; White, C. T. Density Functional Theory Calculations of Solid Nitromethane Under Hydrostatic and Uniaxial Compressions with Empirical Van der Waals Correction. *J. Phys. Chem. A* **2009**, *113*, 3610.
147. Wang, F.; Du, H.-C.; Liu, H.; Gong, X.-D. Density Functional Theory Study of High-Pressure Effect on Crystalline 4,4′,6,6′-Tetra(Azido)Hydrazo-1,3,5-Triazine. *J. Comput. Chem.* **1820**, *2012*, 33.
148. Liu, H.; Wang, F.; Gong, X. DFT Studies on 7-Nitrotetrazolo [1,5]Furazano[4,5-b]Pyridine 1-Oxide: Crystal Structure, Detonation Properties, Sensitivity and Effect of Hydrostatic Compression. *Struct. Chem.* **2013**. http://dx.doi.org/10.1007/s11224-013-0279-0 published online May 11, 2013, Ahead of Print.
149. Wu, Q.; Zhu, W.; Xiao, H. Pressure Effects on Structural, Electronic, Absorption, and Thermodynamic Properties of Crystalline 2,4,6-Triamino-3,5-Dinitropyridine-1-Oxide: A DFT Study. *J. Phys. Org. Chem.* **2013**, *26*, 589.
150. Liu, Y.; Zhang, L.; Wang, G.; Wang, L.; Gong, X. First-Principle Studies on the Pressure-Induced Structural Changes in Energetic Ionic Salt 3-Azido-1,2,4-Triazolium Nitrate Crystal. *J. Phys. Chem. C* **2012**, *116*, 16144.
151. Qiu, L.; Zhu, W.-H.; Xiao, J.-J.; Xiao, H.-M. Theoretical Studies of Solid Bicyclo-HMX: Effects of Hydrostatic Pressure and Temperature. *J. Phys. Chem. B* **2008**, *112*, 3882.
152. Liu, Y.; Gong, X.; Wang, L.; Wang, G. Effect of Hydrostatic Compression on Structure and Properties of 2-Diazo-4,6-Dinitrophenol Crystal: Density Functional Theory Studies. *J. Phys. Chem. C* **2011**, *115*, 11738.
153. Qiu, L.; Xiao, H.-M.; Zhu, W.-H.; Xiao, J.-J.; Zhu, W. Ab Initio and Molecular Dynamics Studies of Crystalline TNAD (trans-1,4,5,8-Tetranitro-1,4,5,8-Tetraazadecalin). *J. Phys. Chem. B* **2006**, *110*, 10651.
154. Zhu, W.; Zhang, X.; Wei, T.; Xiao, H. DFT Studies of Pressure Effects on Structural and Vibrational Properties of Crystalline Octahydro-1,3,5,7-Tetranitro-1,3,5,7-Tetrazocine. *Theor. Chem. Acc.* **2009**, *124*, 179.
155. Wu, C. J.; Yang, L. H.; Fried, L. E.; Quenneville, J.; Martinez, T. J. Electronic Structure of Solid 1,3,5-Triamino-2,4,6-Trinitrobenzene Under Uniaxial Compression: Possible Role of Pressure-Induced Metallization in Energetic Materials. *Phys. Rev. B* **2003**, *67*, 235101.
156. Dong, Y.-J.; Cheng, X.-L. Structural Properties of Nitromethane Molecular Crystal Under High Pressure: An Ab Initio Investigation. *Chin. J. Struct. Chem.* **2009**, *28*, 1105.
157. Budzevich, M. M.; Landerville, A. C.; Conroy, M. W.; Lin, Y.; Oleynik, I. I.; White, C. T. Hydrostatic and Uniaxial Compression Studies of 1,3,5-Triamino-2,4,6-Trinitrobenzene Using Density Functional Theory with Van der Waals Correction. *J. Appl. Phys.* **2010**, *107*, 113524.
158. Bardwell, D. A.; Adjiman, C. S.; Arnautova, Y. A.; Bartashevich, E.; Boerrigter, S. X. M.; Braun, D. E.; Cruz-Cabeza, A. J.; Day, G. M.; Della Valle, R. G.; Desiraju, G. R.; van Eijck, B. P.; Facelli, J. C.; Ferraro, M. B.; Grillo, D.; Habgood, M.; Hofmann, D. W. M.; Hofmann, F.; Jose, K. V. J.; Karamertzanis, P. G.; Kazantsev, A. V.; Kendrick, J.; Kuleshova, L. N.; Leusen, F. J. J.; Maleev, A. V.; Misquitta, A. J.; Mohamed, S.; Needs, R. J.; Neumann, M. A.; Nikylov, D.; Orendt, A. M.; Pal, R.; Pantelides, C. C.;

Pickard, C. J.; Price, L. S.; Price, S. L.; Scheraga, H. A.; van de Streek, J.; Thakur, T. S.; Tiwari, S.; Venuti, E.; Zhitkov, I. K. Towards Crystal Structure Prediction of Complex Organic Compounds—A Report on the Fifth Blind Test. *Acta Crystallogr. B* **2011**, *67*, 535.
159. Lund, A. M.; Orendt, A. M.; Pagola, G. I.; Ferraro, M. B.; Facelli, J. C. Optimization of Crystal Structures of Archetypical Pharmaceutical Compounds: A Plane-Wave DFT-D Study Using Quantum Espresso. *Cryst. Growth Des.* **2013**, *13*, 2181.
160. King, M. D.; Blanton, T. N.; Misture, S. T.; Korter, T. M. Prediction of the Unknown Crystal Structure of Creatine Using Fully Quantum Mechanical Methods. *Cryst. Growth Des.* **2011**, *11*, 5733.
161. Kendrick, J.; Stephenson, G. A.; Neumann, M. A.; Leusen, F. J. J. Crystal Structure Prediction of a Flexible Molecule of Pharmaceutical Interest with Unusual Polymorphic Behavior. *Cryst. Growth Des.* **2013**, *13*, 581.
162. Kendrick, J.; Leusen, F. J. J.; Neumann, M. A.; van de Streek, J. Progress in Crystal Structure Prediction. *Chem. Eur. J.* **2011**, *17*, 10736.
163. Rice, B. M.; Sewell, T. D. Equilibrium Molecular Dynamics Simulations. In *Static Compression of Energetic Materials*; Peiris, S. M.; Piermarini, G. J. Eds.; Springer-Verlag: Berlin, 2008; pp 255–290.
164. Sorescu, D. C.; Rice, B. M.; Thompson, D. L. Theoretical Studies of Solid Nitromethane. *J. Phys. Chem. B* **2000**, *104*, 8406.
165. Smith, G. D.; Bharadwaj, R. K. Quantum Chemistry Based Force Field for Simulations of HMX. *J. Phys. Chem. B* **1999**, *103*, 3570.
166. Sun, H. COMPASS: An Ab Initio Force-Field Optimized for Condensed-Phase Applications—Overview with Details on Alkane and Benzene Compounds. *J. Phys. Chem. B* **1998**, *102*, 7338.
167. Bunte, S. W.; Sun, H. Molecular Modeling of Energetic Materials: The Parameterization and Validation of Nitrate Esters in the COMPASS Force Field. *J. Phys. Chem. B* **2000**, *104*, 2477.
168. Zhao, G.-Z.; Lu, M. Molecular Design of New Nitramine Explosive: 1,3,5,7-Tetraaza-1,2,3,5,6,7-Hexahydros-Indacene Derivatives. *Polycycl. Aromat. Compd.* **2013**, *33*, 297.
169. Li, X.-H.; Zhang, X.-Z. Computational Studies on a High Density Cage Compound Hexanitrohexaazaisowurtzitane Derivative. *Can. J. Chem.* **2013**, *91*, 369.
170. Zhao, G.; Lu, M. Comparative Theoretical Studies of Energetic Dodecahydrodiimidazo[4,5-B:4′,5′-E]Pyrazine Derivatives. *Comput. Theor. Chem.* **2013**, *1007*, 57.
171. Xiao, J. J.; Li, S. Y.; Chen, J.; Ji, G. F.; Zhu, W.; Zhao, F.; Wu, Q.; Xiao, H.-M. Molecular Dynamics Study on the Correlation Between Structure and Sensitivity for Defective RDX Crystals and Their PBXs. *J. Mol. Model.* **2013**, *19*, 803.
172. Zhao, G.; Lu, M. Computational Studies on 3,5,7,10,12,14,15,16-Octanitro-3,5,7,10,12,14,15,16-Octaaza-Pentacyclo[7.5.1.1(2,8).0(4,13).0(6,11)] Hexadecane as Potential High-Energy-Density Compound. *Struct. Chem.* **2013**, *24*, 139.
173. Zhang, Y.; Ji, G.; Gong, Z.; Wei, D.-Q. New Coupling Mechanism of the Silane Coupling Agents in the TATB-Based PBX. *Mol. Simul.* **2013**, *39*, 423.
174. Zhao, G.; Lu, M. Theoretical Studies on the Crystal Structure, Thermodynamic Properties, Detonation Performance and Thermal Stability of Cage-Tetranitrotetraazabicyclooctane as a Novel High Energy Density Compound. *J. Mol. Model.* **2013**, *19*, 57.
175. Zhao, G.; Lu, M. Theoretical Investigations of Pyridine Derivatives as Potential High Energy Density Materials. *J. Phys. Org. Chem.* **2013**, *26*, 211.

176. Zhang, C.; Ji, C.; Li, H.; Zhou, Y.; Xu, J.; Xu, R.; Li, J.; Luo, Y. Occupancy Model for Predicting the Crystal Morphologies Influenced by Solvents and Temperature, and Its Application to Nitroamine Explosives. *Cryst. Growth Des.* **2013**, *13*, 282.
177. Lin, H.; Zhu, S.-G.; Zhang, L.; Peng, X.-H.; Chen, P.-Y.; Li, H.-Z. Intermolecular Interactions, Thermodynamic Properties, Crystal Structure, and Detonation Performance of HMX/NTO Cocrystal Explosive. *Int. J. Quantum Chem.* **2013**, *113*, 1591.
178. Lu, L.-Y.; Zhou, X.-L.; Chen, X.-R. Pressure Effect of Structural and Vibrational Properties of Solid Pentaerythritol by Molecular Dynamics Simulations. *Phys. B Condens. Matter* **2012**, *407*, 3996.
179. Zhang, J.-Y.; Du, H.-C.; Wang, F.; Gong, X.-D.; Ying, S.-J. Crystal Structure, Detonation Performance, and Thermal Stability of a New Polynitro Cage Compound: 2,4,6,8,10,12,13,14,15-Nonanitro-2,4,6,8,10,12,13,14,15-Nonaazaheptacyclo [5.5.1.1 (3,11).1(5,9)] Pentadecane. *J. Mol. Model.* **2012**, *18*, 2369.
180. Xiao, J.; Wang, W.; Chen, J.; Ji, G.; Zhu, W.; Xiao, H. Study on the Relations of Sensitivity with Energy Properties for HMX and HMX-Based PBXs by Molecular Dynamics Simulation. *Phys. B Condens. Matter* **2012**, *407*, 3504.
181. Zhang, C.; Cao, X.; Xiang, B. Understanding the Desensitizing Mechanism of Olefin in Explosives: Shear Slide of Mixed HMX-Olefin Systems. *J. Mol. Model.* **2012**, *18*, 1503.
182. Cui, H.; Chen, X.; Ji, G.; Wei, D. The Mesoscopic Structure of Beta-HMX-Based PBXs by Dissipative Particle Dynamics Simulation. *Adv. Mater. Res. (Durnten-Zurich, Switzerland)* **2012**, 403–408 (Pt. 5, MEMS, NANO and Smart Systems), 4430.
183. Zhang, J.-Y.; Du, H.-C.; Wang, F.; Gong, X.-D.; Huang, Y.-S. Theoretical Investigations of a High Density Cage Compound 10-(1-Nitro-1,2,3,4-Tetraazol-5-yl) Methyl-2,4,6,8,12-Hexanitrohexaazaisowurtzitane. *J. Mol. Model.* **2012**, *18*, 165.
184. Zhou, Y.; Long, X. P.; Wei, X. W. Theoretical Study on the Diffusive Transport of 2,4,6-Trinitrotoluene in Polymer-Bonded Explosive. *J. Mol. Model.* **2011**, *17*, 3015.
185. Wei, C.; Huang, H.; Duan, X.; Pei, C. Structures and Properties Prediction of HMX/TATB Co-Crystal. *Propel. Explos. Pyrotech.* **2011**, *36*, 416.
186. Zhang, C.; Li, Y.; Xiong, Y.; Wang, X.; Zhou, M. Acid and Alkali Effects on the Decomposition of HMX Molecule: A Computational Study. *J. Phys. Chem. A* **2011**, *115*, 11971.
187. Kempa, P. B.; Herrmann, M. Molecular Mechanical Simulations of Nitramines: A Comparison of Force Fields. In *42nd International Annual Conference of ICT (Energetic Materials)*, 2011; p 81/1.
188. Ringuette, S.; Lafleur-Lambert, X.; Stowe, R.; Abou-Rachid, H.; Jaidann, M.; Lima, R. P.; Dubois, C. Molecular Modeling of Hybrid Metal-Polymer Materials. In *42nd International Annual Conference of ICT (Energetic Materials)*, 2011; p 19/1.
189. Tan, J.-J.; Ji, G.-F.; Chen, X.-R.; Li, Z. Structure, Equation of State and Elasticity of Crystalline HNIW by Molecular Dynamics Simulations. *Phys. B Condens. Matter* **2011**, *406*, 2925.
190. Li, M.-M.; Li, F.-S.; Shen, R.-Q. Molecular Dynamics Study of RDX/AMMO Propellant. *Chin. J. Chem. Phys.* **2011**, *24*, 199.
191. Zhang, J.-Y.; Du, H.-C.; Wang, F.; Gong, X.-D.; Huang, Y.-S. DFT Studies on a High Energy Density Cage Compound 4-Trinitroethyl-2,6,8,10,12-Pentanitrohezaazaisowurtzitane. *J. Phys. Chem. A* **2011**, *115*, 6617.
192. Shen, J. P.; Duan, X. H.; Luo, Q. P.; Zhou, Y.; Bao, Q.; Ma, Y. J.; Pei, C. H. Preparation and Characterization of a Novel Cocrystal Explosive. *Cryst. Growth Des.* **2011**, *11*, 1759.
193. Zhang, Y.; Ji, G.; Zhao, F.; Gong, Z.; Wei, D.; Chen, L.; Li, W. Mesoscopic Simulation of Aggregate Behaviour of Fluoropolymers in the TATB-Based PBX. *Mol. Simul.* **2011**, *37*, 237.

194. Li, M.; Li, F.; Shen, R.; Guo, X. Molecular Dynamics Study of the Structures and Properties of RDX/GAP Propellant. *J. Hazard. Mater.* **2011**, *186*, 2031.
195. Zhang, C.; Cao, X.; Xiang, B. Sandwich Complex of TATB/Graphene: An Approach to Molecular Monolayers of Explosives. *J. Phys. Chem. C* **2010**, *114*, 22684.
196. Cui, H.-L.; Ji, G.-F.; Chen, X.-R.; Zhang, Q.-M.; Wei, D.-Q.; Zhao, F. Phase Transitions and Mechanical Properties of Octahydro-1,3,5,7-Tetranitro-1,3,5,7-Tetrazocine in Different Crystal Phases by Molecular Dynamics Simulation. *J. Chem. Eng. Data* **2010**, *55*, 3121.
197. Zhang, C. Understanding the Desensitizing Mechanism of Olefin in Explosives Versus External Mechanical Stimuli. *J. Phys. Chem. C* **2010**, *114*, 5068.
198. Xu, X.; Xiao, J.; Huang, H.; Li, J.; Xiao, H. Molecular Dynamic Simulations on the Structures and Properties of Epsilon-CL-20(001)/F-2314 PBX. *J. Hazard. Mater.* **2010**, *175*, 423.
199. Duan, X.; Wei, C.; Liu, Y.; Pei, C. A Molecular Dynamics Simulation of Solvent Effects on the Crystal Morphology of HMX. *J. Hazard. Mater.* **2010**, *174*, 175.
200. Lu, L.-Y.; Wei, D.-Q.; Chen, X.-R.; Ji, G.-F.; Wang, X.-J.; Chang, J.; Zhang, Q.-M.; Gong, Z.-Z. The Pressure-Induced Phase Transition of the Solid HMX. *Mol. Phys.* **2009**, *107*, 2373.
201. Wang, G.-X.; Shi, C.-H.; Gong, X.-D.; Zhu, W.-H.; Xiao, H.-M. Packing Structures and Periodic Band Calculations on DPO (2,5-Dipicryl-1,3,4-Oxadiazole). *J. Hazard. Mater.* **2009**, *169*, 813.
202. Zhu, W.; Xiao, J.; Zhu, W.; Xiao, H. Molecular Dynamics Simulations of RDX and RDX-Based Plastic-Bonded Explosives. *J. Hazard. Mater.* **2009**, *164*, 1082.
203. Qiu, L.; Xiao, H. Molecular Dynamics Study of Binding Energies, Mechanical Properties, and Detonation Performances of Bicyclo-HMXBased PBXs. *J. Hazard. Mater.* **2009**, *164*, 329.
204. Xiao, J.-J.; Zhang, H.; Huang, H.; Li, J.-S.; Zhu, W.; Xiao, H.-M. NPT Ensemble MD Simulation Investigation on the Mechanical Properties of HMX/F2311 Polymer-Bonded Explosive. *Chin. J. Chem.* **1969**, *2008*, 26.
205. Xiao, J.; Huang, H.; Li, J.; Zhang, H.; Zhu, W.; Xiao, H. Computation of Interface Interactions and Mechanical Properties of HMX-Based PBX with Estane 5703 from Atomic Simulation. *J. Mater. Sci.* **2008**, *43*, 5685.
206. Abou-Rachid, H.; Lussier, L.-S.; Ringuette, S.; Lafleur-Lambert, X.; Jaidann, M.; Brisson, J. On the Correlation Between Miscibility and Solubility Properties of Energetic Plasticizers/Polymer Blends: Modeling and Simulation Studies. *Propel. Explos. Pyrotech.* **2008**, *33*, 301.
207. Zhang, C.; Wang, X.; Huang, H. π-Stacked Interactions in Explosive Crystals: Buffers Against External Mechanical Stimuli. *J. Am. Chem. Soc.* **2008**, *130*, 8359.
208. Xu, X.-J.; Zhu, W.-H.; Gong, X.-D.; Xiao, H.-M. Theoretical Studies on New Potential High Energy Density Compounds (HEDCs) Adamantyl Nitrates from Gas to Solid. *Sci. China Ser. B Chem.* **2008**, *51*, 427.
209. Xu, X.-J.; Zhu, W.-H.; Xiao, H.-M. Theoretical Predictions on the Structures and Properties for Polynitrohexaazaadamantanes (PNHAAs) as Potential High Energy Density Compounds (HEDCs). *J. Mol. Struct.* **2008**, *853*, 1.
210. Xiao, J.; Huang, H.; Li, J.; Zhang, H.; Zhu, W.; Xiao, H. A Molecular Dynamics Study of Interface Interactions and Mechanical Properties of HMX-Based PBXs with PEG and HTPB. *J. Mol. Struct.* **2008**, *851*, 242.
211. Ma, X.; Zhao, F.; Ji, G.; Zhu, W.; Xiao, J.; Xiao, H. Computational Study of Structure and Performance of Four Constituents HMX-Based Composite Material. *J. Mol. Struct.* **2008**, *851*, 22.
212. Xu, X.-J.; Xiao, J.-J.; Huang, H.; Li, J.-S.; Xiao, H.-M. Molecular Dynamics Simulations on the Structures and Properties of ε-CL-20-Based PBXs—Primary Theoretical Studies on HEDM Formulation Design. *Sci. China Ser. B Chem.* **2007**, *50*, 737.

213. Xiao, J.-J.; Ma, X.; Zhu, W.; Huang, Y.; Xiao, H.; Huang, H.; Li, J. Molecular Dynamics Simulations of Polymer-Bonded Explosives (PBXs): Modeling, Mechanical Properties and Their Dependence on Temperatures and Concentrations of Binders. *Propel. Explos. Pyrotech.* **2007**, *32*, 355.
214. Burnett, A. D.; Fan, W. H.; Upadhya, P. C.; Cunningham, J. E.; Edwards, H. G. M.; Kendrick, J.; Munshi, T.; Hargreaves, M.; Linfield, E. H.; Davies, A. G. Broadband Terahertz Time-Domain and Raman Spectroscopy of Explosives. In *Proceedings of SPIE—The International Society for Optical Engineering, Vol. 6549 (Terahertz for Military and Security Applications V)*, 2007; p 654905/1.
215. Zhang, C. Computational Investigation on the Desensitizing Mechanism of Graphite in Explosives Versus Mechanical Stimuli: Compression and Glide. *J. Phys. Chem. B* **2007**, *111*, 6208.
216. Qiu, L.; Zhu, W.-H.; Xiao, J.-J.; Zhu, W.; Xiao, H.-M.; Huang, H.; Li, J.-S. Molecular Dynamics Simulations of trans-1,4,5,8-Tetranitro-1,4,5,8-Tetraazadecan-Based Polymer-Bonded Explosives. *J. Phys. Chem. B* **2007**, *111*, 1559.
217. Yin, K.; Zou, D.; Zhong, J.; Xu, D. A New Method for Calculation of Elastic Properties of Anisotropic Material by Constant Pressure Molecular Dynamics. *Comput. Mater. Sci.* **2007**, *38*, 538.
218. Ma, X.-F.; Xiao, J.-J.; Huang, H.; Ju, X.-H.; Li, J.-S.; Xiao, H.-M. Simulative Calculation of Mechanical Property, Binding Energy and Detonation Property of TATB/Fluorine-Polymer PBX. *Chin. J. Chem.* **2006**, *24*, 473.
219. Xu, X.-J.; Xiao, H.-M.; Xiao, J.-J.; Zhu, W.; Huang, H.; Li, J.-S. Molecular Dynamics Simulations for Pure ε-CL-20 and ε-CL-20-Based PBXs. *J. Phys. Chem. B* **2006**, *110*, 7203.
220. Gee, R. H.; Roszak, S.; Balasubramanian, K.; Fried, L. E. Molecular Dynamics Simulations for Pure ε-CL-20 and ε-CL-20-Based PBXs. *J. Chem. Phys.* **2004**, *120*, 7059.
221. Podeszwa, R.; Bukowski, R.; Rice, B. M.; Szalewicz, K. Potential Energy Surface for Cyclotrimethylenetrinitramine Dimer from Symmetry-Adapted Perturbation Theory. *Phys. Chem. Chem. Phys.* **2007**, *9*, 5561–5569.
222. Taylor, D. E.; Rob, F.; Rice, B. M.; Podeszwa, R.; Szalewicz, K. A Molecular Dynamics Study of 1,1-Diamino-2,2-Dinitroethylene (FOX-7) Crystal Using a Symmetry Adapted Perturbation Theory-Based Intermolecular Force Field. *Phys. Chem. Chem. Phys.* **2011**, *13*, 16629–16636.
223. Taylor, D. E. Intermolecular Forces and Molecular Dynamics Simulation of 1,3,5-Triamino-2,4,6-Trinitrobenzene (TATB) Using Symmetry Adapted Perturbation Theory. *J. Phys. Chem. A* **2013**, *117*, 3507–3520.
224. Sorescu, D.; Boatz, J. A.; Thompson, D. L. Classical and Quantum-Mechanical Studies of Crystalline FOX-7 (1,1-Diamino-2,2-Dinitroethylene). *J. Phys. Chem. A* **2001**, *105*, 5010–5021.
225. Bedrov, D.; Borodin, O.; Smith, G. D.; Sewell, T. D.; Dattelbaum, D. M.; Stevens, L. L. A Molecular Dynamics Simulation Study of Crystalline 1,3,5-Triamino-2,4,6-Trinitrobenzene as a Function of Pressure and Temperature. *J. Chem. Phys.* **2009**, *131*, 224703.
226. Van Duin, A. C. T.; Dasgupta, S.; Lorant, F.; Goddard, W. A. ReaxFF: A Reactive Force Field for Hydrocarbons. *J. Phys. Chem. A* **2001**, *105*, 9396–9409.
227. Zhang, L.; Zybin, S. V.; van Duin, A. C. T.; Dasgupta, S.; Goddard, W. A.; Kober, E. M. Carbon Cluster Formation During Thermal Decomposition of Octahydro-1,3,5,7-Tetranitro-1,3,5,7-Tetrazocine and 1,3,5-Triamino-2,4,6-Trinitrobenzene High Explosives from ReaxFF Reactive Molecular Dynamics Simulations. *J. Phys. Chem. A* **2009**, *113*, 10619.
228. Guo, F.; Zhang, H.; Cheng, X. Molecular Dynamic Simulations of Solid Nitromethane Under High Pressures. *J. Theor. Comput. Chem.* **2010**, *9*, 315.

229. van Duin, A. C. T.; Zeiri, Y.; Dubnikova, F.; Kosloff, R.; Goddard, W. A. Atomistic-Scale Simulations of the Initial Chemical Events in the Thermal Initiation of Triacetonetriperoxide. *J. Am. Chem. Soc.* **2005**, *127*, 11053.
230. Liu, L.; Zybin, S.; Sun, H.; Goddard, W. A. ReaxFF-lg: Correction of the ReaxFF Reactive Force Field for London Dispersion, with Applications to the Equations of State for Energetic Materials. *J. Phys. Chem. A* **2011**, *115*, 11016–11022.
231. Hui, C. Y.; Ruina, A.; Long, R.; Jagota, A. Cohesive Zone Models and Fracture. *J. Adhesion* **2011**, *87*, 1–52.
232. Clayton, J. D.; Becker, R. Elastic-Plastic Behavior of Cyclotrimethylenetrinitramine Single Crystals under Spherical Indentation: Modeling and Simulation. *J. Appl. Phys.* **2012**, *111*, 063512.
233. Peter, C.; Kremer, K. Multiscale Simulation of Soft Matter Systems. *Faraday Discuss.* **2010**, *144*, 9, and references therein.
234. Bhatnagara, N.; Kamathb, G.; Potoff, J. J. Prediction of 1-Octanol-Water and Air-Water Partition Coefficients for Nitro-Aromatic Compounds from Molecular Dynamics Simulations. *Phys. Chem. Chem. Phys.* **2013**, *15*, 6467.
235. Hervouet, A.; Desbiens, N.; Bourasseau, E.; Maillet, J.-B. Microscopic Approaches to Liquid Nitromethane Detonation Properties. *J. Phys. Chem. B* **2008**, *112*, 5070.
236. Desbiens, N.; Bourasseau, E.; Maillet, J.-B. Potential Optimization for the Calculation of Shocked Liquid Nitromethane Properties. *Mol. Simul.* **2007**, *33*, 1061.
237. Desbiens, N.; Bourasseau, E.; Maillet, J.-B.; Soulard, L. Molecular Based Equation of State for Shocked Liquid Nitromethane. *J. Hazard. Mater.* **2009**, *166*, 1120.
238. Gee, R. H.; Wu, C.; Maiti, A. Coarse-Grained Model for a Molecular Crystal. *Appl. Phys. Lett.* **2006**, *89*, 021919.
239. Lin, P.-H.; Khare, R.; Weeks, B. L.; Gee, R. H. Molecular Modeling of Diffusion on a Crystalline Pentaerythritol Tetranitrate Surface. *Appl. Phys. Lett.* **2007**, *91*, 104107.
240. Maiti, A.; Gee, R. H. Modeling Growth, Surface Kinetics, and Morphology Evolution in PETN. *Propel. Explos. Pyrotech.* **2009**, *34*, 489.
241. Izvekov, S.; Voth, G. A. A Multiscale Coarse-Graining Method for Biomolecular Systems. *J. Phys. Chem. B* **2005**, *109*, 2469.
242. Izvekov, S.; Chung, P. W.; Rice, B. M. The Multiscale Coarse-Graining Method: Assessing Its Accuracy and Introducing Density Dependent Coarse-Grain Potentials. *J. Chem. Phys.* **2010**, *133*, 064109.
243. Izvekov, S.; Chung, P. W.; Rice, B. M. Particle-Based Multiscale Coarse Graining with Density-Dependent Potentials: Application to Molecular Crystals (Hexahydro-1,3,5-Trinitro-s-Triazine). *J. Chem. Phys.* **2011**, *135*, 044112.
244. Depa, P. K.; Maranas, J. K. Speed Up of Dynamic Observables in Coarse-Grained Molecular-Dynamics Simulations of Unentangled Polymers. *J. Chem. Phys.* **2005**, *123*, 094901.
245. Lopez, C. F.; Nielsen, S. O.; Moore, P. B.; Shelley, J. C.; Klein, M. L. Self-Assembly of a Phospholipid Langmuir Monolayer using Coarse-Grained Molecular Dynamics Simulations. *J. Phys. Condens. Matter* **2002**, *14*, 9431.
246. Milano, G.; Müller-Plathe, F. Mapping Atomistic Simulations to Mesoscopic Models: A Systematic Coarse-Graining Procedure for Vinyl Polymer Chains. *J. Phys. Chem. B* **2005**, *109*, 18609.
247. Chen, X.; Carbone, P.; Cavalcanti, W. L.; Milano, G.; Müller-Plathe, F. Viscosity and Structural Alteration of a Coarse-Grained Model of Polystyrene Under Steady Shear Flow Studied by Reverse Nonequilibrium Molecular Dynamics. *Macromolecules* **2007**, *40*, 8087.
248. Moore, J. D.; Izvekov, S.; Lisal, M.; Brennan, J. K. Particle Based Multiscale Modeling of the Dynamic Response of RDX. *AIP Conf. Proc.* **2012**, *1426*, 1237.

249. Hoogerbrugge, P. J.; Koelman, J. M. V. A. Simulating Microscopic Hydrodynamic Phenomena with Dissipative Particle Dynamics. *Europhys. Lett* **1992**, *19*, 155.
250. Avalos, J. B.; Mackie, A. D. Dissipative Particle Dynamics with Energy Conservation. *Europhys. Lett.* **1997**, *40*, 141.
251. Español, P. Dissipative Particle Dynamics with Energy Conservation. *Europhys. Lett.* **1997**, *40*, 631.
252. Moore, J. D.; Larentzos, J. P.; Izvekov, S.; Lísal, M.; Brennan, J. K. Comprehensive Suite of Coarse-Grain Modeling Tools Formulated Upon Dissipative Particle Dynamics. In Abstracts of Papers, 246th ACS National Meeting & Exposition, Indianapolis, IN, United States, September 8–12, 2013, 2013, and PHYS-151.
253. Brennan, J. K.; Tran, L; Fermen-Coker, M. Physics-Based Multi-Scale Modeling of Shear Initiated Reactions in Energetic and Reactive Materials. *US Army Research Laboratory, ARL-TR-4753*, 2009.
254. Moore, J. D.; Izvekov, S.; Lísal, M.; Brennan, J. K. Particle Based Multiscale Modeling of the Dynamic Response of RDX. In *AIChE Annual Meeting, Conference Proceedings, Pittsburgh, PA, USA, October 28–November 2*, 2012.
255. Brennan, J. K.; Moore, J. D. U.S. Army Research Laboratory. Private Communication.
256. Brennan, J. K.; Lísal, M. Dissipative Particle Dynamics: Addressing Deficiencies and Establishing New Frontiers. *Mol. Simul.* **2009**, *35*, 766.
257. Brennan, J. K.; Fermen-Coker, M.; Tran, L. Physics-Based Multi-Scale Modeling of Shear Initiated Reactions in Energetic and Reactive Materials. *ARL-MR-0744*, 2010.
258. Silling, S. A. Unpublished.
259. Singh, U. C.; Kollman, P. A. A Combined Ab Initio Quantum-Mechanical and Molecular Mechanical Method for Carrying Out Simulations on Complex Molecular-Systems—Applications to the $CH_3Cl + Cl^-$ Exchange-Reaction and Gas-Phase Protonation of Polyethers. *J. Comput. Chem.* **1986**, *7*, 718.
260. Poteau, R.; Ortega, I.; Alary, F.; Solis, A. R.; Barthelat, J. C.; Daudey, J. P. Effective Group Potentials. 1. Method. *J. Phys. Chem.* **2001**, *105*, 198.
261. Ferre, N.; Olivucci, M. The Amide Bond: Pitfalls and Drawbacks of the Link Atom Scheme. *J. Mol. Struct.* **2003**, *632*, 71.
262. Mallik, A.; Taylor, D. E.; Runge, K.; Dufty, J. W.; Cheng, H. P. Procedure for Building a Consistent Embedding at the QM–CM Interface. *J. Comput. Aided Mater. Des.* **2006**, *13*, 45.
263. Mallik, A.; Runge, K.; Dufty, J. W.; Cheng, H. P. Multiscale Modeling of Materials Based on Force and Charge Density Fidelity. *J. Chem. Phys.* **2007**, *127*, 224707.
264. Reed, E. J.; Fried, L. E.; Joannopoulos, J. D. A Method for Tractable Dynamical Studies of Single and Double Shock Compression. *Phys. Rev. Lett.* **2003**, *90*, 235503.
265. Reed, E. J.; Manaa, M. R.; Fried, L. E.; Glaesemann, K. R.; Joannopoulos, J. D. A Transient Semimetallic Layer in Detonating Nitromethane. *Nat. Phys.* **2008**, *4*, 72.
266. Manaa, M. R.; Reed, E. J.; Fried, L. E.; Goldman, N. Nitrogen-Rich Heterocycles as Reactivity Retardants in Shocked Insensitive Explosives. *J. Am. Chem. Soc.* **2009**, *131*, 5483.
267. Reed, E. J.; Rodriguez, A. W.; Manaa, M. R.; Fried, L. E.; Tarver, C. M. Ultrafast Detonation of Hydrazoic Acid (HN_3). *Phys. Rev. Lett.* **2012**, *109*, 038301.
268. Zhu, W.; Huang, H.; Huang, H.; Xiao, H. Initial Chemical Events in Shocked Octahydro-1,3,5,7-Tetranitro-1,3,5,7-Tetrazocine: A New Initiation Decomposition Mechanism. *J. Chem. Phys.* **2012**, *136*, 044516.
269. Ge, N.-N.; Wei, Y.-K.; Ji, G.-F.; Chen, X.-R.; Zhao, F.; Wei, D.-Q. Initial Decomposition of the Condensed-Phase Beta-HMX Under Shock Waves: Molecular Dynamics Simulations. *J. Phys. Chem. B* **2012**, *116*, 13696.

270. Shan, T.-R.; Wixom, R. R.; Mattsson, A. E.; Thompson, A. P. Atomistic Simulation of Orientation Dependence in Shock-Induced Initiation of Pentaerythritol Tetranitrate. *J. Phys. Chem. B* **2013**, *117*, 928.
271. Reed, E. J. Electron-Ion Coupling in Shocked Energetic Materials. *J. Phys. Chem. C* **2012**, *116*, 2205.
272. Delpuech, A.; Cherville, J. Relation Between Shock Sensitiveness of Secondary Explosives and Their Molecular Electronic-Structure. 1. Nitroaromatics and Nitramines. *Propel. Explos. Pyrotech.* **1978**, *3*, 169.
273. Delpuech, A.; Cherville, J. Relation Between Shock Sensitiveness of Secondary Explosives and Their Molecular Electronic-Structure. 2. Nitrate Esters. *Propel. Explos. Pyrotech.* **1979**, *4*, 121.
274. Xiao, H.-M.; Fan, J.-F.; Gu, Z.-M.; Dong, H.-S. Theoretical Study on Pyrolysis and Sensitivity of Energetic Compounds—(3)-Nitro Derivatives of Aminobenzenes. *Chem. Phys.* **1998**, *226*, 15.
275. Fan, J.; Gu, Z.; Xiao, H.; Dong, H. Theoretical Study on Pyrolysis and Sensitivity of Energetic Compounds—(3)-Nitro Derivatives of Aminobenzenes. *J. Phys. Org. Chem.* **1998**, *11*, 177.
276. Wu, C. J.; Fried, L. E. First-Principles Study of High Explosive Decomposition Energetics. In *Proceedings of the Eleventh Symposium (International) on Detonation*, 1998; p 490.
277. Rice, B. M.; Hare, J. A Quantum Mechanical Investigation of the Relation Between Impact Sensitivity and the Charge Distribution in Energetic Molecules. *J. Phys. Chem. A* **2002**, *106*, 1770.
278. Yau, A.; Byrd, E. F. C.; Rice, B. M. An Investigation of KS-DFT Electron Densities Used in Atoms-in-Molecules Studies of Energetic Molecules. *J. Phys. Chem. A* **2009**, *113*, 6166.
279. Jones, T. Role of Inter- and Intramolecular Bonding on Impact Sensitivity. *J. Phys. Chem. A* **2012**, *116*, 11008–11014.
280. Taylor, D. E. *Prediction of the Impact Sensitivity of Energetic Materials Using Symmetry Adapted Perturbation Theory:Army Research Laboratory Technical Report*; 2011.
281. Ye, S.; Tonokura, K.; Koshi, M. Theoretical Calculations of Lattice Properties of Secondary Explosives. *J. Jpn. Explos. Soc.* **2002**, *63*, 104.

CHAPTER SIX

The Reactivity of Energetic Materials Under High Pressure and Temperature

M. Riad Manaa, Laurence E. Fried[1]
Energetic Materials Center, Lawrence Livermore National Laboratory, Livermore, California, USA
[1]Corresponding author: e-mail address: fried1@llnl.gov

Contents

Abstract

Chemical transformations that occur at the reactive shock front of energetic materials determine many aspects of material properties and performance. One major shortcoming of current explosive models is the lack of chemical kinetics data of the reacting explosive at the high pressures and temperatures experienced under detonation conditions. In the absence of experimental data, long time-scale atomistic molecular dynamics simulations with reactive chemistry provide insight into the decomposition mechanisms of explosives and allow us to obtain effective reaction rates. These rates can then be incorporated into a thermochemical continuum code for accurate and predictive description of grain- and continuum-scale dynamics of reacting explosives. During the course of the past decade, we have examined the chemistry of several reacting explosive materials, such as the high-performing HMX and the very insensitive TATB explosives, both of which are organic molecular solids at ambient conditions. We have used quantum-based, self-consistent charge density functional tight-binding method to calculate the interatomic forces in molecular dynamics simulations either for thermal decomposition studies (constant volume–temperature) or dynamical shock studies using the multiscale shock simulation technique (MSST). These studies allow us to investigate the chemical reactivity of explosives and to examine electronic properties at extreme conditions of temperature and pressure for a relatively long timescale on the order of several hundreds of picoseconds. In this chapter, we discuss challenges in simulating the reactions of shocked energetic materials and review specific examples of our recent simulations on HMX, PETN, and shocked TATB. Each of these studies revealed interesting aspects

Advances in Quantum Chemistry, Volume 69
ISSN 0065-3276
http://dx.doi.org/10.1016/B978-0-12-800345-9.00006-4

associated with known macroscopic properties of these materials. We also discuss simulations on nonenergetic materials such as shocked carbon and methane.

In the brief instant of a high-explosive detonation, the shock wave produces a pressure 500,000 times that of the Earth's atmosphere and internal temperatures soar up to 3500 K. The detonation wave travels as fast as 10 km/s. As the shock propagates through the explosive, the rapid heating coupled with compression that results in an almost 30% volume reduction initiates complex chemical reactions. A dense, highly reactive supercritical fluid is established behind the propagating detonation front. The energy release from the exothermic chemical reactions drives and sustains the detonation process until complete reactivity is achieved. Several experimental results suggest the existence of strong correlations between the applied mechanical stress, the local heterogeneity (dislocations, vacancies, cracks, impurities, etc.), and the onset of chemical reactions.[1] The reaction chemistry of energetic materials at high pressure and temperature is, therefore, of considerable importance in understanding processes that these materials experience under impact and detonation conditions. Chemical decomposition models are critical ingredients in order to predict, for example, the measured times to explosion and the conditions for ignition of hot spots,[2] which are the localized regions of highly concentrated energy associated with defects.[3] From a more applied perspective, understanding reaction mechanisms has important ramifications in disposing of energetic materials safely and cheaply, since there exist vast stockpiles of high-explosive (HE) materials with corresponding environmental hazards related to earth and groundwater contamination.[4]

Detailed investigation of the kinetics and decomposition mechanisms and general reactivity in explosive materials is critical in order to improve the accuracy and reliability of micromechanical models that are implemented in computational continuum codes. The time frame associated with explosive reactions and the highly exothermic conditions associated with an explosion make experimental investigation of the decomposition pathways difficult at best. Elucidating decomposition pathways is further complicated due to the increased reactivity of explosive materials under high temperature and pressure regimes. At present, computer simulations at the atomistic level provide the best access to the short time-scale processes (~100 femtoseconds (fs) to a few tens of picoseconds (ps)) occurring in regions of extreme conditions. Computational investigations aspire to unravel at the molecular level the detailed decomposition pathways and the kinetic rate laws at high pressure and temperature. The ultimate objective of such studies is to construct accurate and predictive models of performance and sensitivity for the development of new energetic materials.[5]

In what follows, we briefly discuss methods used to simulate the chemistry of shocked and thermally heated energetic materials, with some emphasis on methods employed in our studies. We next examine the chemical reactivity of the two explosives HMX and TATB from condensed-phase atomistic simulations using the quantum-based density functional tight-binding (DFTB) method.[6]

1. METHODS TO SIMULATE CHEMISTRY AT EXTREME CONDITIONS

The chemical reactions of energetic materials at extreme conditions present significant challenges to simulation methodologies. Energetic materials release energy through complex multistep pathways. The measured reaction time of a detonating HE called the "reaction zone length" is inferred from hydrodynamic measurements to be on the ns–μs timescale, depending on the material under study.[7] These times are exceedingly long when compared to a typical molecular dynamics time step (0.5 fs).

HEs are compressed by the shock wave before they react, with a typical pressure before reaction of 40 GPa (400 kbar). These pressures are sufficient to fundamentally change the nature of chemical bonding. As matter is compressed, the mean electron kinetic energy increases. The increase in kinetic energy leads to weakening of chemical bonds and possibly ionization.[8] The extreme environment of a detonation demands that chemistry be modeled in the condensed phase. Condensed-phase electronic structure approaches based on Kohn–Sham density-functional theory (DFT),[9] are typically used to simulate systems of up to several hundred atoms for times on the order of 10 ps.[10] A system of several hundred atoms would represent roughly 10 energetic molecules (e.g., TATB has 24 atoms). Therefore, standard DFT approaches are very limited both in terms of system size and simulation duration when applied to condensed-phase energetic molecular chemistry.

One approach to extend both time and spatial scales is to employ reactive empirical force fields to represent intermolecular interactions. While many physical attributes of energetic materials can be studied with classical, nonreactive force fields,[11] the ReaxFF force field enables insights into the chemical transformations occurring at the extreme conditions of pressures and temperatures.[12] The ReaxFF approach is scalable for systems with thousands of atoms, and timescales roughly up to a few ns. ReaxFF has been used to study thermal and shock decomposition processes of various energetic systems.[13] One of the difficulties of the ReaxFF approach, however, is the many parameters used in the force field. The determination of optimal parameters is not unique, and parameters are typically refined for particular applications.[13d] The ReaxFF method scales almost linearly with the number of atoms,[14] and therefore is potentially applicable to very large system sizes. The wide variety of chemical environments encountered during the complex chemistry of HE decomposition could be difficult to represent adequately with a force field approach.

A second approach is to employ more approximate forms of quantum simulation that could enable longer simulation times and larger system sizes. Unlike reactive force fields, quantum-based methods permit us for the first time to examine electronic properties at detonation conditions. Semi-empirical, quantum-based methods such as tight binding are less costly, extending the time profile of chemical events to be investigated for a few hundred ps with computational cells of several hundred atoms. Recent studies with variants of tight-binding methods include shocked benzene,[15] shocked methane,[16] the chemistry of dense HMX explosive,[17] shocked nitromethane,[18] shocked and dense TATB explosive,[19] and complete reactivity of detonating HN_3.[20]

Our group has applied the DFTB with self-consistent charges to HE reaction chemistry.[10d,18] DFTB is an approximate quantum method that uses a minimal basis set to represent covalent interactions. Unlike DFT, DFTB contains an empirical repulsive force field that is fitted to experimental data and DFT simulations. The two-body DFTB repulsive force field, however, is much simpler in form than the many-body interactions represented in ReaxFF. A DFTB calculation is 10–100 times faster than a DFT calculation. Since the DFTB method requires a matrix diagonalization, the technique has an N^3 asymptotic scaling with the number of atoms. Linear scaling methods can be adapted to DFTB,[21] but most implementations to date have not done so. In practice, this limits the system size of a DFTB molecular dynamics simulation to several hundred atoms. The increased efficiency of DFTB has allowed simulations to be conducted for as long as several hundreds of ps.[19]

Since DFTB the method is semi-empirical, its accuracy for energetic materials should be verified through comparison to experiment and other simulations. DFTB has been benchmarked against DFT for the static compression[22] and the reactivity of nitromethane.[10d] These comparisons showed that qualitatively similar results were obtained with DFT generalized gradient approximation[23] (GGA) and DFTB. In benchmarking against the reaction energy of small organic molecules, DFTB was found to have a mean absolute error of 12.5 kcal/mol when compared to the experiment. Although the DFTB method is not as accurate as more elaborate *ab initio* methods, the high temperature condition of the simulation ameliorates this difficulty. For instance, at 3500 K, a 10 kcal/mol error in a reaction barrier leads to a factor of 4 errors in the reaction rate. This is an acceptable error for a qualitative study. At 600 K, however, a comparable error in the barrier would lead to a factor of 4000 in the reaction rate.

Recent work has been undertaken to improve the accuracy of DFTB by adding higher-order corrections to charge fluctuations,[24] extending the repulsive energy terms,[25] and by using nonminimal basis sets.[25b] Goldman et al. showed that DFTB simulations with a nonminimal basis could be successfully applied to shocked carbon at pressures in excess of 10 Mbar, while the standard DFTB implementation showed significant errors when compared to the experiment. In Fig. 6.1, we show a comparison of the improved DFTB method of Goldman et al.[25b] to DFT GGA calculations. The DFT GGA calculations are in good agreement with experiment (not shown), although the spread of results from different experimental studies is large.

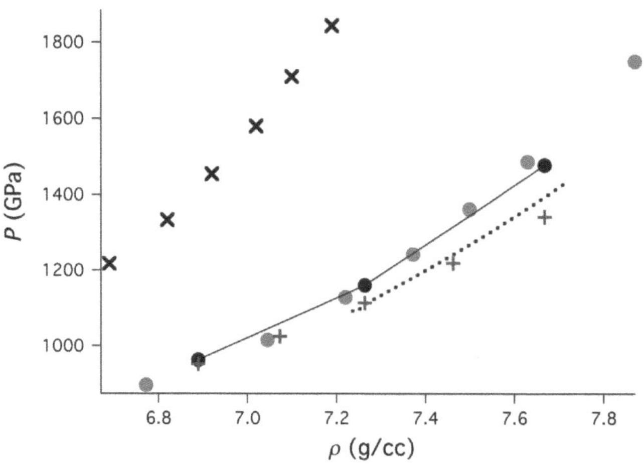

Figure 6.1 Comparison of simulation and experimental results for the pressure and density of shocked carbon. The black "x" symbols are standard DFTB. The filled red symbols are DFTB including a polarized basis set. The closed black and blue "+" symbols are GGA DFT calculations. (For interpretation of the references to color in this figure legend, the reader is referred to the online version of this chapter.)

Energetic material detonation involves the propagation of a supersonic shock wave. Therefore, the study of the chemistry of detonation requires that some sort of approximation to the shock be made. One of the simplest approximations is to simply study the system at a fixed density and temperature. While this is a well-defined statistical mechanics problem, it does not correspond closely to an actual shock wave. The exothermic nature of energetic material reactivity necessarily leads to a large change in the temperature upon reaction. This change is suppressed when the temperature is fixed in a molecular dynamics simulation. Simulations at fixed volume and energy do not have this difficulty, but still miss important features of the shock wave.

The detonation shock wave itself is a traveling mechanical excitation in the material, with a typical velocity of 8 km/s. This corresponds to 8 nm/ps when converted into typical molecular dynamics units. Therefore, the wave propagating through the material will move 800 nm in a 100-ps simulation. This is much larger than the system size in most reactive molecular dynamics simulations, especially those using quantum methods (where a simulation cell size of 5–10 nm is typical). There are several approaches to overcome this problem. One approach is to employ direct numerical simulations for shorter timescales of several ps.[26] Another is to use a moving frame of reference in simulating the shock wave.

Reed et al.[27] developed a method with a moving frame of reference called the multiscale shock simulation technique (MSST). This method simulates a notional Lagrangian element (essentially, a fixed set of atoms) as it propagates through a steady shock wave. MSST allows both the system density and temperature to vary in a way that conserves mass, momentum, and energy during the shock wave evolution. It has been known since the 1940s that HEs expand and move toward lower pressure states as they react in the shock front. The macroscopic theory describing this is called the Zeldovich–von Neumann–Doering (ZND) theory.[28] The MSST can be thought of as coupling the macroscopic ZND theory to the MD simulation cell. There are alternative approaches to implementing molecular dynamics in a moving frame. One such technique is to inject particles on one side of the simulation cell and remove particles on the other side.[29] A second approach is to employ a fixed density but constrain the temperature according to macroscopic hydrodynamics, which is implemented in the Hugoniotstat technique by Maillet et al.[30]

The MSST has been successfully compared to both continuum direct numerical simulations and molecular dynamics direct numerical simulations.[31] In Fig. 6.2, we show a comparison between the MSST and direct numerical simulation for a shock wave in amorphous solid carbon.[32]

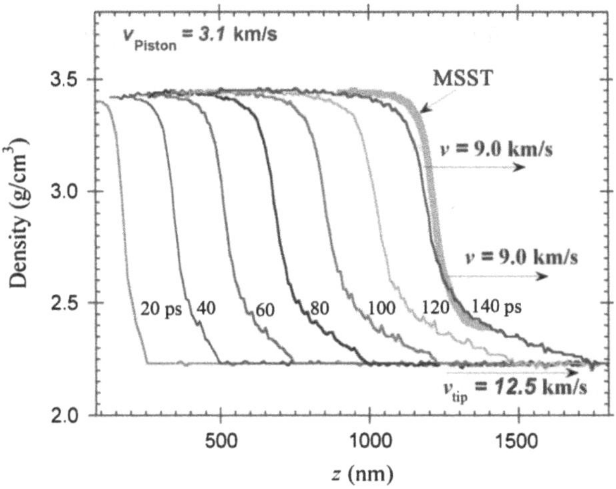

Figure 6.2 Comparison of a shock wave profiles in amorphous carbon (thin solid lines) with a velocity of 9.0 km/s compared to the MSST method (thick gold line), also at a velocity of 9.0 km/s. The MSST method does not include the nonsteady foot in the shock wave profile, which propagates at the velocity v_{tip}. (For interpretation of the references to color in this figure legend, the reader is referred to the online version of this chapter.)

Nonetheless, it is important to remember that the MSST does not treat density gradients and therefore is less accurate very near the shock front where density gradients are very high. Recently, the MSST has been extended to include electronic excitations[33] and the self-consistent incorporation of quantum nuclear effects.[16b]

It is known that the shock initiation of energetic materials is sensitive to micron-scale material structure, such as the presence of voids, particle size, and plastic-bonded binder. It is difficult to treat these phenomena at the atomic scale. Instead, we have focused on overdriven shock waves that propagate at velocities greater than that of a steady detonation, or on simulations under a fixed temperature above 2000 K. Under these conditions, the HE transforms in a ps or less to a highly reactive dense fluid phase. There are important differences between the dense fluid (supercritical) phase and the solid phase, which is stable at standard conditions. Namely, the dense fluid phase cannot accommodate long-lived voids, bubbles, or other static defects. Instead, numerous fluctuations in the local environment occur within a timescale of 10 s of fs. The fast reactivity of the dense fluid phase and the short spatial coherence length make it well suited for molecular dynamics study with a finite system for a limited period of time, although it should be remembered that overdriven shock waves do not correspond to the most commonly performed experiments for energetic materials.

The quantum nature of molecular vibrations poses an additional challenge to simulations of reactive chemistry at extreme conditions. Molecular dynamics simulations predict a classical heat capacity regardless of whether the force field is empirical or calculated via quantum simulations. The classical heat capacity can be significantly greater than the quantum (experimental) heat capacity for materials that are rich in hydrogen. Goldman et al.[34] have developed a method to correct shock temperatures obtained through MD simulations for quantum vibrational effects. In Fig. 6.3, we show a comparison of simulated and experimental temperatures for shocked methane. Although methane is not an energetic material, it is an expected detonation product. One potential route to developing insight into the chemistry of shocked energetic materials is to study the simpler problem of detonation product mixtures.[35]

Recently, the MSST has been modified to include a vibrational quantum correction through the use of a colored noise Langevin thermostat.[16b]

We mention a final difficulty in simulations of chemistry at extreme conditions, which is the chemical interpretation of results. Under the high temperature and pressure conditions of a detonation, reaction rates are predicted

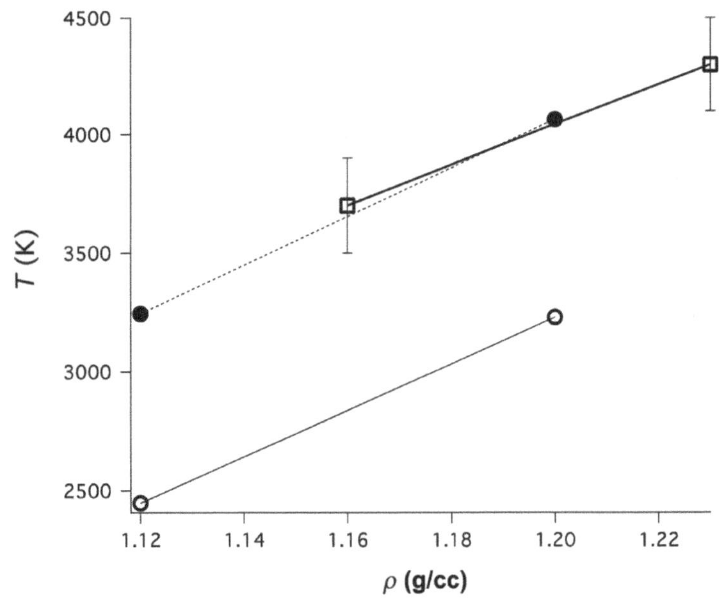

Figure 6.3 Plot of the temperature versus density for shocked CH_4. The open circles correspond to the classical ionic temperatures from a DFT calculation, the solid circles are values corrected for quantum effects as in Goldman et al.[34] The open squares with error bars show experimental results.

by simulation to be comparable to vibrational frequencies. This makes it difficult to unambiguously determine molecular species and chemical reactions at extreme conditions. The physical basis of this issue is that as matter is compressed, it becomes increasingly nonmolecular. A good example of this phenomenon is found in water, where increasing compression has been shown to lead to hydrogen bond symmetrization and nonmolecular phases.[36] In simulations of the HE PETN at high density and temperature, Wu et al.[37] found that hydrogen diffused at a much higher rate than other atoms. An example is shown in Fig. 6.4, where we show the motion of hydrogen and other elements in "unwrapped" coordinates.

Nonetheless, we find the identification of species to be a useful interpretive aid in analyzing simulations at extreme conditions. A scheme to identify molecular species in our simulations is as follows: a bond distance and bond duration are both used to determine whether two molecules are bonded. This approach is independent of the method used to calculate forces. For instance, reactive force fields often have a distance-dependent bond order

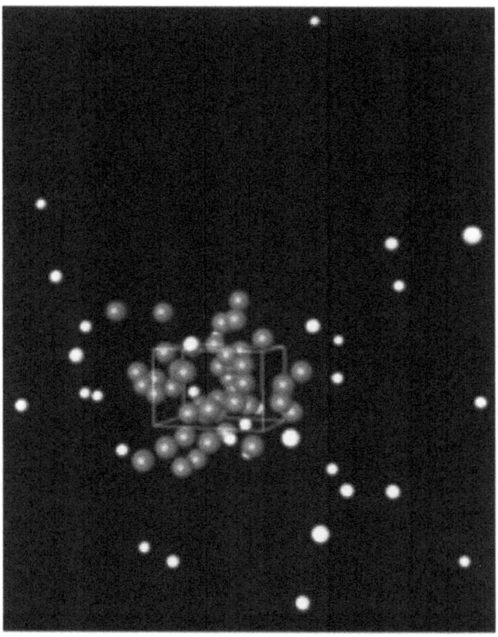

Figure 6.4 The simulated positions of hydrogen atoms (white) and other elements (C,N,O) (orange) are shown relative to the initial simulation cell (yellow box). The hydrogen atom diffuses much further than the other elements, indicating that hydrogen has taken on a nonmolecular character. (For interpretation of the references to color in this figure legend, the reader is referred to the online version of this chapter.)

parameter.[12] Bond order parameters, however, are more difficult to generate in quantum simulations. We determine the bond distance by identifying the first minimum in the atom–atom radial distribution function.[36] This corresponds to the maximum of the potential of mean force, which is a nature effective interaction potential in a many-body condensed-phase environment. We next determine the bond duration to be equal to a several times a typical vibrational period. This enforces a loose time-scale separation between molecular lifetimes and vibrational periods. We find that using these criteria leads to reasonable (though not unique) identifications of molecular species in complex reactive environments.

We now discuss our application of the DFTB methodology to conduct relatively large time-scale simulations of thermally activated or shocked explosives using MSST. We consider two markedly different explosive materials in terms of sensitivity and performance: HMX is a sensitive and a high-performing explosive, while TATB is an insensitive and a low-performing explosive. The chemistry revealed from these simulations provided us with a better understanding with regard to reactivity of these two explosives. As we will see, HMX decomposition produces mostly small gas species such as water and N_2. In contrast, the insensitive TATB explosive produces large, stable, high-nitrogen, carbon-rich heterocycles, which retard TATB's complete reactivity toward a final form of carbon (graphite or diamond).

2. CHEMISTRY OF HMX

HMX (1,3,5,7-tetranitro-1, 3,5,7-tetraazacyclooctane; Fig. 6.5) is widely used as an ingredient in various explosives and propellants. A molecular solid at standard state, it has four known polymorphs. We have conducted a quantum-based molecular dynamics simulation of the chemistry of HMX under extreme conditions, similar to those encountered at the Chapman–Jouguet detonation state.[17b] The simulation studied the reactivity of dense (1.9 g/cm^3) fluid HMX at 3500 K for reaction times of up to 55 ps, thus allowing the formation of stable product molecules. There are numerous experimental characterizations at low temperatures (i.e., <1000 K, well below detonation temperature) of decomposition products of condensed-phase HMX.[38] These studies tend to identify final gas products (such as H_2O, N_2, H_2, CO, and CO_2) from the surface reactions and aspire to establish a global decomposition mechanism. The early thermal decomposition study using mass spectrometry at $T=503$, 527, and 553 K of Suryanarayana et al.[38j] identified a

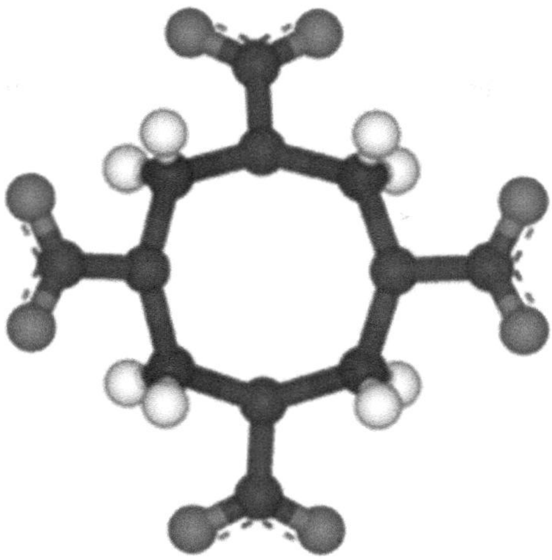

Figure 6.5 Molecular HMX. (For color version of this figure, the reader is referred to the online version of this chapter.)

concerted decomposition into four methylenenitramine ($CH_2N_2O_2$), which can further decompose into CH_2O and N_2O. At 448–548 K, Farber and Srivastava[39] identified a major decomposition product with $m/e = 148$ and proposed a homolytic cleavage of HMX to two $C_2H_4N_4O_4$ fragments. CH_2N and NO_2 were later detected as later stage decomposition products from an $CH_2N_2O_2$ intermediate in an electron spin resonance pyrolysis study.[38h] Experiments using thermogravimetric modulated beam mass spectrometry and isotope scrambling identified gaseous pyrolysis products such as H_2O, HCN, CO, CH_2O, NO, and N_2O between 483 and 508 K.[38a–c] Brill et al. have analyzed rate measurements for the early stage of HMX thermal decomposition,[38e] revealing the existence of an approximate linear relationship between the Arrhenius prefactor, ln A, and the apparent activation energy, E_a. Brill later suggested two competing global mechanisms for thermal decomposition, the first leading to 4 HONO and 4 HCN, while the second leads to the formation of 4 CH_2O and 4 N_2O.[38d]

The above experimental work on the thermal decomposition of condensed phase HMX is largely restricted to relatively low temperature (~550 K) and pressure (0.1 GPa) regimes. A recent study examined the effect of pressure on the global thermal decomposition rate of HMX up to 3.6 GPa.[40] Similar experimental observations at detonation conditions (temperatures 2000–5000 K and pressure 10–30 GPa), however, have not been realized to date. At present, computer simulations provide the best access to the short time-scale processes occurring in these regions of extreme conditions of pressure and temperature.[41] Many computational studies have included electronic structure calculations of various decomposition channels of the gas-phase HMX molecule. For example, Melius determined decomposition pathways for nitramine compounds HMX and RDX at the MP4 level of theory. He found that the initial step in decomposition is N—NO_2 bond breaking, which subsequently causes a significant weakness in the second-nearest-neighbor bond breaking energies (18 kcal/mol for the C—N bond dissociation), leading to HCN, NO_2, and H as the net products for rapid thermal heating. In the condensed phase, however, Melius made the observation that alternative decomposition mechanisms can occur. The free NO_2 fragment can recombine as a nitride, which then decomposes to form NO, or attract weakly bound hydrogen atoms and form HONO. The HONO molecules can then rapidly equilibrate to form water via the reaction $2HONO \rightarrow H_2O + NO_2 + NO$.[42]

Lewis et al. calculated four possible decomposition pathways of the α-HMX polymorph: N—NO_2 bond dissociation, HONO elimination, C—N bond scission, and the concerted ring fission. Based on the energetics, it was determined that N—NO_2 dissociation was the initial mechanism of decomposition in the gas phase, while they proposed HONO elimination and C—N bond scission to be favorable in the condensed phase.[43] The more recent study of Chakraborty et al., using the DFT (B3LYP) method, reported detailed decomposition pathways of β-HMX, the stable polymorph at room temperature. It was concluded that consecutive HONO elimination (4 HONO) and subsequent decomposition into HCN, OH, and NO are energetically the most favorable pathways in the gas phase.[44] Zhang et al. have concluded from their ReaxFF study that only small gas fragments occur during the thermal decomposition of HMX, in contrast to the formation of large clusters obtained from decomposing TATB.[45] To date, however, our study remains the only quantum-based study of the fast chemistry of condensed-phase HMX at conditions similar to those encountered under detonation.

In our molecular dynamics simulations, we used DFTB to calculate the forces on atoms.[6] The initial condition of the simulation included six HMX molecules in a cell, corresponding to the unit cell of the δ phase of HMX, with a total of 168 atoms. It is well known that HMX undergoes a phase transition at 436 K from the β phase (two molecules per unit cell with a chair molecular conformation, density = 1.89 g/cm^3) to the δ phase (with boat molecular conformation, density = 1.50 g/cm^3).[46] The δ phase was chosen as the initial starting structure to include all the relevant physical attributes of the system prior to chemical decomposition. The calculation started with the experimental unit cell parameters and atomic positions of δ HMX. The atomic positions were optimized in an energy minimization procedure.

The cell was then compressed to the final density of the simulation. The atomic structure was subsequently fully optimized at the corresponding cell volume. Since the purpose of the simulation is to study the high-pressure and high-temperature chemistry of HMX in general, the exact density and temperature used in our simulation are somewhat arbitrary. The density of 1.9 g/cm^3 and a temperature of 3500 K were chosen. This state is in the neighborhood of the Chapman–Jouguet state of β-HMX (3500 K, 2.1 g/cm^3) as predicted through thermochemical calculations described later. The closest experimental condition corresponding to this simulation would be a sample of HMX, which is suddenly heated under constant volume conditions, such as in a diamond anvil cell.

The molecular dynamics simulation was conducted at constant volume and constant temperature. Periodic boundary conditions were imposed. Constant temperature conditions were implemented through velocity rescaling. A time step of 0.5 fs was used.

Under the simulation conditions, chemical reactions occurred within 50 fs. Stable molecular species were formed in less than 1 ps. Figure 6.6A–D displays the product formation of H_2O, N_2, CO_2, and CO, respectively. The concentration, $C(t)$, is represented by the actual number of product molecules formed at the corresponding time t. Each point on the graphs (open circles) represents a 250 fs averaged interval. The number of the molecules in the simulation was sufficient to capture clear trends in the chemical composition of the species studied. These concentrations were in turn fit to an expression of the form: $C(t)=C_\infty(1-e^{-bt})$, where C_∞ is the equilibrium concentration and b is the effective rate constant. From this fit to the data, we estimate effective reaction rates for the formation of H_2O, N_2, CO_2, and CO to be 0.48, 0.08, 0.05, and 0.11 ps^{-1}, respectively.

It is not surprising that the rate of H_2O formation is much faster than that of N_2. Fewer reaction steps are required to produce a triatomic species like water, while the formation of N_2 involves a much more complicated mechanism.[42] The formation of water (Fig. 6.6A) starts around 0.5 ps and reaches a steady state at 10 ps, with oscillatory behavior of decomposition and formation clearly visible. The formation of N_2 (Fig. 6.6B), on the other hand, starts around 1.5 ps and is still progressing (slope of the graph is slightly positive) after 55 ps of simulation time.

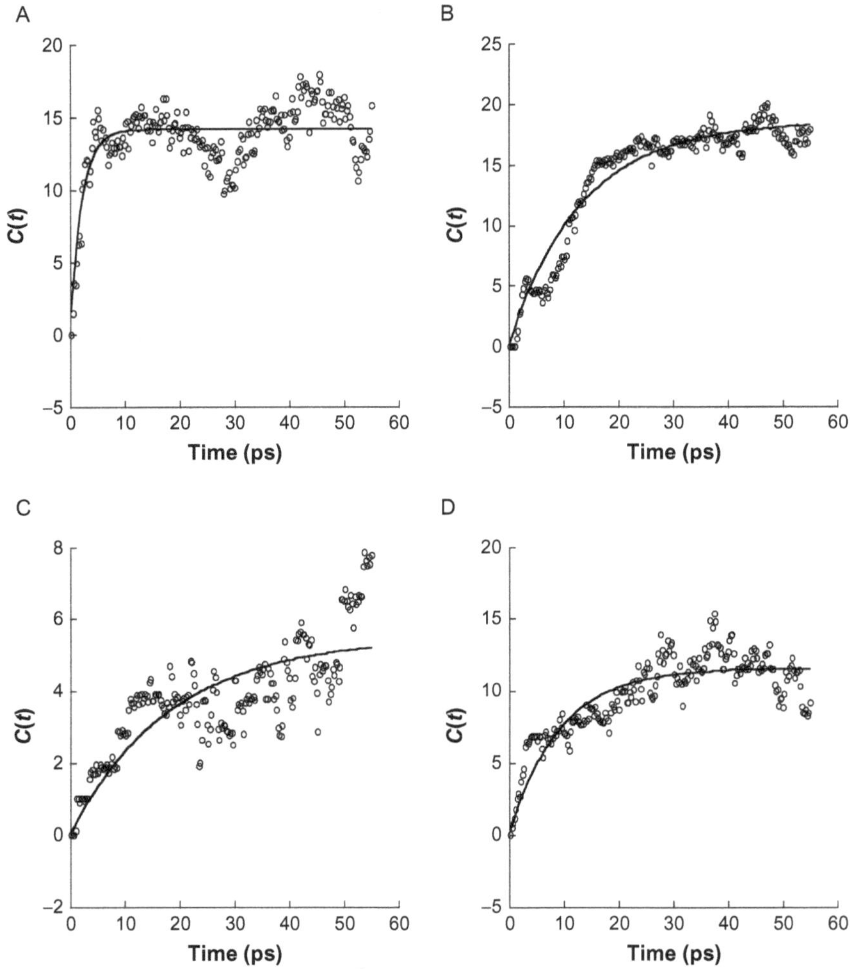

Figure 6.6 Product particle number as a function of time for (A) H_2O, (B) N_2, (C) CO_2, and (D) CO.

In Fig. 6.7, we show a comparison of dominant species formation for decomposing HMX obtained from an entirely different theoretical approach. The concentration of species at chemical equilibrium was calculated through thermodynamic calculations, as implemented within the Cheetah thermochemical code.[47] For HMX, the molecules N_2, H_2O, CO_2, HNCO, and CO were predicted to be present in quantities greater than 1 mol/kg HMX. The species CO, NH_3, H_2, CH_4, H, CH_3OH, NO, and C_2H_4 were also predicted to be present in quantities greater than

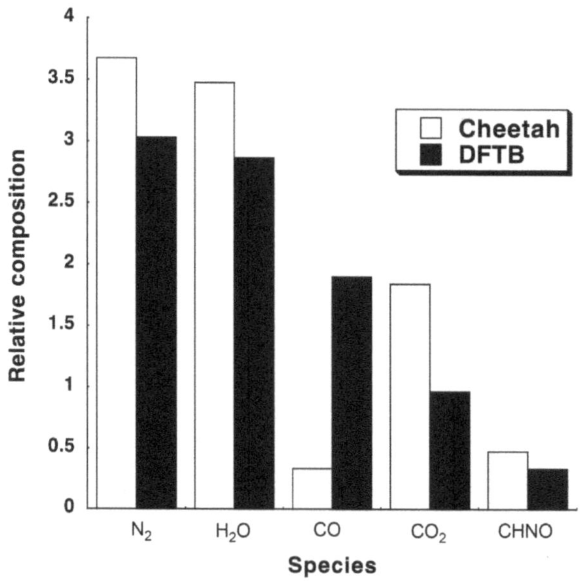

Figure 6.7 Comparison of relative composition of dominant species determined from current DFTB simulation and from a thermodynamical calculation.

0.0001 mol/kg HMX. The species N_2O, C_2H_2, N, O, O_2, NO_2, HCN, atomic C, and O_3 were not predicted to have significant concentrations. Carbon in the diamond phase was predicted to be in equilibrium with the other species at a concentration of 4.9 mol/kg HMX. The thermochemical calculations predict a pressure for fully reacted HMX of 16 GPa, or 160 kbar.

As can be noticed in Fig. 6.7, the results of the MD simulation compare very well with the formation of H_2O, N_2, and HNCO. The relative concentration of CO and CO_2, however, is reversed at the end of the simulation. This could be due to the limited simulation time. No condensed carbon was found in the current simulation.

One expects more CO_2 than CO as final products, as predicted by Cheetah (Fig. 6.7). The results displayed in Fig. 6.6C and D show that, at simulation time of 40 ps, the system is still in the second stage of reaction chemistry. At this stage, the CO concentration is rising and has not yet undergone the water–gas shift reaction ($CO + H_2O \rightarrow CO_2 + H_2$) conversion. Interestingly, this shift seems to occur at around 50 ps of the simulation, with CO_2 molecules being formed while CO concentration is correspondingly diminishing.

Here, the initial steps of the decomposition pathways are considered. The first chemical event in the simulation is the breaking of the N—NO_2 bond and the dissociation of NO_2 fragments. At 200 fs of simulation time, the number of NO_2 fragments is 10, out of a possible total of 24, with some being successive elimination from the same HMX molecule. This preference to bond rupture is consistent with the recent observation that the energetic barriers for the cleavage of N—NO_2 bond in the solid phase of the nitramine RDX vary depending on the location of the molecule in the crystal.[48] At this stage of the simulation, the C—N bond breaking is also exhibited, occurring in two ways: the first (1) producing methylenenitramine ($CH_2N_2O_2$), while the symmetric breaking (2) leads to the formation of two $C_2H_4N_4O_4$ moieties, as shown below.

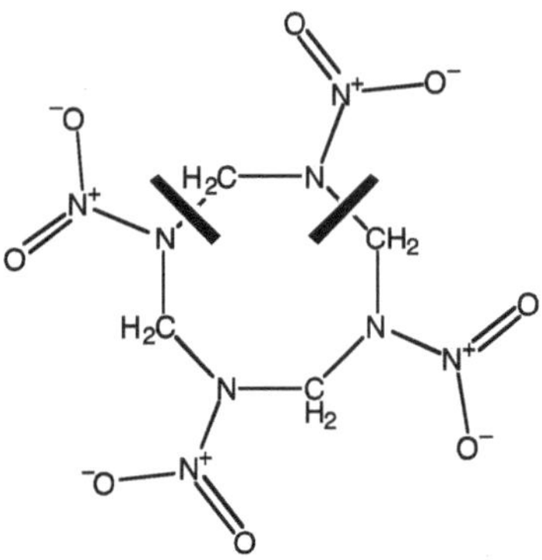

(1)

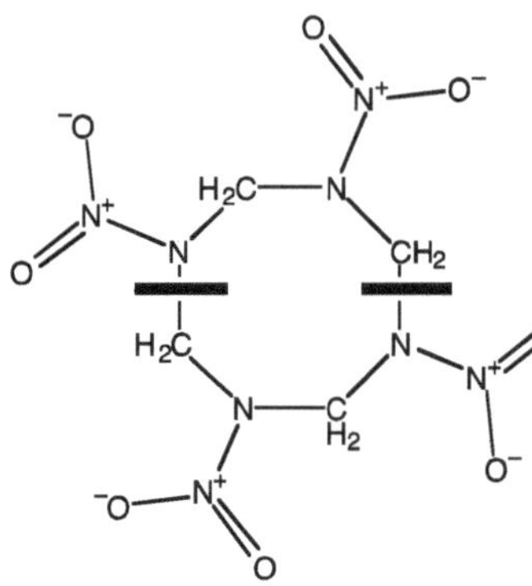

(2)

Seven $CH_2N_2O_2$ species have been formed at around 200 fs of simulation time. These results are similar to those identified in thermal decomposition experiments.[38j,39] A further N—NO_2 bond breaking then follows the decomposition (1) and (2) above. From (1), this leads to the formation of CH_2N and NO_2. These pathways are remarkably similar to those predicted previously by Melius from the decomposition of nitramines at fast heating rates.[42]

As the radical CH_2N is formed, the production of HCN occurs via the reaction:

$$CH_2N + CH_2N \rightarrow C_2H_4N_2 \rightarrow HCN + H_2NH.$$

Another source for the formation of HCN follows from a series of complex reactions that also produce nitric acid, HNO_3:

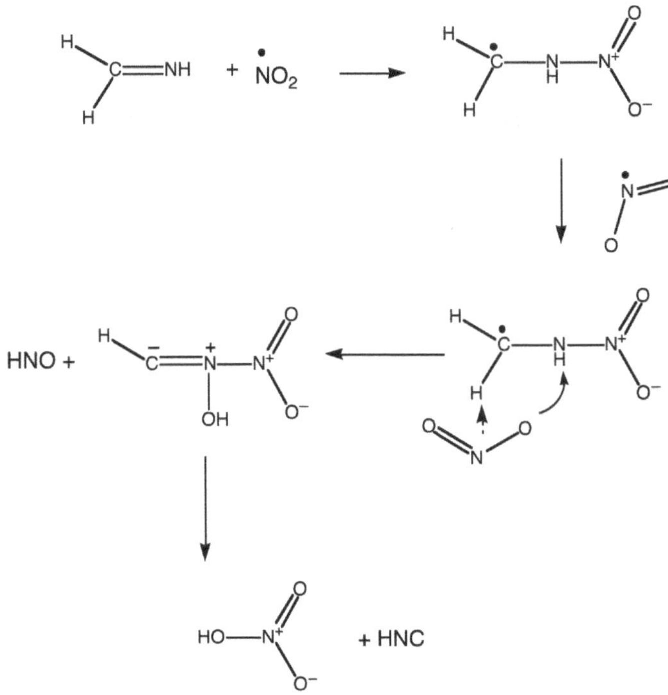

Figure 6.8 Reaction scheme for the formation of HCN.

$$CH_2NH + NO_2 \rightarrow CH_3N_2O_2$$
$$CH_3N_2O_2 + NO_2 \rightarrow CH_2N_2O_3 + HNO$$
$$CH_2N_2O_3 \rightarrow HCN + HNO_3$$

The schematic mechanism for these reactions is illustrated in Fig. 6.8. It should be noted that formaldehyde, CH_2O, is first formed from a reaction involving large intermediate fragments. The formation occurs from the reaction of the $C_2H_4N_4O_4$ moiety, which is produced from the symmetric bond scission of the HMX molecule as in (2) above, with HNO. The reaction leads to the production of CH_2O and a larger intermediate fragment that undergoes further decomposition.

3. CHEMISTRY OF TATB

It has been noted that explosives rich in carbon tend to have much longer reaction zones than those that do not. Such explosives form graphitic or diamond-like carbon particles during detonation.[49] Formation of the

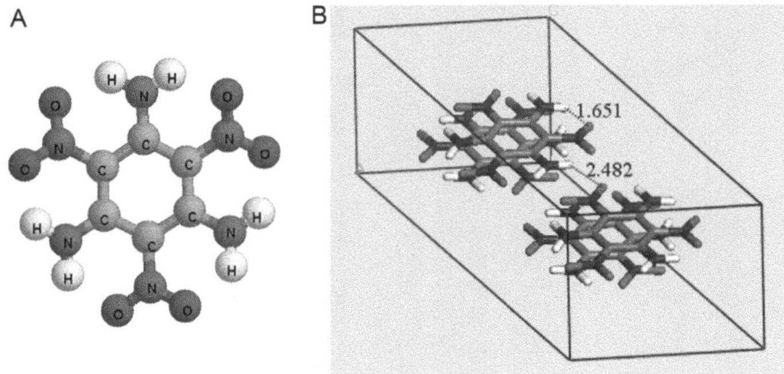

Figure 6.9 (A) Molecular TATB ($C_6H_6N_6O_6$). (B) Graphite-like sheets in a $1 \times 2 \times 2$ supercell of nonorthorhombic crystals of TATB solid phase. The X-ray crystal structure of TATB includes 2 molecules per unit cell (triclinic) with parameters $a = 9.010$, $b = 9.028$, $c = 6.812$ Å, and $\alpha = 108.59°$, $\beta = 91.82°$, and $\gamma = 119.97°$.[51] (For color version of this figure, the reader is referred to the online version of this chapter.)

bulk solid carbon from the clusters is believed to play a key role in determining the size of the reaction zone of a given explosive.[49,50] The latter process is thought to be initiated by collecting carbon atoms from a relatively large volume and governed by diffusion-limited growth of these clusters. TATB ($C_6H_6N_6O_6$, Fig. 6.9) is an insensitive explosive with an estimated reaction zone in the microsecond regime. It has been found to be remarkably stable under extreme pressure, persisting metallization or chemical decomposition.[52] It can be described as markedly under oxidized or, alternatively, carbon rich. As a consequence, significant carbon formation in some form (graphite, clusters, or diamond) is expected to be part of its final products under detonation reactions.

We used DFTB molecular dynamics to model the reactivity of TATB under thermal and shock conditions. Relatively few studies have been performed on the chemistry of stronger shocks due to the requirement for accurate methods of simulating a variety of chemical reactions. In addition, our simulations have revealed an important feature of TATB decomposition: the first evidence for the formation of an extended region of nitrogen-rich heterocyclic clusters, the formation of which inhibits reactivity toward final decomposition products of fluid N_2 and solid carbon.[19] This result suggests a new mechanism for carbon-rich explosive materials that precede the slow diffusion-limited process of forming the bulk solid from carbon clusters. Another important feature undisclosed previously in the reactivity of energetic materials at extreme conditions is the prevailing persistence of small charged species throughout thermal and shock simulations.

Our thermal decomposition simulations were conducted at constant volume and temperature. Periodic boundary conditions were imposed in all directions, while constant temperature conditions were implemented through simple velocity rescaling. The probability to rescale atom velocities was chosen to be 0.1 per time step. A dynamic time step of 0.48 fs was used. Several simulations with various density, size, and temperature were carried out: (1) four simulations used a $1\times1\times3$ supercell (144 atoms) at density $\rho=2.87\ \mathrm{g/cm^3}$, corresponding to an initial pressure of 29.7 GPa, which was thermally heated at various temperatures of $T=3500$, 3000, 2500, and 1500 K, and (2) two simulations used a $2\times2\times1$ supercell (192 atoms) at two densities $\rho=2.72$ and $2.9\ \mathrm{g/cm^3}$, corresponding to initial pressures of about 22 and 34.5 GPa, respectively, which was thermally heated at $T=3000$ K. These simulations varied in duration from 400 ps to 2.0 ns.

We also simulated crystal TATB's reactivity under steady overdriven shock compression using the MSST.[27,53] We performed MD simulations with a time step of 0.5 fs for up to 430 ps with shock speeds of 8 and 9 km/s, and 202 ps for a shock speed of 10 km/s on the smaller $1\times1\times2$ computational cell. For the larger $1\times2\times2$ supercell, the simulation was conducted up to 140 ps. In all simulations, initial position of TATB molecules was taken from the experimental X-ray structure and then optimized with the DFTB method. Initial velocities were randomly chosen, with an initial temperature of 300 K. We then performed an equilibration for a period of up to 2 ps at temperature of 300 K. Shock compression occurred along the *c* lattice direction in all simulations.

In our shock simulations, TATB must be substantially overdriven to observe chemistry on sub-ns timescales. The experimental TATB detonation velocity is 7.7 km/s at $\rho=1.88\ \mathrm{g/cm^3}$ and a pressure of 29 GPa. Figure 6.10 displays the time evolution of the average temperature, stress, and volume for all three initial shock speeds. For the 9 km/s shock speed simulation, the temperature gradually increases from around 1900 to 3500 K, while the stress increases from 58 to 70 GPa. This is due to exothermic chemical reactions. More dramatic conditions are noted for the 10 km/s shock speed simulation, where the temperature increases from 2500 to 3500 K and the stress reaches 85 GPa in less than 50 ps of simulation time. For the 8 km/s simulation, the temperature and stress have nearly constant values of 1500 K and 42 GPa throughout the simulation period. The only observed reactivity at this shock speed is protons transfer between amino and nitro groups up to 0.43 ns of simulation time. A similar observation is obtained from thermal heating at $T=1500$ K up to 2 ns.

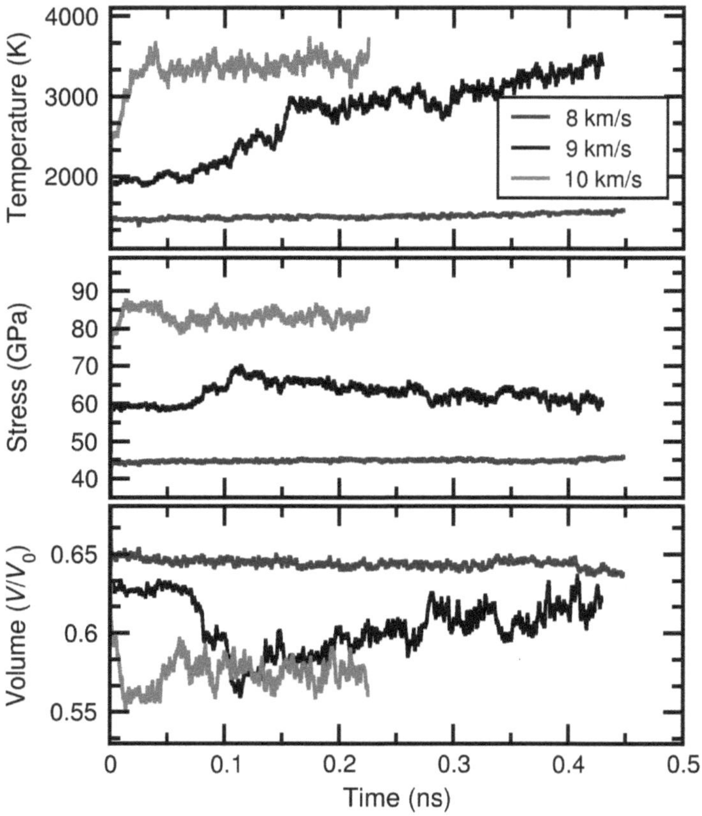

Figure 6.10 Time evolution of average temperature, shock-propagation-direction stress, and volume of $1 \times 1 \times 2$ computational cell for simulations at shock speeds of 8, 9, and 10 km/s. (For color version of this figure, the reader is referred to the online version of this chapter.)

These results clearly indicate that chemical transformations occur faster as the temperature increases (or shock speed increases).

The use of quantum-based methods permits us for the first time to examine electronic properties at detonation conditions. Our simulations of shocked nitromethane showed that a transient band-gap reduction to a semi-metallic state occurred during shock compression with a speed of 6.5 km/s.[18] Our shock simulations of TATB also show a nonmonotonic change in electronic properties. Figure 6.11 compares the dimensionless electronic state overlap parameter, which is often used to characterize metallization,[54] of TATB and detonating nitromethane. With an overdriven shock speed of 9 km/s, TATB is in a metallic state throughout

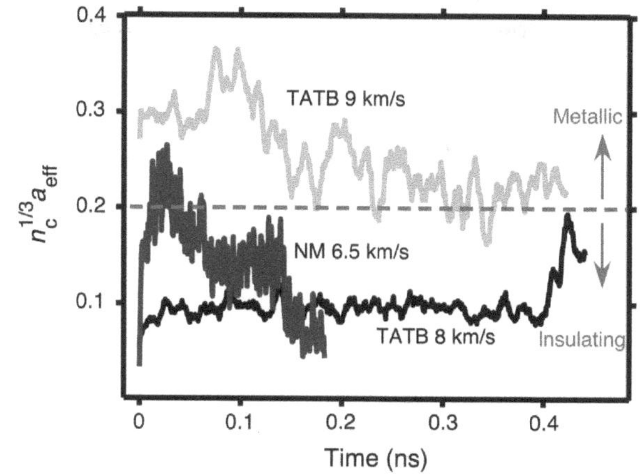

Figure 6.11 Time profile of the electronic state overlap parameter $n_c^{1/3}a_{eff}$ for TATB and nitromethane. n_c is the charge-carrier concentration, and a_{eff} is the effective Bohr radius of electronic states. (For color version of this figure, the reader is referred to the online version of this chapter.)

most of the simulation. For a shock speed of 8 km/s, it stays mostly as an insulator, which is consistent with its lack of reactivity.

Global features of TATB decomposition obtained from both thermal and shock simulations revealed similar trends. The shock simulation with a speed of 9 km/s has a temperature range of about 1900–3500 K, allowing for comparison with thermal decomposition simulations within $T=2500$–3500 K. Both simulation types have fast chemical transformations, particularly at high T (e.g., $T=3500$ K). At this temperature, H_2O and N_2-dominant gas formation occurred in less than 100 ps, with water being first to appear. At the high T condition, dissociation–recombination reactions of the type $H_2O \leftrightarrow H + OH$ are very prevalent. Furazan-like structures seem to accompany H_2O formation. Once hydrogen is eliminated from amine groups to form water, nitrogen of adjacent molecules joins in to form N_2.

Unlike water, however, the formation of N_2 does not proceed before a polymerization process is underway in which the heavy fragments of the parent TATB have bonded together, as illustrated in Fig. 6.12. Nitrogen formation can thus be described as having several growth steps: (1) elimination of H_2O and NO with furazan-like structures as intermediates, (2) formation of extended ring network (condensation reactions), and (3) later steps are always preceded by ring formation with N—N or N—N—N. It should be noted that experimental evidence for benzofurazan formation have been

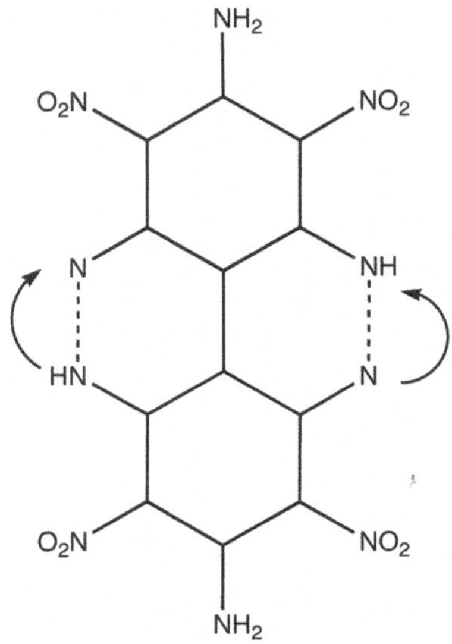

Figure 6.12 Condensation reactions (e.g., bimolecular) to form N—N bonds from parent fragments.

reported for decomposing TATB subjected to various stimuli such as heat, impact, or UV photolysis.[55]

Fast chemistry in our thermal and shock simulations is also exhibited by the fact that both H_2O and N_2 appear to reach steady state of concentration at about 100 and 250 ps, respectively. Figure 6.13 displays the concentration profile of TATB and most dominant, stable products from the shock simulation with a speed of 9 km/s. Figure 6.14 shows the concentration profile of water from thermal decomposition simulations of two temperatures, $T=2500$, and 3500 K at $\rho=2.87$ g/cm^3, along with an exponential fit to obtain reaction rates of this product.

We found that concentration profiles of stable products such as H_2O and N_2 can be fit to an expression of the form: $C(t)=C_\infty(1-\mathrm{e}^{-bt})$, where C_∞ is the equilibrium concentration and b is the effective rate constant. From this fit to the data in Fig. 6.14, we estimate effective reaction rates for the formation of H_2O to be 0.04 and 0.44 ps^{-1} for $T=2500$ and 3500 K, respectively. The shock simulation with a speed of 9 km/s also provides a similar fit for H_2O, and an effective rate of 0.02 ps^{-1}. A similar fit for N_2, as shown in Fig. 6.15, at $T=3500°$ yielded an effective rate of 0.11 ps^{-1}. The figure presents the N_2 concentration profile from thermal simulations at $T=2500$ and 3500 K, and from the shock simulation (inset). We note the similarities in N_2 delayed formation at $T=2500$ K

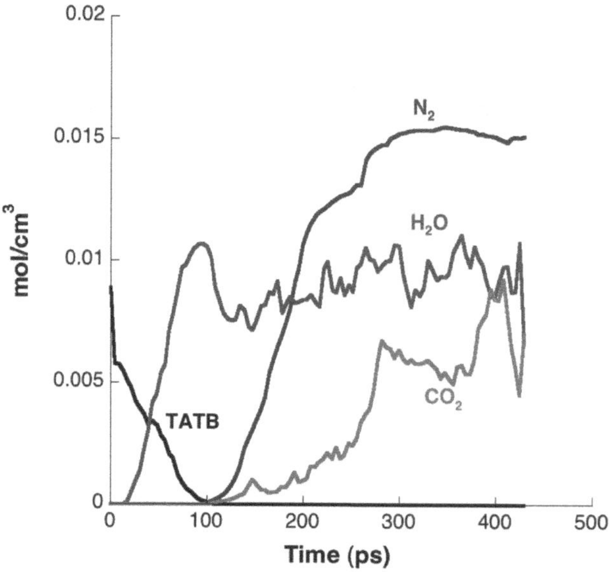

Figure 6.13 Concentration–time profile of TATB and dominant gas products N_2, H_2O, and CO_2, from the simulation with shock speed of 9 km/s. (For color version of this figure, the reader is referred to the online version of this chapter.)

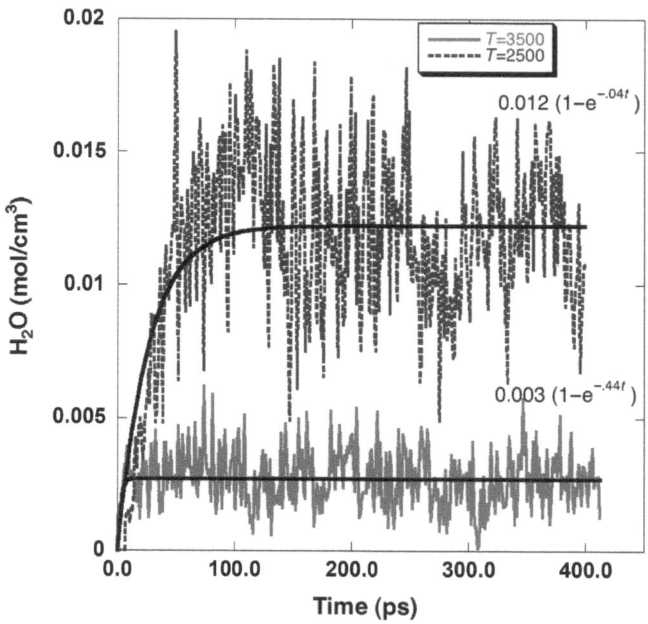

Figure 6.14 Water concentration profile from thermal decomposition simulations at $T=2500$ and 3500 K. (For color version of this figure, the reader is referred to the online version of this chapter.)

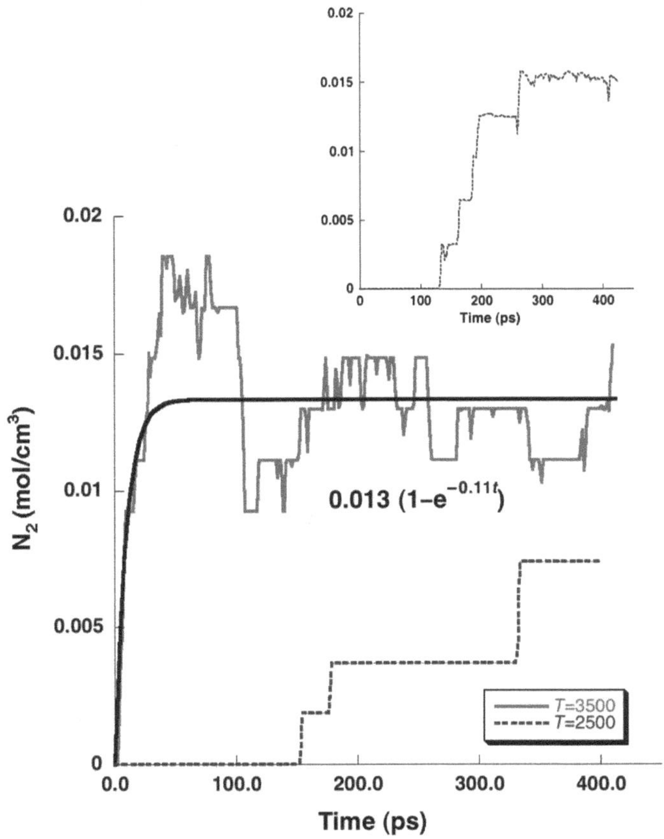

Figure 6.15 N_2 concentration profile from thermal decomposition simulations at $T=2500$ and 3500 K at $\rho=2.87$ g/cm^3, and shock simulation with speed of 9 km/s (inset). (For color version of this figure, the reader is referred to the online version of this chapter.)

and the shock simulation, along with its step-type formation due to "condensation reactions" of the type shown in Fig. 6.12.

Our thermal and shock simulations revealed a very important step in TATB decomposition process: the formation of carbon-rich, nitrogen-rich heterocyclic clusters of various sizes. In Fig. 6.16, we compare the results obtained from simulations of cell sizes of $1 \times 1 \times 2$ and $1 \times 2 \times 2$ for the shock simulation with a speed of 10 km/s. The plot shows the time profile of the average nitrogen to carbon ratio in carbon-containing clusters. As shown, both simulations confirm the formation of nitrogen-rich heterocycles, albeit at different time due to increased reactivity with increase in the shock speed. Figure 6.16 also shows the nitrogen content variation in these

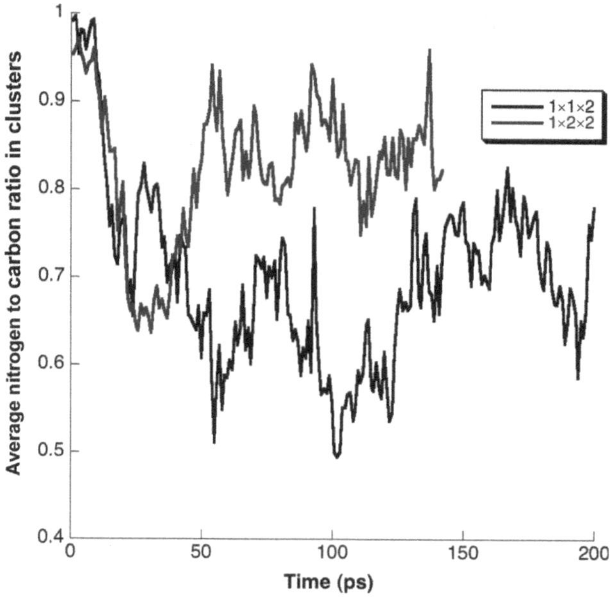

Figure 6.16 Time evolution of the average nitrogen to carbon ratio in carbon-containing clusters obtained from simulated shocked TATB experiencing a shock wave speed of 10 km/s. (For color version of this figure, the reader is referred to the online version of this chapter.)

clusters, ranging between 50% and 100% N in the smaller cell simulation, and between 63% and 96% N in the larger $1 \times 2 \times 2$ simulation cell. In all of our thermal simulations, we found that the formation of these heterocycles is persistent so much that no further decomposition occurred, even at the highest temperature. For example, the formation of high N clusters persists from ~50 ps to 0.42 ns at $T=3500$ K. This is a testament to the stability of these clusters with respect to further decomposition and further slowing of the chemical reactivity of TATB. A representative heterocycle is shown in Fig. 6.17.

Finally, our MD simulations are providing very detailed decomposition mechanism for TATB.[56] We have been able to construct a preliminary four-step decomposition mechanism for use in hydrodynamic grain-scale simulations (e.g., micron-size pore collapse in TATB). The first step involves the decomposition of TATB into mono-furazan and water, followed by the formation of di-furazan and water. The third step accounts for the formation of high-nitrogen heterocycles and other gaseous products such as water, N_2, and CO from condensation reactions. The last step involves further decomposition of these heterocycles into N_2 and the final formation of graphite.

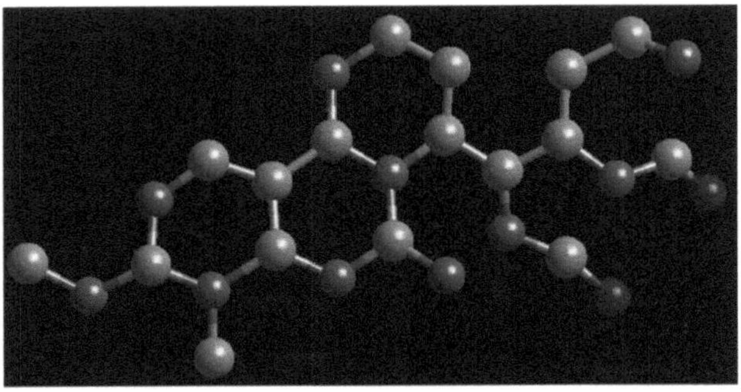

Figure 6.17 A representative nitrogen-rich heterocycle near the end of simulated TATB experiencing a steady shock wave with a speed of 9 km/s. (For color version of this figure, the reader is referred to the online version of this chapter.)

Table 4.1 Arrhenius parameters for a reduced four-step decomposition mechanism of TATB

Mechanistic step	Prefactor (in μs^{-1})	Activation energy (in K)
TATB → mono-furazan + H_2O	2.0×10^6	3.1×10^4
Mono-furazan → di-furazan + H_2O	3.2×10^3	1.74×10^4
2 (Di-furazan) → $4N_2+2H_2O+6CO+2C_3N_2$	0.75×10^6	3.0×10^4
C_3N_2 → graphite + N_2	2.0×10^7	4.0×10^4

We obtained Arrhenius parameters for the expression $K=Ae^{-Ea/RT}$, with A being the prefactor and Ea is the activation energy, for each mechanistic step from fit of rates at different temperatures. These parameters are shown in Table 4.1 for the four-step mechanism. Using these kinetics parameters in the thermochemical equilibrium code, Cheetah provided good agreement with our MD-based decomposition kinetics at these various T, with an example shown in Fig. 6.18. Refinement of this overall decomposition mechanism to a desired accuracy is always possible through inclusion of more detailed steps from our atomistic simulations.

4. CONCLUSIONS

Our thermal and shock simulations allow us to (a) incorporate high-pressure reaction rates in thermochemical hydrocodes to simulate HE reactivity under various scenarios, including grain-scale dynamics, (b) enhance

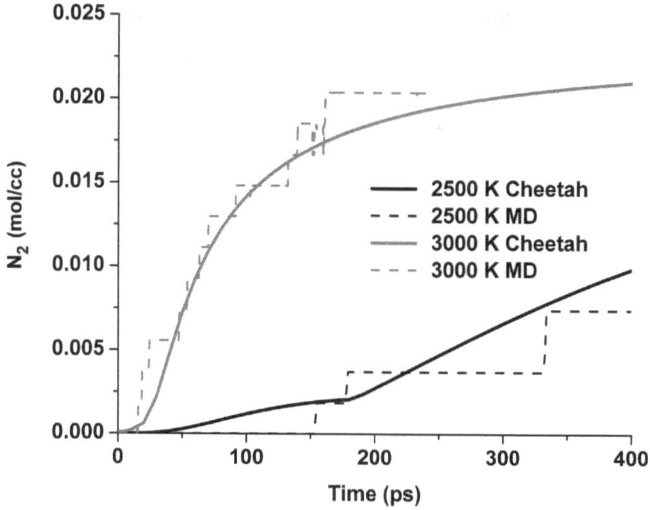

Figure 6.18 A comparison between the thermochemical prediction of the formation of N_2 based on the four steps reduced TATB decomposition mechanism, and DFTB-MD thermal decomposition calculations at $T = 2500$ and 3000 K. (For color version of this figure, the reader is referred to the online version of this chapter.)

thermochemical modeling capabilities by providing a complete account of relevant species, and (c) obtain "reduced" reaction sets from insight into detailed decomposition pathways.

Our results are useful in determining the nature of the Chapman–Jouguet detonation state.[28] Thermodynamic models of detonation have long assumed that the Chapman–Jouguet state was composed primarily of small gas species and condensed solid particles.[57] Our studies on oxygen-rich explosives, such as PETN,[37] HMX,[17b] and nitromethane[10d] confirm this picture, with the strong caveat that hydrogen becomes highly diffusive and nonmolecular. We find that the more sensitive, high-performing HMX explosive produces mostly small gas species such as water and N_2. In contrast, the low-performing, insensitive TATB explosive produces large, stable, high-nitrogen, carbon-rich heterocycles, which retard TATB's complete reactivity toward a final form of carbon (graphite or diamond). Our results on TATB indicate that the inclusion of nanoparticles in thermodynamic models[58] could be required to accurately describe speciation at extreme conditions.

The methods used to study the complex chemistry under extreme conditions are also applicable to other instances where organic materials are subjected to strong shock waves. A good example of this is the impact of comets upon the earth and the possible formation of precursor molecules to life.[59] We believe that knowledge of the chemistry of shocked organic matter will continue to develop rapidly as new computational and experimental tools become available.[60]

ACKNOWLEDGMENTS

The authors thank Nir Goldman, I-F Will Kuo, Amitesh Maiti, Evan J. Reed, and Christine J. Wu for many invaluable contributions to the work discussed in this review. We particularly remember the many insights and contributions of our recently deceased colleague, W. Michael Howard. We dedicate this article to his memory. This research was performed under the auspices of the U.S. Department of Energy by Lawrence Livermore National Laboratory under Contract No. DE-AC52-07NA27344.

REFERENCES

1. (a) Campbell, A. W.; Davis, W. C.; Ramsay, J. B.; Travis, J. R. Shock Initiation of Solid Explosives. *Phys. Fluids* **1961**, *4*, 511; (b) Dremin, A. N.; Savrov, S. D.; Trofimov, V. S.; Shvedov, K. K. *Detonation Waves in Condensed Matter*; Nauka: Moscow, 1970; (c) Elban, W. L.; Armstrong, R. W.; Yoo, K. C.; Rosemeier, R. G.; Yee, R. Y. X-Ray Reflection Topographic Study of Growth Defect and Microindentation Strain Fields in an RDX Explosive Crystal. *J. Mater. Sci.* **1989**, *24*, 1273; (d) Kanel, G. I.; Razorenov, S. V.; Utkin, A. V.; Fortov, V. E. *Impact-Wave Phenomena in Condensed Matter*. Yanus-K: Moscow, 1996; (e) Miles, M. H.; Dickinson, J. T. Fracto-Emission from Pentaerythritol-Tetranitrate and Cyclotetramethylene Tetranitramine Single-Crystals. *Appl. Phys. Lett.* **1982**, *41*, 924.
2. Tarver, C. M.; Chidester, S. K.; Nichols, A. L., III Critical Conditions for Impact and Shock Hot Spots in Solid Explosives. *J. Phys. Chem.* **1996**, *100*, 5794.
3. (a) Bowden, F. P.; Yoffe, A. D. *Initiation and Growth of Explosion in Liquids and Solids*; Cambridge University Press: London, 1952; (b) Coffey, C. S. Phonon Generation and Energy Localization by Moving Edge Dislocations. *Phys. Rev. B* **1981**, *24*, 6984.
4. Hawthorne, S. B.; Lagadec, A. J. M.; Kalderis, D.; Like, A. V.; Miller, D. J. Pilot-Scale Destruction of TNT, RDX, and HMX on Contaminated Soils Using Subcritical Water. *Environ. Sci. Technol.* **2000**, *34*(15), 3224.
5. (a) Fried, L. E.; Manaa, M. R.; Pagoria, P. F.; Simpson, R. L. Design and Synthesis of Energetic Materials. *Annu. Rev. Mater. Res.* **2001**, *31*, 291; (b) Sorescu, D. C.; Rice, B. M. Theoretical Predictions of Energetic Molecular Crystals at Ambient and Hydrostatic Compression Using Dispersion Corrections to DFT (DFT-D). *J. Phys. Chem. C* **2010**, *114*, 6734–6748.
6. Elstner, M.; Porezag, D.; Jungnickel, G.; Elsner, J.; Hauk, M.; Frauenheim, T.; Suhai, S.; Seifert, G. Self-Consistent-Charge Density-Functional Tight-Binding Method for Simulations of Complex Materials Properties. *Phys. Rev. B* **1998**, *58*, 7260.
7. Tarver, C. M.; Breithaupt, R. D.; Kury, J. W. Detonation Waves in Pentaerythritol Tetranitrate. *J. Appl. Phys.* **1997**, *81*(11), 7193–7202.
8. Hemley, R. J. Effects of High Pressure on Molecules. *Annu. Rev. Phys. Chem.* **2000**, *51*, 763–800.
9. Kohn, W.; Sham, L. J. Self-Consistent Equations Including Exchange and Correlation Effects. *Phys. Rev.* **1965**, *140*, A1133.
10. (a) Ancilotto, F.; Chiarotti, G. L.; Scandolo, S.; Tosatti, E. Dissociation of Methane at Extreme Pressure and Temperature. *Science* **1997**, *275*, 1288; (b) Cavazzoni, C.;

Chiarotti, G. L.; Scandolo, S.; Tosatti, E.; Bernasconi, M.; Parrinello, M. Superionic and Metallic States of Water and Ammonia at Giant Planet Conditions. *Science* **1999**, *283*, 44; (c) Goldman, N.; Reed, E. J.; Fried, L. E.; Kuo, I. F. W.; Maiti, A. Synthesis of Glycine-Containing Complexes in Impacts of Comets on Early Earth. *Nat. Chem.* **2010**, *2*(11), 949–954; (d) Manaa, M. R.; Reed, E. J.; Fried, L. E.; Galli, G.; Gygi, F. Early Chemistry in Hot Dense Nitromethane. *J. Chem. Phys.* **2004**, *120*, 10146; (e) Mundy, C. J.; Curioni, A.; Goldman, N.; Kuo, I. F. W.; Reed, E. J.; Fried, L. E.; Ianuzzi, M. Ultrafast Transformation of Graphite to Diamond: An Ab Initio Study of Graphite Under Shock Compression. *J. Chem. Phys.* **2008**, *128*(18), 184701.

11. (a) Kabadi, V. N.; Rice, B. M. Molecular Dynamics Simulations of Normal Mode Vibrational Energy Transfer in Liquid Nitromethane. *J. Phys. Chem. A* **2004**, *108*(4), 532–540; (b) Rivera-Rivera, L. A.; Sewell, T. D.; Thompson, D. L. Post-Shock Relaxation in Crystalline Nitromethane. *J. Chem. Phys.* **2013**, *138*(8), 084512; (c) Siavosh-Haghighi, A.; Dawes, R.; Sewell, T. D.; Thompson, D. L. A Molecular Dynamics Study of Classical Vibrational Spectra in Hydrostatically Compressed Crystalline Nitromethane. *J. Phys. Chem. B* **2010**, *114*(51), 17177–17186; (d) Smith, G. D.; Bharadwaj, R. K. Quantum Chemistry Based Force Field for Simulations of HMX. *J. Phys. Chem. B* **1999**, *103*, 3570; (e) Sorescu, D. C.; Rice, B. M.; Thompson, D. L. Theoretical Studies of the Hydrostatic Compression of RDX, HMX, HNIW, and PETN Crystals. *J. Phys. Chem. B* **1999**, *103*, 6783; (f) Sorescu, D. C.; Rice, B. M.; Thompson, D. L. Molecular Dynamics Simulations of Nitromethane. *J. Phys. Chem. A* **2001**, *105*, 9336.
12. van Duin, A. C. T.; Dasgupta, S.; Lorant, F.; Goddard, W. A., III ReaxFF: A Reactive Force Field for Hydrocarbons. *J. Phys. Chem. A* **2001**, *105*, 9396.
13. (a) Budzien, J.; Thompson, A. P.; Zybin, S. V. Reactive Molecular Dynamics Simulations of Shock Through a Single Crystal of Pentaerythritol Tetranitrate. *J. Phys. Chem. B* **2009**, *113*(40), 13142–13151; (b) Han, S. P.; van Duin, A. C. T.; Goddard, W. A.; Strachan, A. Thermal Decomposition of Condensed-Phase Nitromethane from Molecular Dynamics from ReaxFF Reactive Dynamics. *J. Phys. Chem. B* **2011**, *115*(20), 6534–6540; (c) Strachan, A.; Kober, E. M.; van Duin, A. C. T.; Oxgaard, J.; Goddard, W. A. Thermal Decomposition of RDX from Reactive Molecular Dynamics. *J. Chem. Phys.* **2005**, *122*(5), 054502; (d) Strachan, A.; van Duin, A. C. T.; Chakraborty, D.; Dasgupta, S.; Goddard, W. A. Shock Waves in High-Enegy Materials (RDX). *Phys. Rev. Lett.* **2003**, *91*, 098301; (e) Zhang, L. Z.; van Duin, A. C. T.; Zybin, S. V.; Goddard, W. A. Thermal Decomposition of Hydrazines from Reactive Dynamics Using the ReaxFF Reactive Force Field. *J. Phys. Chem. B* **2009**, *113*(31), 10770–10778; (f) Zhou, T. T.; Zybin, S. V.; Liu, Y.; Huang, F. L.; Goddard, W. A. Anisotropic Shock Sensitivity for Beta-Octahydro-1,3,5,7-Tetranitro-1,3,5,7-Tetrazocine Energetic Material Under Compressive-Shear Loading from ReaxFF-lg Reactive Dynamics Simulations. *J. Appl. Phys.* **2012**, *111*(12), 124904; (g) Zybin, S. V.; Goddard, W. A.; Xu, P.; van Duin, A. C. T.; Thompson, A. P. Physical Mechanism of Anisotropic Sensitivity in Pentaerythritol Tetranitrate from Compressive-Shear Reaction Dynamics Simulations. *Appl. Phys. Lett.* **2010**, *96*(8), 081918.
14. Aktulga, H. M.; Fogarty, J. C.; Pandit, S. A.; Grama, A. Y. Parallel Reactive Molecular Dynamics: Numerical Methods and Algorithmic Techniques. *Parallel Comput.* **2012**, *38*, 245–259.
15. Bickham, S. R.; Kress, J. D.; Collins, L. A. Molecular Dynamics Simulations of Shocked Benzene. *J. Chem. Phys.* **2000**, *112*, 9695.
16. (a) Kress, J. D.; Bickham, S. R.; Collins, L. A.; Holian, B. L.; Goedecker, S. Tight-Binding Molecular Dynamics of Shock Waves in Methane. *Phys. Rev. Lett.* **1999**, *83*, 3896; (b) Qi, T. T.; Reed, E. J. Simulations of Shocked Methane Including

Self-Consistent Semiclassical Quantum Nuclear Effects. *J. Phys. Chem. A* **2012**, *116*(42), 10451–10459.

17. (a) Ge, N. N.; Wei, Y. K.; Ji, G. F.; Chen, X. R.; Zhao, F.; Wei, D. Q. Initial Decomposition of the Condensed-Phase Beta-HMX Under Shock Waves: Molecular Dynamics Simulations. *J. Phys. Chem. B* **2012**, *116*(46), 13696–13704; (b) Manaa, M. R.; Fried, L. E.; Melius, C. F.; Elstner, M.; Frauenheim, T. Decomposition of HMX at Extreme Conditions: A Molecular Dynamics Simulations. *J. Phys. Chem. A* **2002**, *106*, 9024.
18. Reed, E. J.; Manaa, M. R.; Fried, L. E.; Glaesemann, K. R.; Joannopoulos, J. D. A Transient Semimetallic Layer in Detonating Nitromethane. *Nat. Phys.* **2008**, *4*(1), 72–76.
19. Manaa, M. R.; Reed, E. J.; Fried, L. E.; Goldman, N. Nitrogen-Rich Heterocycles as Reactivity Retardants in Shocked Insensitive Explosives. *J. Am. Chem. Soc.* **2009**, *131*(15), 5483–5487.
20. Reed, E. J.; Rodriguez, A. W.; Manaa, M. R.; Fried, L. E.; Tarver, C. M. Ultrafast Detonation of Hydrazoic Acid (HN3). *Phys. Rev. Lett.* **2012**, *109*(3), 038301.
21. Hu, H.; Lu, Z.; Elstner, M.; Hermans, J.; Yang, W. Simulating Water with the Self-Consistent-Charge Density Functional Tight Binding Method: From Molecular Clusters to the Liquid State. *J. Phys. Chem. A* **2007**, *111*, 5685–5691.
22. Margetis, D.; Kaxiras, E.; Elstner, M.; Frauenheim, T.; Manaa, M. R. Electronic Structure of Solid Nitromethane: Effects of High Pressure and Molecular Vacancies. *J. Chem. Phys.* **2002**, *117*, 788.
23. Perdew, J. P.; Burke, K.; Ernzerhof, M. Generalized Gradient Approximation Made Simple. *Phys. Rev. Lett.* **1997**, *78*, 1396.
24. Yang, Y.; Yu, H.; York, D.; Cui, Q.; Elstner, M. Extension of the Self-Consistent-Charge Density-Functional Tight-Binding Method: Third-Order Expansion of the Density Functional Theory Total Energy and Introduction of a Modified Effective Coulomb Interaction. *J. Phys. Chem. A* **2007**, *111*, 10861–10873.
25. (a) Goldman, N.; Fried, L. E. Extending the Density Functional Tight Binding Method to Carbon Under Extreme Conditions. *J. Phys. Chem. C* **2011**, *116*, 2198–2204; (b) Goldman, N.; Goverapet Srinivasan, S.; Hamel, S.; Fried, L. E.; Gaus, M.; Elstner, M. Determination of a Density Functional Tight Binding Model with an Extended Basis Set and Three-Body Repulsion for Carbon Under Extreme Pressures and Temperatures. *J. Phys. Chem. C* **2013**, *117*, 7885–7894.
26. Holian, B. L. Modeling Shock-Wave Deformation via Molecular-Dynamics. *Phys. Rev. A* **1988**, *37*, 2562.
27. Reed, E. J.; Fried, L. E.; Joannopoulos, J. D. A Method for Tractable Dynamical Studies of Single Shock Compression. *Phys. Rev. Lett.* **2003**, *90*, 2355031.
28. Fickett, W.; Davis, W. C. *Detonation*. University of California Press: Berkeley, 1979.
29. Budzevich, M. M.; Zhakovsky, V. V.; White, C. T.; Oleynik, I. I. Evolution of Shock-Induced Orientation-Dependent Metastable States in Crystalline Aluminum. *Phys. Rev. Lett.* **2012**, *109*(12), 125505.
30. Maillet, J. B.; Mareschal, M.; Soulard, L.; Ravelo, R.; Lomdahl, P. S.; Germann, T. C.; Holian, B. L. Uniaxial Hugoniotstat: A Method for Atomistic Simulations of Shocked Materials. *Phys. Rev. E.* **2001**, *63*, 016121.
31. Reed, E. J.; Fried, L. E.; Henshaw, W. D.; Tarver, C. M. Analysis of Simulation Technique for Steady Shock Waves in Materials with Analytical Equations of State. *Phys. Rev. E.* **2006**, *74*(5), 056706.
32. Reed, E. J.; Maiti, A.; Fried, L. E.; Reed, E. J.; Maiti, A.; Fried, L. E. Anomalous Sound Propagation and Slow Kinetics in Dynamically Compressed Amorphous Carbon. *Phys. Rev. E.* **2010**, *81*(1), 016607.

33. Reed, E. J. Electron-Ion Coupling in Shocked Energetic Materials. *J. Phys. Chem. C* **2012**, *116*(3), 2205–2211.
34. Goldman, N.; Reed, E. J.; Fried, L. E. Quantum Mechanical Corrections to Simulated Shock Hugoniot Temperatures. *J. Chem. Phys.* **2009**, *131*(20), 204103.
35. Maillet, J. B.; Bourasseau, E. Ab Initio Simulations of Thermodynamic and Chemical Properties of Detonation Product Mixtures. *J. Chem. Phys.* **2009**, *131*(8), 084107.
36. Goldman, N.; Fried, L. E.; Kuo, I. F. W.; Mundy, C. J. Bonding in the Superionic Phase of Water. *Phys. Rev. Lett.* **2005**, *94*, 217801.
37. Wu, C. J.; Fried, L. E.; Yang, L. H.; Goldman, N.; Bastea, S. Catalytic Behavior of Dense Hot Water. *Nat. Chem.* **2009**, *1*, 57–62.
38. (a) Behrens, R. Identification of Octahydro-1,3,5,7-Tetranitro-1,3,5,7-Tetrazocine (HMX) Pyrolysis Products by Simulataneous Thermogravimetric Modulated Beam Mass-Spectrometry and Time-of-Flight Velocity-Spectra Measurements. *Int. J. Chem. Kinet.* **1990**, *22*, 135; (b) Behrens, R. Thermal-Decomposition of Energetic Materials—Temporal Behaviors of the Rate of Formation of the Gaseous Pyrolysis Products from Condensed-Phase Decomposition of Octahydro-1,3,5,7-Tetranitro-1,3,5,7-Tetrazocine. *J. Phys. Chem.* **1990**, *94*, 6706; (c) Behrens, R.; Bulusu, S. Thermal-Decompsition of Energetic Materials. 2. Deuterium-Isotope Effects and Isotopic Scrambling in Condensed-Phase Decomposition of Octahydro-1,3,5,7-Tetranitro-1,3,5,7-Tetrazocine. *J. Phys. Chem.* **1991**, *95*, 5838; (d) Brill, T. B. Multiphase Chemistry Considerations at the Surface of Burning Nitramine Monopropellants. *J. Prop. Power* **1995**, *11*, 740; (e) Brill, T. B.; Gongwer, P. E.; Williams, G. K. Thermal-Decomposition of Energetic Materials. 66. Kinetic Compensation Effects in HMX, RDX, and NTO. *J. Phys. Chem.* **1994**, *98*, 12242; (f) Bulusu, S.; Axenrod, T.; Milne, G. W. A. Electron-Impact Fragmentation of Some Secondary Aliphatic Nitramines. Migration of Nitro Group in Heterocyclic Nitramines. *Org. Mass Spectrom.* **1970**, *3*, 13; (g) Fifer, R. A. Fundamentals of Solid Propellant Combustion. In *Progress in Astronautics and Aeronautics*; Kuo, K. K.; Summerfield, M. Eds.; Vol. 90, . AIAA Inc.: New York, 1984; p 177; (h) Morgan, C. V.; Bayer, R. A. Electron-Spin-Resonance Studies of HMX Pyrolysis Products. *Combust. Flame* **1979**, *36*, 99; (i) Oxley, J. C.; Kooh, A. B.; Szekers, R.; Zhang, W. Mechanisms of Nitramine Thermolysis. *J. Phys. Chem.* **1994**, *98*, 7004; (j) Suryanarayana, B.; Graybush, R. J.; Autera, J. R. Thermal Degradation of Secondary Nitramines: A Nitrogen-15 Tracer Study of HMX (1,3,5,7-Tetranitro-1,3,5,7-Tetrazacyclooctane). *Chem. Ind.* **1967**, *52*, 2177; (k) Tang, C.-J.; Lee, Y. J.; Kudva, G.; Litzinger, T. A. A Study of the Gas-Phase Chemical Structure During CO2 Laser Assisted Combustion of HMX. *Combust. Flame* **1999**, *117*, 170; (l) Tang, C.-J.; Lee, Y. J.; Litzinger, T. A. Simultaneous Temperature and Species Measurements During Self-Oscillating Burning of HMX. *J. Prop. Power* **1999**, *15*, 296
39. Farber, M.; Srivastava, R. D. HMX Decomposition Mechanisms. In 16th JANNAF Combus. Meeting, CPIA Pub., 1979; p 59.
40. Glascoe, E. A.; Zaug, J. M.; Burnham, A. K. Pressure-Dependent Decomposition Kinetics of the Energetic Material HMX up to 3.6 GPa. *J. Phys. Chem. A* **2009**, *113*(48), 13548–13555.
41. Politzer, P.; Boyd, S. Molecular Dynamics Simulations of Energetic Solids. *Struct. Chem.* **2002**, *13*, 105.
42. Melius, C. F. Decomposition Mechanisms of HMX. In *Chemistry and Physics of Energetic Materials*; Bulusu, D. N. Ed.; Kluwer: Dordrecht, 1990.
43. Lewis, J. P.; Glaesemann, K. R.; Van Opdorp, K.; Voth, G. A. Ab Initio Calculations of Reactive Pathways for Alpha-Octahydro-1,3,5,7-Tetranitro-1,3,5,7-Tetrazocine (Alpha-HMX). *J. Phys. Chem. A* **2000**, *104*, 11384.

44. Chakraborty, D.; Muller, R. P.; Dasgupta, S.; Goddard, W. A., III Mechanism for Unimolecular Decomposition of HMX (1,3,5,7-Tetranitro-1,3,5,7-Tetrazocine) an Ab Initio Study. *J. Phys. Chem. A* **2001**, *105*, 1302.
45. Zhang, L. Z.; Zybin, S. V.; van Duin, A. C. T.; Dasgupta, S.; Goddard, W. A.; Kober, E. M. Carbon Cluster Formation During Thermal Decomposition of Octahydro-1,3,5,7-Tetranitro-1,3,5,7-Tetrazocine and 1,3,5-Triamino-2,4,6-Trinitrobenzene High Explosives from ReaxFF Reactive Molecular Dynamics Simulations. *J. Phys. Chem. A* **2009**, *113*(40), 10619–10640.
46. Landers, A. G.; Brill, T. B. Pressure-Temperature Dependence of the Beta-Delta-Polymorph Interconversion in Octahydro-1,3,5,7-Tetranitro-1,3,5,7-Tetrazocine. *J. Phys. Chem.* **1980**, *84*, 3573.
47. (a) Fried, L. E.; Howard, W. M. An Accurate Equation of State for the Exponential-6 Fluid Applied to Supercritical Dense Nitrogen. *J. Chem. Phys.* **1998**, *109*, 7338; (b) Fried, L. E.; Howard, W. M. Explicit Gibbs Free Energy Equation of State Applied to the Carbon Phase Diagram. *Phys. Rev. B* **2000**, *61*, 8734.
48. Kuklja, M. M. Thermal Decomposition of Solid Cyclotrimethylene Trinitramine. *J. Phys. Chem. B* **2001**, *105*, 10159.
49. (a) Greiner, N. R.; Phillips, D. S.; Johnson, J. D.; Volk, F. Diamonds in Detonation Soot. *Nature* **1988**, *333*(6172), 440–442; (b) Shaw, M. S.; Johnson, J. D. Carbon Clustering in Detonations. *J. Appl. Phys.* **1987**, *62*(5), 2080–2085.
50. Viecelli, J. A.; Ree, F. H. Carbon Clustering Kinetics in Detonation Wave Propagation. *J. Appl. Phys.* **1999**, *86*(1), 237–248.
51. Cady, H. H.; Larson, A. C. Crystal Structure of 1,3,5-Triamino-2,4,6-Trinitrobenzene. *Acta Crystallogr.* **1965**, *18*, 485.
52. (a) Davidson, A. J.; Dias, R. P.; Dattelbaum, D. M.; Yoo, C. S. "Stubborn" Triaminotrinitrobenzene: Unusually High Chemical Stability of a Molecular Solid to 150 GPa. *J. Chem. Phys.* **2011**, *135*(17), 174507; (b) Manaa, M. R.; Fried, L. E. Nearly Equivalent Inter- and Intramolecular Hydrogen Bonding in 1,3,5-Triamino-2,4,6-Trinitrobenzene at High Pressure. *J. Phys. Chem. C* **2012**, *116*(3), 2116–2122.
53. Reed, E. J.; Fried, L. E.; Manaa, M. R.; Joannopoulos, J. D. A Multi-scale Approach to Molecular Dynamics Simulations of Shock Waves. In *Chemistry at Extreme Conditions*; Manaa, M. R. Ed.; Elsevier B.V: Amsterdam, 2005; pp 297–326.
54. Mott, N. F. *Metal-Insulator Transitions*. Taylor & Francis: Bristol, 1990.
55. Sharma, J.; Forbes, J. W.; Coffey, C. S.; Liddiard, T. P. The Physical and Chemical Nature of Sensitization Centers Left from Hot Spots Caused in Triaminotrinitrobenzene by Shock or Impact. *J. Phys. Chem.* **1987**, *91*, 5139–5144.
56. Fried, L. E.; Najjar, F.; Howard, W. M.; Manaa, M. R.; Reed, E. J.; Goldman, N.; Bastea, S.; Nichols, A. L. I. In *Multiscale Simulation of Hot Spot Ignition*, 14th International Detonation Symposium, Couer d'Alene, Idaho, 2011; Pereis, S.; Boswell, C.; Asay, B. Eds.; Office of Naval Research: Couer d'Alene, Idaho, 2010; pp 1412–1420.
57. Bastea, S.; Fried, L. E. Chemical Equilibrium Detonation. In *Shock Wave Science and Technology Reference Library*; Zhang, F. Ed.; Vol. 6, Springer-Verlag: Berlin Heidelberg, 2012; pp 1–27.
58. Bourasseau, E.; Maillet, J. B. Coupling Microscopic and Mesoscopic Scales to Simulate Chemical Equilibrium Between a Nanometric Carbon Cluster and Detonation Products Fluid. *Phys. Chem. Chem. Phys.* **2011**, *13*(15), 7019–7039.
59. Goldman, N.; Reed, E. J.; Fried, L. E.; Kuo, I. F. W.; Maiti, A. Synthesis of Glycine-Containing Complexes in Impacts of Comets on Early Earth. *Nat. Chem.* **2010**, *2*, 949–954.
60. Dang, N. C.; Bolme, C. A.; Moore, D. S.; McGrane, S. D. Shock Induced Chemistry in Liquids Studied with Ultrafast Dynamic Ellipsometry and Visible Transient Absorption Spectroscopy. *J. Phys. Chem. A* **2012**, *116*(42), 10301–10309.

CHAPTER SEVEN

Ab Initio Chemical Kinetics of Key Processes in the Hypergolic Ignition of Hydrazine and Nitrogen Tetroxide

Putikam Raghunath[*], **N.T. Nghia**[*,†], **Ming-Chang Lin**[*,1]

[*]Center for Interdisciplinary Molecular Science, Department of Applied Chemistry, National Chiao Tung University, Hsinchu, Taiwan

[†]Department of Physical Chemistry, Hanoi University of Science and Technology, Hanoi, Vietnam

[1]Corresponding author: e-mail address: chemmcl@emory.edu

Contents

Abstract

Kinetics and mechanisms of key processes leading to the hypergolic ignition of N_2H_4 and N_2O_4 have been studied by high-level *ab initio* molecular orbital theory in conjunction with statistical theory calculations. The results of this work suggest that the explosion following the contact of N_2H_4 and N_2O_4 in liquid or gaseous state at low temperatures can be initiated by the rapid reaction of N_2H_4 with $ONONO_2$ isomers of N_2O_4 producing $H_2NN(H)NO + HNO_3$ without thermal activation accompanying with 21 kcal mol^{-1} exothermicity. The HNO_3 thus formed can instantaneously produce $[N_2H_5^+][NO_3^-]$ salt releasing twice as much acid–base neutralization energy to

Advances in Quantum Chemistry, Volume 69
ISSN 0065-3276
http://dx.doi.org/10.1016/B978-0-12-800345-9.00007-6

enhance NO_2 and N_2H_3 radical production. The thermochemistry and rate constants for reactions of species involved (NO_2, NO_3, N_2O_4, and its isomers, OH and N_2H_x ($x=2, 3$, and 4)) in the chain initiation and propagation have been predicted and compared with available data in the literature.

1. INTRODUCTION

There has been much interest in the mechanism for the autoignition of hydrazine (N_2H_4) and nitrogen tetroxide (N_2O_4, NTO) upon mixing since the 1960s when the phenomenon of the hypergolic reaction was investigated intensively by the US Defense industry and the NASA.[1–6] Since then, there have been many attempts to study the mechanism of the ignition process using a variety of techniques. Two types of studies are worth mentioning. In 1962, Friedman et al.[1] investigated the explosion induced by the contact of a falling droplet of N_2H_4 on a pool of liquid N_2O_4 using a high-speed camera. A similar study by Kimura and coworkers[7] compared the explosion data obtained by falling droplets of N_2O_4 and N_2H_4 in contact with the surfaces of liquid N_2H_4 and N_2O_4, respectively. Although the direct comparison of the two versions of the falling droplet experiments showed a distinct difference in the splashing of the liquids upon contact, the probability of occurrences and the strength of the explosion appeared to be similar between the two experiments. This suggests that the initiation reaction involved in the explosion observed in the two different experimental configurations is essentially the same.

Another kind of studies centered on the kinetics of the reaction of N_2H_4 with N_2O_4 in the gas phase, in which the latter reagent is essentially in the form of the NO_2 monomer because of the low dissociation energy of N_2O_4. Sawyer and Glassman[8] studied the reactions of N_2H_4 with NO_2, NO, and O_2 in the temperature range 800–1000 K in a quartz adiabatic flow reactor at atmospheric pressure. For the $N_2H_4+NO_2$ reaction, they reported the second-order global reaction kinetics, $d[N_2H_4]/dt=-k[N_2H_4][NO_2]$ with the rate constant $k=10^{15.83}\exp[-26700/RT]\,cm^3\,mol^{-1}\,s^{-1}$ (where $R=1.987\,cal\,mol^{-1}\,K^{-1}$) based on the rate of temperature change measured. The reaction was believed to be initiated by the bimolecular metathetical process $N_2H_4+NO_2\rightarrow N_2H_3+HONO$. The N_2H_3 radical thus formed could then set off the chain reactions producing H_2O and N_2 as final products, releasing a large amount of energy. Miyajima and Sakamoto[9] carried out a series of studies attempting to correlate the ignition-delay time with the temperature of the system upon mixing of N_2H_4 and NO_2; they reported the existence of the linear correlation between $\ln(\tau P^3/T^4)$ with $1/T$ according to the thermal theory of explosion (where τ is the ignition delay time, P and T are pressure and temperature of the system, respectively),[10] but no kinetic data were extracted from the study.

More recently, Koshi and coworkers[11] investigated the mechanism and the dependence of the ignition-delay time on temperature aided by extensive quantum chemical and statistical reaction rate theory calculations for some selected elementary reactions which were believed to play a key role in the initiation and chain propagation processes. This work is so far a more extensive and useful attempt to elucidate the complex chemistry involved in the combustion of N_2H_4 and NTO system in the gas phase. The reaction responsible for the autoignition of the NO_2-rich N_2O_4–N_2H_4 system was attributed to the $NO_2 + N_2H_4$ bimolecular reaction as concluded by Sawyer and Glassman in 1967.[8]

To date, the mechanism responsible for the ignition process, particularly at low temperatures involving the reactions in the liquid phase, remains uncertain due in part to the complexity of the system and in part to the lack of experiments with better product diagnostics at the molecular level. A cross molecular-beam study for the direct interaction of N_2H_4 and N_2O_4 using a universal, angularly resolved product analysis with a mass spectrometer, for example, should provide detailed reaction dynamics data for mechanistic elucidation. In view of the absence of such a refined reaction dynamics study under secondary collision free conditions, we have recourse to the current refined state of *ab initio* molecular orbital theory calculations to search for potential exothermic reaction paths involving N_2H_4 and N_2O_4, including its known isomers, *cis*- and *trans*-$ONONO_2$, with high molecular dipoles which may be responsible for the hypergolic molecular interaction in the condensed phase, giving rise to the explosion in the falling droplet experiments cited above. Our extensive search has led to the discovery of the highly exothermic reaction of N_2H_4 with *cis*- and *trans*-$ONONO_2$, which can occur without an intrinsic activation barrier producing $H_2NN(H)NO + HNO_3$.[12] The energy released in the exothermic initiation reaction as well as the acid–base neutralization reaction of N_2H_4 and HNO_3 formed in the initiation reaction producing $[N_2H_5^+][NO_3^-]$ should enhance greatly the dissociation of N_2O_4 as well as the excited $H_2NN(H)NO$ product formed in the gas phase and at the droplet-liquid pool interface generating more reactive radical species such as NO_2 and N_2H_3 which play a major role in the chain propagation process.

2. COMPUTATIONAL METHODS

2.1. *Ab initio* MO calculations

The development and successful applications of *ab initio* molecular orbital methods for quantitative prediction of potential energy surfaces (PESs) of complex molecular systems, together with the recent refinement of statistical reaction rate theories to be described below, have allowed us to directly compute rate constants for processes with multireaction channels reaching chemical accuracies (± 2 kcal mol^{-1} for heats of formation and $\pm 20\%$ for rate constant prediction). These computational techniques have been tested in numerous cases by comparing the predicted results with available data for known reactions. The Gaussian-X (GX) (X = 1, 2, 3) series of methods and the most recent new family of the G3 method, referred to as G3X, such as G3X(MP3) and G3X(MP2) developed by Pople coworkers[13] and Curtiss et al.[14–16] and have been widely employed to predict PESs of practically important reactions. To alleviate the problem from CPU limitation, in 1995, Mebel et al.[17] proposed several modified GAUSSIAN-2 (G2M) schemes for different molecular sizes based on geometry optimization and vibrational frequency calculations with the hybrid density functional B3LYP/6-311G(d,p) approach (i.e., Becke's three-parameter nonlocal exchange functional[18,19] with the nonlocal correlation functional of Lee et al.[20]); higher level electron correlations were accomplished with stepwise expansions of basis sets using the MP2 and MP4 methods, and the final single point energy calculation with the coupled cluster technique to approximate the RCCSD(T)/6-311+G(3df,2p) level of theory. For the calculation of the mid-size N_2H_4–N_2O_4 system, we employed the G2M(CC3) method,[17] which approximates RCCSD(T)/6-311+G(3df,2p) using a series of single-point calculations by MP2 and MP4 with the B3LYP-optimized geometries. The total energy in G2M(CC3) with the ZPE correction is calculated as follows:

$$E[\mathrm{G2M(CC3)}] = E_{\mathrm{bas}} + \Delta E(+) + \Delta E(2df) + \Delta E(\mathrm{cc}) + \Delta' + \Delta E(\mathrm{HLC}) + \mathrm{ZPE}.$$

where

$$E_{\mathrm{bas}} = E[\mathrm{PMP4/6\text{-}311G(d,p)}],$$

$$\Delta E(+) = E[\mathrm{PMP4/6\text{-}311+G(d,p)}] - E[\mathrm{PMP4/6\text{-}311G(d,p)}],$$

$$\begin{aligned}
\Delta E(+) &= E[\text{PMP4/6-311+G(d, p)}] - E[\text{PMP4/6-311G(d, p)}],\\
\Delta E(2df) &= E[\text{PMP2/6-311G(2df,p)}] - E[\text{PMP2/6-311G(d, p)}],\\
\Delta E(\text{cc}) &= E[\text{CCSD}(T)\text{/6-311G(d, p)}] - E[\text{PMP4/6-311G(d, p)}],\\
\Delta' &= E[\text{MP2/6-311+G(3df, 2p)}] - E[\text{MP2/6-311G(2df, p)}]\\
&\quad - E[\text{MP2/6-311+G(d, p)}] + E[\text{MP2/6-311G(d, p)}].
\end{aligned}$$

The empirical "higher level correction" is given by

$$\Delta E(\text{HLC}) = 0.001\left(-5.63n_\beta - 0.19n_\alpha\right) \text{ for open shell}$$

and

$$\Delta E(\text{HLC}) = 0.001\left(-5.45n_\beta - 0.19n_\alpha\right) \text{ for closed shell,}$$

where n_α and n_β are the numbers of α and β valence electrons, respectively. The G2M method has been successfully employed for interpretation of combustion-related processes with >10 heavy atoms.[21–25] Similar successful applications have been made for Cl_xO_y involving reactions of interest to ammonium perchlorate propulsion processes at high temperatures[26–29] as well as the stratospheric ozone destruction chemistry at low temperatures.[30–37] In the Cl-atom containing cases, the geometry optimization was made with a larger basis set using the B3LYP/6-311+G(3df,2p) method; similar optimizations using the smaller 6-311G(d,p) basis set as proposed in the original G2M schemes failed to predict heats of formation of Cl_xO_y species reliably.[30]

To improve energy prediction, Dunning and coworkers have developed the extrapolation method to the complete basis set (CBS) limit.[38,39] The energies of all the species of some reactions presented in this chapter were calculated by the CCSD(T) method with correlation-consistent polarized split-valence multiple-z basis sets, cc-pVXZ,[38] extrapolated to the CBS limit.[39] The CBS energies were evaluated with these geometries as follows. The total energies $E(X)$ computed with the cc-pVXZ basis sets were extrapolated to the CBS limit, E_{CBS}, employing a mixed Gaussian/exponential form:

$$E(X) = E_{\text{CBS}} + b\exp[-(X-1)] + c\exp\left[-(X-1)^2\right]$$

where X is the cardinal number associated with each basis set, $X=2$, 3, 4 for D, T, and Q. This method has been utilized for refinement of energies of several smaller reaction systems to be discussed below.

In the present calculations, the equilibrium geometries, including reactants, intermediates (IMs or LMs), transition states (TSs), and products, were optimized by using B3LYP/6-311++G(3df,2p) with polarized and diffuse functions. Previous studies show that the widely used B3LYP method with split valence and Dunning correlation basis sets is quite suitable for geometry and property predictions.[40,41] The vibrational frequencies were determined at the same levels of theories for ZPE corrections and for characterizing the stationary points. The intrinsic reaction coordinate calculations[42] were carried out to confirm that transition states connect the designed reactants and products. For more accurate evaluation of energies, higher-level single-point energy calculations of all the species and transition states have been carried out at the CCSD(T)/6-311++G(3df,2p) level[43] based on the optimized geometries. All of the calculations have been carried out with the Gaussian 03 program.[44] Details of the methods we used in various reactions will be briefly referred to in the following sections.

2.2. Rate constant calculations

Rate coefficients for various reaction channels were calculated with a microcanonical transition state theory (TST) and microcanonical variational RRKM (Rice–Ramsperger–Kassel–Marcus) theory[45] using the Variflex Program[46] which solves the 1D master equation involving multistep vibrational energy transfers for reactions occurring via long-lived intermediates. For reactions with multiple intermediates and product channels, we have employed the CHEMRATE code[47] to couple all low-lying reaction channels including isomerization processes by solving the T,P-dependent master equation. Rate constants for the low-energy channels and products were calculated by using the PESs of all the reactions computed at the G2M(CC3) and/or CCSD(T)//B3LYP/6-311++G(3df,2p) levels.

For the T,P-dependent rate constant calculation using Variflex, the specific rate constant $k_{E,J}$ of a unimolecular reaction is predicted with the RRKM theory as a function of the energy E of the molecule and of the total angular momentum J,

$$k_{(E,J)} = \frac{N^{\dagger}_{(E,J)}}{h\rho_{(E,J)}}$$

where $\rho_{E,J}$ is the ro-vibrational density of states for the reactants, $N^{\dagger}_{(E,J)}$ is the transition states which is given by the variationally determined minimum in $N_{E,J}(R)$, as a function of the reaction coordinate R and h is Planck's constant. The method was developed by Wardlaw and Marcus[48] and refined by Klippenstein.[49] The basis of the approach involves a separation of the vibrational modes into conserved and transitional modes. With this separation, one can convolute the sum of vibrational quantum states for the conserved modes with the classical phase space density of states for the transitional modes by Monte Carlo integration.[50,51] The pressure dependence was evaluated with 1D master equation calculations using the Boltzmann probability of the reaction intermediate for the J distribution. The master equation was solved by an inversion-based approach.[48,50,51] The total angular momentum J covered the range from 1 to 241 in steps of 10 for the E,J-resolved calculation. A detailed review on VTST has been made by Truhlar et al.[52]

For a barrierless dissociation or its reverse association process, the fitted Morse function, $V(R) = D_e\{1 - \exp[-\beta(R - R_e)]\}^2$, was used to represent the minimum potential energy path (MEP) obtained by full geometric optimization along the dissociation coordinate. Here, D_e is the binding energy excluding zero-point vibrational energy for an association reaction, R is the reaction coordinate (i.e., the distance between the two bonding atoms), and R_e is the equilibrium value of R at the stable intermediate structure. To evaluate the variational potential energy curves, the reaction pathways have been mapped out along the MEP with the tight convergence criterion were performed at the B3LYP/6-311++G(3df,2p) and/or CASPT2(8,8)/6-311+G(3df)//CASSCF(8,8)/6-311+G(3df) level, which will be discussed later. The CASSCF calculations are performed with the MOLPRO code.[53] For the tight transition states, the numbers of states were evaluated according to the rigid-rotor harmonic-oscillator approximation. In addition, in some cases, the CVTST approach[54] has been utilized to locate transition states for the key steps which occur without well-defined transition states and decomposition processes were evaluated canonically for each temperature and critical separation, $r^{\#}(T)$, based on the maximum Gibbs free energy criterion as described previously for radical–radical reactions.[55,56]

For our one-well system assumed in the present calculation, the master equation takes the form

$$\frac{d\rho_i(t)}{dt} = \phi_i + \omega \sum_{j=1}^{m} P_{ij}\rho_j(t) - \omega\rho_i(t) - (k_{i1} + k_{i2})\rho_i(t),$$

where ϕ_i represents the capturing rate of the incoming two fragments, m is the number of grains chosen such that the population of the mth grain contributes negligibly to the bimolecular rate constant, ω is the collision frequency, $k_{i1}(E)$ and $k_{i2}(E)$ are the microcanonical specific rate constants for decomposition and redissociation, and P_{ij} the probability of energy transfer from grain j to grain i upon collision. A simple exponential down model[57] was employed for P_{ij}

$$P_{ij} = A_j \exp\left[-\alpha\left(E_j - E_i\right)\right], \quad j \geq i,$$

where α is a parameter related to energy transfer efficiency; α^{-1} corresponds to the average energy transferred per collision for the down collision, $\langle \Delta E \rangle_{\text{down}}$. A_j are normalization constants obtained from the normalizing condition:

$$\sum_i p_{ij} = 1.$$

The master equation was solved approximately with the weak collision model.[57] The values of $\langle \Delta E \rangle_{\text{down}}$ used depend on the third bodies involved in the collisional deactivation or activation process; they will be specified for the individual reaction studied below.

3. RESULTS AND DISCUSSION

3.1. Unimolecular decomposition of N_2H_4 and N_2H_3 radical

The thermal unimolecular decomposition of N_2H_4 cannot occur at low temperatures, and the process is therefore unimportant to the initial hypergolic ignition of N_2H_4–N_2O_4 mixtures. It is, however, expected to be very relevant to the chain propagation in the explosion regime at high temperatures producing NH_2 and N_2H_3 by breaking its N—N and N—H bonds. The N_2H_3 radical can also be generated by radical attacks on N_2H_4 in the chain reactions. We have accordingly computed the PESs for the decomposition of N_2H_4 and N_2H_3 at the CCSD(T)/CBS//B3LYP/6-311++G (3df,2p) level of theory as presented in Fig. 7.1.

3.1.1 N_2H_4 decomposition

The kinetics and mechanisms for the thermal decomposition and the pure flame of N_2H_4 have been modeled in detail by Konnov and Ruyck who provided much literature data for the N_2H_4 system.[61] The result of their simulation and others' kinetic data for the unimolecular decomposition reaction will be compared with our predicted values.

The PES shown in Fig. 7.1A indicates that N_2H_4 can decompose by several pathways as described below. Breaking N—N and N—H bonds to produce two NH_2 radicals and $N_2H_3 + H$, respectively, occur without intrinsic

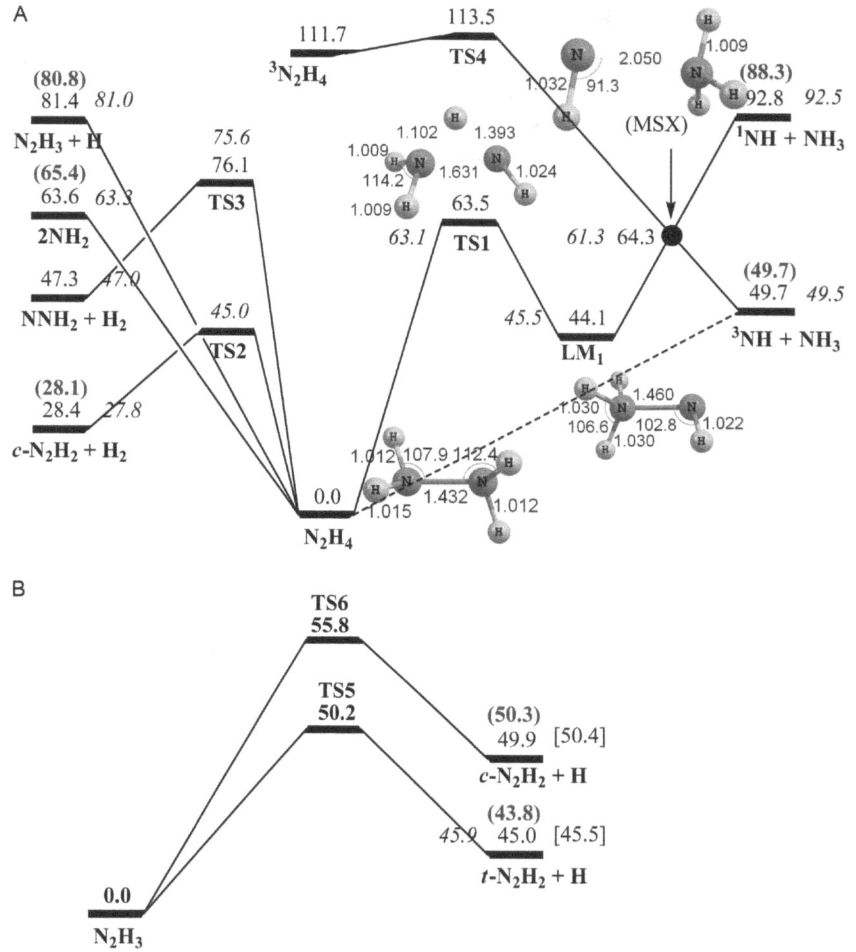

Figure 7.1 (A) Potential energy profile for the N_2H_4 decomposition based on the CCSD(T)/CBS//B3LYP/6-311++G(3df,2p) calculation. Experimental values are in blue from Refs. 58,59 and CCSD(T)/CBS+ZPE (CCSD(T)/aug-pVTZ) energies are in italics from Ref. 60. The bond length in italics values are from Refs. 58,59. (B) N_2H_3 decomposition based on the CCSD(T)/CBS//MP2/6-311++G(3df,2p) calculation. Experimental values are in blue from Refs. 58,59. The values in the brackets are calculated at CCSD(T)/CBS// B3LYP/6-311++G(3df,2p). The italics values are from Refs. 58,59. (For interpretation of the references to color in this figure legend, the reader is referred to the online version of this chapter.)

transition states as confirmed by variational transition state search along their minimum energy paths. The energies for production of NH_2+NH_2 and N_2H_3+H are 63.6 and 81.4 kcal mol^{-1}, respectively, which are in excellent agreement with the values 64.2 and 82.0 kcal mol^{-1}, as well as with 65.4 and 80.8 kcal mol^{-1}, respectively, reported in the literatures.[58–60]

N_2H_4 can also isomerize by migration of an H atom to the neighboring N atom via three-center transition state TS1 (63.3 kcal mol^{-1}) to $HNNH_3$ (43.5 kcal mol^{-1}), which can then dissociate to produce $NH_3 + {}^1NH$ (92.8 kcal mol^{-1}). TS1 has a unique imaginary frequency of 1497*i* with the bond lengths N—N, H—N^1, and H—N^2 being 1.631, 1.102, and 1.393 Å, respectively, corresponding to a structure with the H atom locating between the two N atoms above the N—N bond as indicated in Fig. 7.1A. In addition, N_2H_4 can also decompose by H_2 elimination giving *trans*-N_2H_2 and H_2NN via TS2 and TS3, respectively. The geometries and the related energies of species are noted to be in good agreement with those reported by a previous study.[60]

The rate constant for the unimolecular decomposition reaction was computed by the Variflex code.[46] Because of the absence of well-defined transition states for the fragmentation reactions producing $2NH_2$ and $N_2H_3 + H$, the decomposition potential functions of N_2H_4 were computed variationally to cover the range of N—N separations from the equilibrium value 1.432 to 5.432 Å for the former and 1.012 to 5.012 Å for the latter with the interval step size of 0.1 Å. The dissociation curves were fitted to the Morse potential function with the parameters $\beta = 2.819$ and 2.773 $Å^{-1}$, respectively. The *L–J* parameters for N_2H_4 were calculated using $\varepsilon/K = 0.897^* T_c$ and $\sigma = 0.785^* V_c^{1/3}$, where $T_c = 653.15$ K and $V_c = 158.00\ cm^3\ mol^{-1}$[62,63]; the *L–J* parameters for N_2H_3 and Ar are taken from the literature.[62,63]

The results of the calculation by Variflex gave the rate constant for each decomposition channel as well as the total rate constants at both the high-pressure and low-pressure limits and at the atmospheric pressure of Ar for the temperature range 700–2000 K as shown in Fig. 7.2. The rate constant for $N_2H_4 \rightarrow 2NH_2$ at 760 Torr Ar can be represented by $k_1 = 2.21 \times 10^{50}\ T^{-11.2} \exp(-3678/T)\ s^{-1}$ which is in good agreement with experimental results of McHale et al.[65] studied by mass spectrometry in the temperature range 970–1550 K and Kerr et al.[68] by gas chromatography at 887–1030 K. The high-pressure limit k^{∞} is higher than the value reported by Genich et al.[69] at 100–310 atm. and the low-pressure limit k^0 is in agreement with that of Genich and Szwarc obtained at 6.66×10^{-3}–2.11×10^{-2} atm.[69,70] On account of the lower dissociation energy, the rate constant for NH_2 radical production is much higher than that for N_2H_3 formation at 760 Torr as well as at both pressure limits. The total rate constant at 760 Torr can be presented by $k_{tot,760} = 5.89 \times 10^{42}\ T^{-8.89} \exp(-34{,}737/T)\ s^{-1}$. The total rate constants at the high- and low-pressure limits can be presented by: $k_{tot}^{\infty} = 1.15 \times 10^{17}\ T^{0.15} \exp(-32{,}302/T)\ s^{-1}$ and

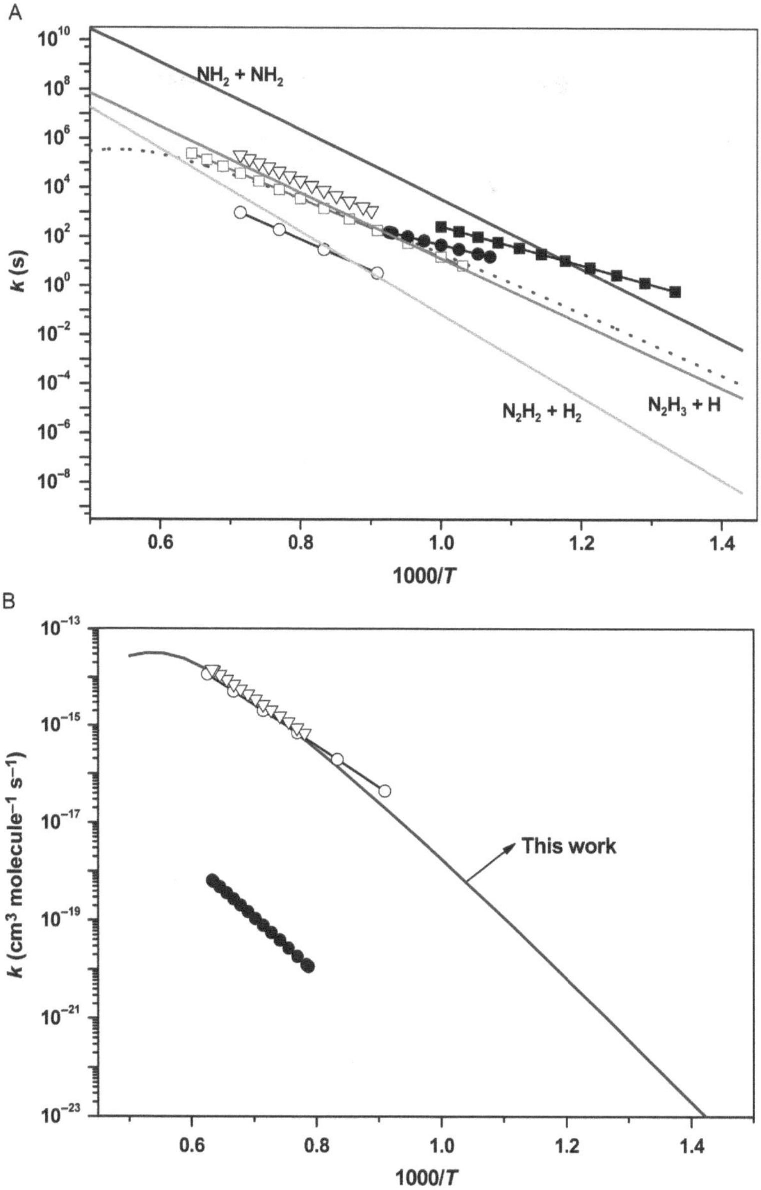

Figure 7.2 (A) The predicted high-pressure limit, Ar (k^{∞}) rate constants for unimolecular dissociation of N_2H_4 reaction products calculated at the CCSD(T)/CBS/B3LYP/6-311++G (3df,2p) level, for different product formation as labeled as NH_2+NH_2 (blue –), N_2H_3+H (red –), and $N_2H_2+H_2$ (green –). 3 atm. result for NH_2+NH_2 given in (blue . . . .). Experimental data (▽) are taken from Ref. 64 at 3.72×10^4 Torr pressure; (□) is taken from Ref. 65 at 1750–6540 Torr pressure; (▪) is taken from Ref. 66 at high-pressure limit; (o) data from Ref. 61; and (•) from Ref. 67. The other two channels are calculated at 760 Torr Ar pressure. (B) The predicted low-pressure limit, Ar rate constant (k^0) data for $N_2H_4 \rightarrow NH_2+NH_2$ channel with respect to temperature. Experimental data (▽) is taken from Ref. 64, (o) data from Ref. 61, and (•) from Ref. 67. (For interpretation of the references to color in this figure legend, the reader is referred to the online version of this chapter.)

$k^0_{tot} = 2.01 \times 10^{39}\ T^{-10.29}\ \exp(-33{,}556/T)\,cm^3\ molecule^{-1}\ s^{-1}$. The predicted 1 atm., low-, and high-pressure rate constants for the N_2H_4 dissociation process at 300–2000 K temperature range can be presented as

$$N_2H_4 \rightarrow NH_2 + NH_2 \tag{7.1}$$

$$k_1 = 2.21 \times 10^{50}\, T^{-11.2} \exp(-3678/T)\, s^{-1} \text{ at } 760\,\text{Torr}$$
$$k_1{}^{\infty} = 1.57 \times 10^{21}\, T^{-1.04} \exp(-33{,}497/T)\, s^{-1}$$
$$k_1{}^{0} = 3.25 \times 10^{28}\, T^{-10.2} \exp(-36{,}069/T)\, cm^3\, molecule^{-1}\, s^{-1}$$

$$N_2H_4 \rightarrow N_2H_3 + H \tag{7.2}$$

$$k_2 = 2.74 \times 10^{39}\, T^{-7.69} \exp(-42{,}131/T)\, s^{-1} \text{ at } 760\,\text{Torr}$$
$$k_2{}^{\infty} = 5.69 \times 10^{14}\, T^{-0.28} \exp(-38{,}586/T)\, s^{-1}$$
$$k_2{}^{0} = 3.24 \times 10^{23}\, T^{-8.5} \exp(-41{,}457/T)\, cm^3\, molecule^{-1}\, s^{-1}$$

$$N_2H_4 \rightarrow t\text{-}N_2H_2 + H_2 \tag{7.3}$$

$$k_3 = 5.19 \times 10^{38}\, T^{-7.84} \exp(-33{,}761/T)\, s^{-1} \text{ at } 760\,\text{Torr}$$
$$k_3{}^{\infty} = 1.63 \times 10^{10}\, T^{1.22} \exp(-29{,}367/T)\, s^{-1}$$
$$k_3{}^{0} = 4.62 \times 10^{28}\, T^{-10.4} \exp(-34{,}036/T)\, cm^3\, molecule^{-1}\, s^{-1}$$

3.1.2 N_2H_3 decomposition

Because of asymmetry of N_2H_3, the two N—H bonds in the NH_2 group can rupture via two transition state TS1 and TS2 located 50.2 and 55.8 kcal mol^{-1} to produce H + *trans*- and *cis*-N_2H_2, respectively. The calculated relative energies for the formation of *cis- and trans*-N_2H_2 isomers are 49.9 and 45.0 kcal mol^{-1}, respectively, which are consistent with the values 50.4 and 45.5 kcal mol^{-1} calculated at the CCSD(T)/CBS//B3LYP/6-311++G(3df,2p) and are also in good agreement with results from the literature, 50.3 and 45.9 kcal mol^{-1}, as well as the value 43.8 kcal mol^{-1} for the formation of the more stable product *trans*-N_2H_2 based on the heats of formation of N_2H_3, *cis*- and *trans*-N_2H_2, and H species.[58–60]

The rate constants for the two product channels and the total rate constant at 760 Torr Ar in the temperature range 300–2000 K have been predicted by using Variflex. As the H + *trans*-N_2H_2 channel is more stable and has a lower barrier, the production of H + *trans*-N_2H_2 is dominant at medium high temperatures at 760 Torr Ar and at the low- and high-pressure limits. At 760 Torr Ar pressure, the rate constants for H + *trans*-N_2H_2, H + *cis*-N_2H_2 production and the total rate constant can be presented by

$k_{trans} = 4.96 \times 10^{20}\ T^{-2.34}\ \exp(-49{,}631/T)\,s^{-1}$; $k_{cis} = 5.75 \times 10^{18}\ T^{-1.72}\ \exp(-54{,}618/T)\,s^{-1}$ and $k_{tot} = 1.66 \times 10^{20}\ T^{-2.18}\ \exp(-49{,}553/T)\,s^{-1}$, respectively. The total rate constants at the low- and high-pressure limits can be given by $k^{\infty} = 3.53 \times 10^{10}\ T^{0.86}\ \exp(-48{,}008/T)\,s^{-1}$; $k^{0} = 9.81 \times 10^{13}\ T^{-5.62}\ \exp(-50{,}340/T)\,cm^3\ molecule^{-1}\ s^{-1}$.

3.2. Unimolecular reactions of N_2O_4 and $ONONO_2$

There have been many experimental and computational studies on the unimolecular decomposition of N_2O_4 (D_{2h}) and its reverse NO_2 recombination reactions. Most of their kinetic data and references can be found at the NIST chemical kinetics database (http://kinetics.nist.gov/kinetics/index.jsp) except some of the recent results related to N_2O_4 and its reactions.[12,71] Our predicted heats of reaction forming $2NO_2$ and $ONONO_2$ isomers at the UCCSD(T)/CBS//UB3LYP/6-311++G(3df,2p) level of theory are in good agreement with available data.[71–73]

The $ONONO_2$ isomers existing at low temperatures expected to be indistinguishable above room temperature through internal rotation with a small energy barrier (4.2 kcal mol^{-1}, predicted at the UCCSD(T)/CBS level of theory). On account of the weak ONO—NO_2 and ON—ONO_2 bonds, the $ONONO_2$ molecule can fragment readily into $2NO_2$ and $NO+NO_3$ radical product pairs without intrinsic barriers. On the basis of the computed MEPs for these fragmentation processes, their variational transition states could be fitted to the Morse function with $\beta = 3.70$ and 2.42 Å^{-1}, respectively; these values were used for rate constant calculations.

The predicted rate constants for the unimolecular decomposition of $ONONO_2$ producing $2NO_2$ and $NO+NO_3$ over the temperature range of 300–2000 K at high- and low-pressure limits using Ar bath gas are given below:

$$ONONO_2 \rightarrow 2NO_2$$

$$k^{\infty} = 1.69 \times 10^{23}\,T^{-2.43}\exp(-4100/T)\,s^{-1}$$

$$k^{0} = 1.02 \times 10^{-3}\,T^{-0.63}\exp(-1974/T)\,cm^3\,molecule^{-1}\,s^{-1}$$

$$ONONO_2 \rightarrow NO + NO_3$$

$$k^{\infty} = 2.41 \times 10^{21}\,T^{-1.76}\exp(-15869/T)\,s^{-1}$$

$$k^{0} = 3.93 \times 10^{17}\,T^{-7.36}\exp(-15954/T)\,cm^3\,molecule^{-1}\,s^{-1}$$

3.3. Reactions of N_2H_4 with NO_2, NO_3, and N_2O_4 isomers

At ambient temperature and pressure, NO_2, N_2O_4, and its structural isomers (*cis-* and *trans*-$ONONO_2$) coexist. The relative concentration of NO_2 and N_2O_4 can be quantitatively calculated from their well-known equilibrium constant. The existence of the asymmetric $ONONO_2$ isomers has been known from various experimental studies[74–84]; their concentration has been estimated to be about 0.05% of N_2O_4 at room temperature, which was shown to be close to the predicted value of 0.058% by Lai et al.[12] The possible geometries of *cis-* and *trans*-$ONONO_2$ (C_s symmetry)[85–87] and that of *cis*-$ONONO_2$ with C_1 symmetry[88] have been studied by various theoretical groups. Based on these studies, *sym*-N_2O_4 and *trans*-$ONONO_2$ are most stable isomers. On account of their coexistence, in this work, the reactions of N_2H_4 with NO_2 and all isomers of N_2O_4 have been investigated for their potential contributions to the hypergolic initiation reaction. The kinetics and mechanisms of these reactions summarized below are those of Lai et al.[12]; their results will be compared with available kinetic data in the literature.[8,11]

3.3.1 Reaction of N_2H_4 with NO_2

The major low energy product channel of the $N_2H_4 + NO_2$ reaction potential energy profile, shown in Fig. 7.3, was calculated with the G2M(CC2)//B3LYP/6-311++G(3df,2p) method.[12,17] The reaction takes place via two pre-reaction van der Waals complexes with differing stabilities. These complexes have no influences on the kinetics of the reaction, which is controlled by $TS2_a$ with 7.6 kcal mol^{-1} barrier above the reactants for H-abstraction giving the product complex, $IM3_a$ ($H_2NN(H)...HONO$), by hydrogen bonding with 5.0 kcal mol^{-1} binding energy. The reaction producing $N_2H_3 +$ *cis*-HONO is predicted with a small endothermicity of 3.4 kcal mol^{-1}. The other higher energy pathways of $N_2H_4 + NO_2$ producing $N_2H_3 +$ *cis*-HONO and $N_2H_3 + HNO_2$ by H-abstraction were found to have 9.4 and 13.0 kcal mol^{-1} barriers, respectively.[12,17] It should be mentioned that the 7.6 kcal mol^{-1} G2M barrier at $TS2_a$ was found to be in close agreement with the 7.8 kcal mol^{-1} value directly computed at the CCSD(T)/6-311++G(3df,2p)//B3LYP/6-311++G(3df,2p) level. Both values are noticeably lower than that reported by Koshi and coworkers, 9.8 kcal mol^{-1}, computed at the CBS-QB3 level of theory.[11]

The rate constant for $N_2H_4 + NO_2 \rightarrow N_2H_3 +$ *cis*-HONO via $TS2_a$ with the 7.6 kcal mol^{-1} barrier has been calculated for the temperature range 250–2500 K with Eckart tunneling corrections; the least-squares-fitted expression can be represented by

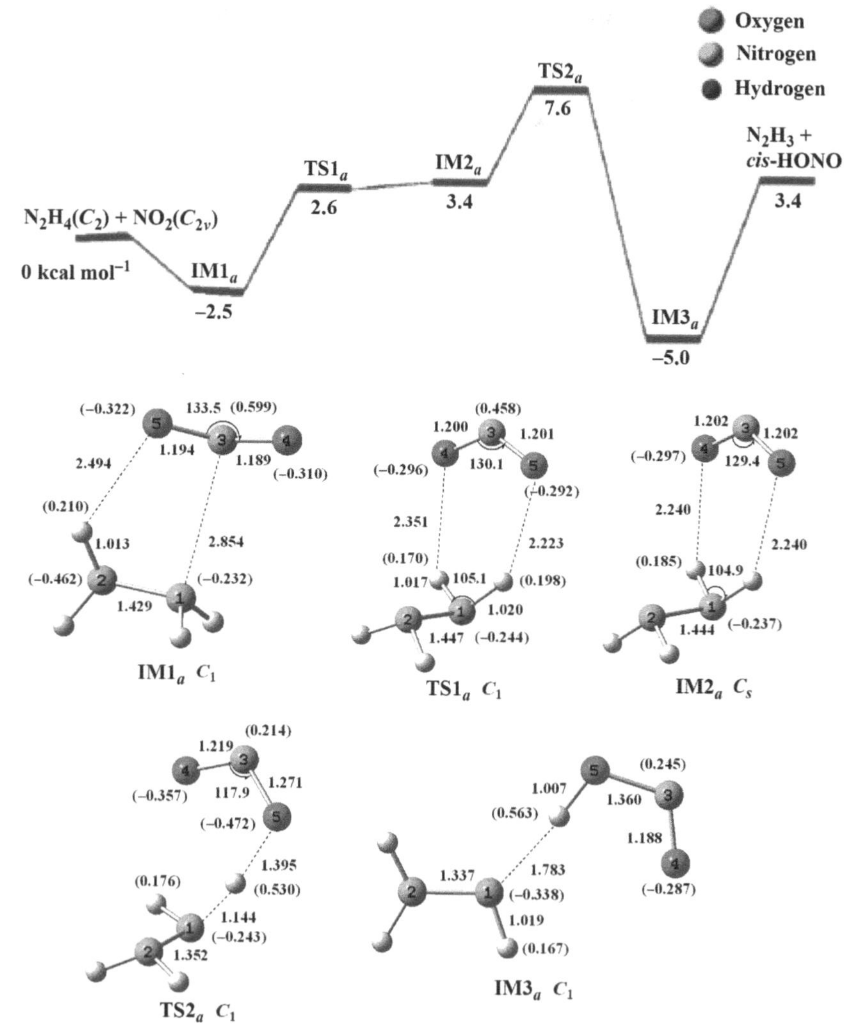

Figure 7.3 The major reaction pathway of potential energy profile and schematic geometry of species for $N_2H_4+NO_2$ (C_{2v}) are calculated with G2M(CC3) based on B3LYP geometries. The optimized geometries of the intermediated complexes and transition states calculated at the B3LYP/6-311++G(3df,2p) level. (For color version of this figure, the reader is referred to the online version of this chapter.)

$$k_{1a} = 5.36 \times 10^{-24} T^{3.56} \exp(384/T)\,\text{cm}^3\,\text{molecule}^{-1}\,\text{s}^{-1}$$

Our result at 800 K, 1.87×10^{-13} cm^3 molecule^{-1} s^{-1}, may be compared with the second-order global rate constant reported by Sawyer and Glassman[8] for the 800–1200 K range based on the rate of temperature change upon mixing N_2H_4 with NO_2, 5.70×10^{-16} cm^3 $molecule^{-1}$ s^{-1}, which is not believed to be trust worthy.

3.3.2 Reaction of N_2H_4 with NO_3

Under combustion conditions, NO_3 can be generated by the well-known bimolecular reaction, $2NO_2 \rightarrow NO + NO_3$ with appreciable rates (http://kinetics.nist.gov/kinetics/index.jsp). It can also be produced, in principle, by the decomposition of $ONONO_2$, which coexist with N_2O_4. We have accordingly computed the PES for the reaction of NO_3 with N_2H_4, which is unavailable in the literature. In the low energy PES of the $N_2H_4 + NO_3$ reaction, including pre-reaction complex computed with the BHandHLYP/6-311++G(3df,2p) method, the energies of all species were refined by CCSD(T) single point calculations with the same basis set as shown in Fig. 7.4. Recently, Wille et al.[89] studied nitrate radical reactions with alkynes using the various DFT methods: B3LYP, BHandHLYP, mPW1PW91, and mPW1K. Their results showed that BHandHLYP/cc-pVDZ gave the best agreement with those by the QCISD and CCSD(T) methods.

The optimized geometries of the reactants, transition states, and products at the BHandHLYP level of theory are shown in Fig. 7.4. The initial reaction of N_2H_4 and NO_3, proceeding via the van der Waals complex, N_2H_4:NO_3 (LM1) with 4.2 kcal mol^{-1} binding energy, can readily occur by H-abstraction reaction from one of the N—H bonds in N_2H_4 with a 1.3 kcal mol^{-1} barrier via TS1 forming the post-reaction complex, N_3H_3:HNO_3 (LM2); the process is exothermic by −38.4 kcal mol^{-1} at the CCSD(T)//BHandHLYP/6-311++G(3df,2p) level. Finally, LM2 can dissociate to produce the products $HNO_3 + N_2H_3$ barrierlessly with the predicted exothermicity, 23.3 kcal mol^{-1}. The computed potential energies using the DFT method for decomposition to $LM1 \rightarrow N_2H_4 + NO_3$, $LM2 \rightarrow HNO_3 + N_3H_3$, and $LM3 \rightarrow N_2H_4 + NO_3$ along their MEPs could be fitted to the Morse function with $\beta = 1.40$ Å^{-1}, $\beta = 1.35$ Å^{-1} and $\beta = 1.82$ Å^{-1}, respectively.

In the initial step of the second mechanism, one of the O atoms of NO_3 interacting with the one of N atoms in N_2H_4 by a barrierless process forming LM3 7.1 kcal mol^{-1} binding energy calculated at CCSD(T)/6-311++G(3df,2p)//BHandHLYP/6-31+G(d,p) level. As the reaction proceeds, LM3 undergoes further transformation by the transfer of an H atom from N_2H_4 to a neighboring O atom in NO_3 with the concerted transfer of another O atom to N_2H_3 via the five-membered ring transition state TS2 (−7.4 kcal mol^{-1}) resulting in the production of $N_2H_3O + HONO$.

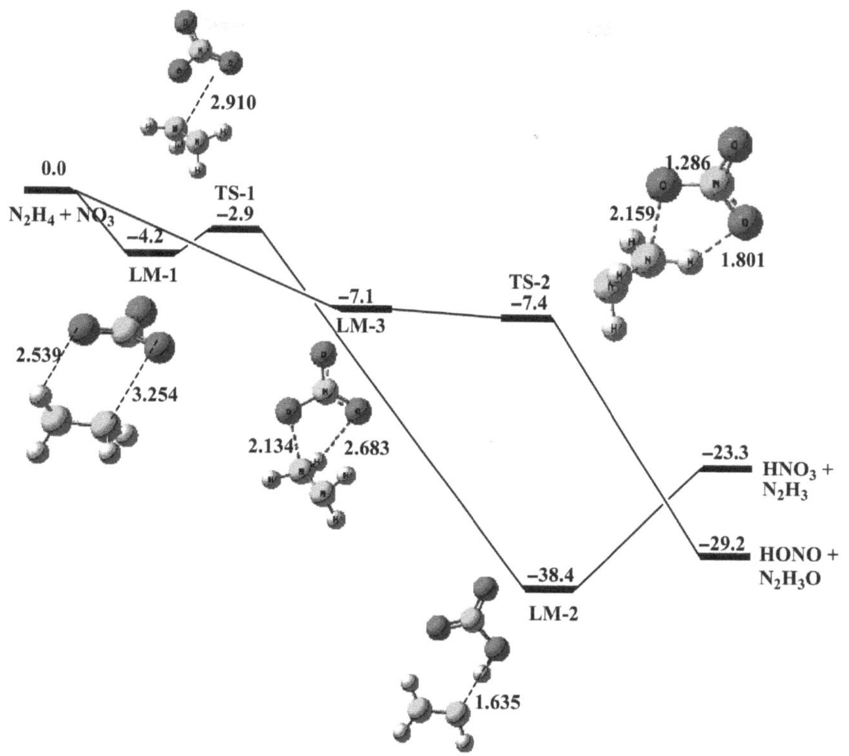

Figure 7.4 The major reaction pathway of potential energy profile and schematic geometry of species for $N_2H_4+NO_3$ (C_{2v}) are calculated with CCSD(T) based on BHandHLYP geometries. (For color version of this figure, the reader is referred to the online version of this chapter.)

The rate constants for the forward reaction of $N_2H_4+NO_3$ have been computed in the temperature range of 300–3000 K; the pressure-independent results at $P < 100$ atm. can be expressed by

$$N_2H_4+NO_3 \rightarrow N_2H_3+HNO_3$$
$$k=2.12\times10^{-20}T^{2.53}\exp(1483/T)\,\mathrm{cm^3\,molecule^{-1}\,s^{-1}}$$
$$N_2H_4+NO_3 \rightarrow HONO+N_2H_3O$$
$$k=1.82\times10^{-6}T^{-1.84}\exp(323/T)\,\mathrm{cm^3\,molecule^{-1}\,s^{-1}}\quad(300-1000\mathrm{K})$$
$$k=5.74\times10^{-21}T^{2.51}\exp(3750/T)\,\mathrm{cm^3\,molecule^{-1}\,s^{-1}}\quad(1000-3000\mathrm{K})$$

3.3.3 Reactions of N_2H_4 with N_2O_4

3.3.3.1 Reactions of N_2H_4 with N_2O_4 (D_{2h})

We have carried out an exhaustive search for the potential reaction paths and mechanism of the redox process involving N_2H_4 and N_2O_4 including H-abstraction and association reactions. Interestingly, as shown in

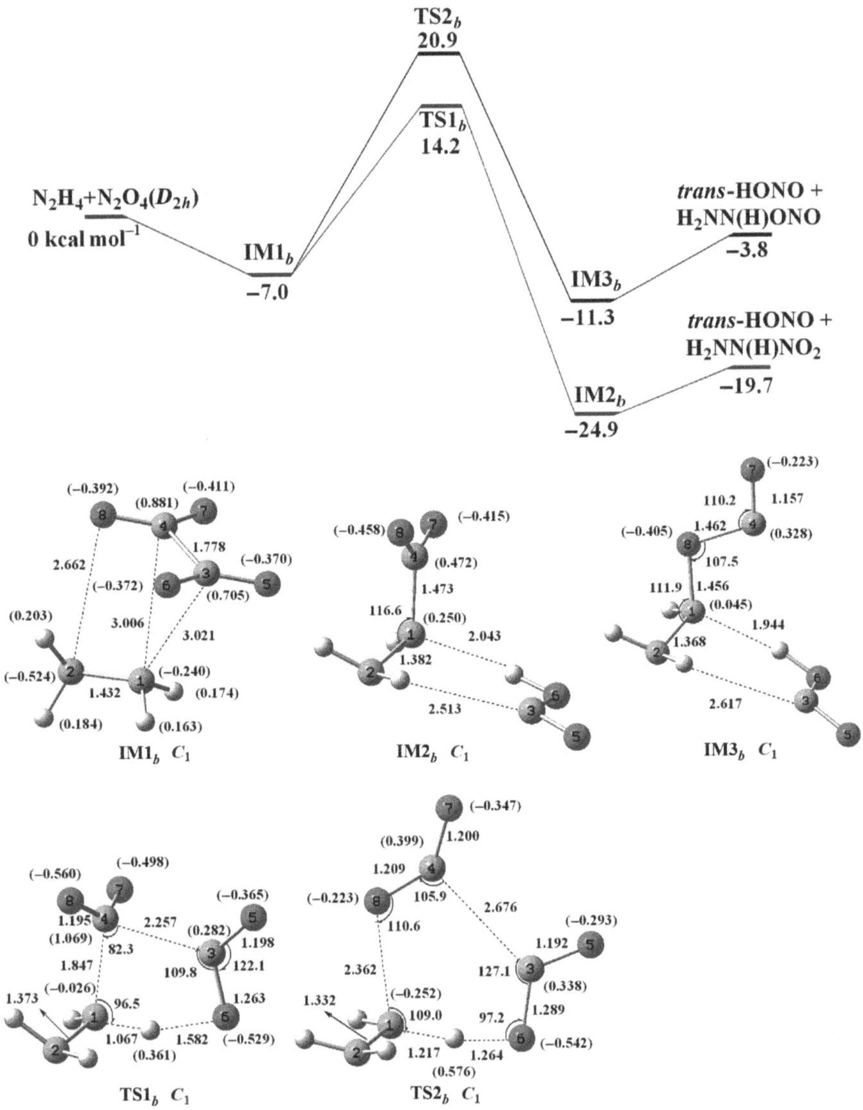

Figure 7.5 The major reaction pathway of potential energy profile and schematic geometry of species for $N_2H_4+N_2O_4$ (D_{2h}). Relative energies (kcal mol^{-1}) are calculated with G2M(CC3) based on B3LYP geometries. (For color version of this figure, the reader is referred to the online version of this chapter.)

Fig. 7.5, the major product channel proceeding via a loose pre-reaction complex LM1b with 7.0 kcal mol^{-1} binding energy occurs by the transfer of an H atom from N_2H_4 to one of the NO_2 groups with the concerted transfer of the second NO_2 group to N_2H_3 via the five-membered ring transition state $TS1_b$ with 14.2 kcal mol^{-1} barrier; the reaction produces a more stable postreaction complex $IM2_b$, *trans*-HONO. . .$H_2NN(H)NO_2$. A similar redox reaction producing other products *t*-HONO + $H_2NN(H)$ONO via $TS2_b$ has to overcome a 20.9 kcal mol^{-1} energy barrier.

The rate constant for the $N_2H_4 + N_2O_4 \rightarrow$ *t*-HONO + $H_2NN(H)NO_2$ reaction calculated with Eckart tunneling corrections for the temperature range 200–2500 K can be expressed as

$$k_{1b} = 2.30 \times 10^{-22} T^{2.62} \exp(-6598/T)\,\text{cm}^3\,\text{molecule}^{-1}\,\text{s}^{-1}$$

No experimental data exist in the literature for comparison.

3.3.3.2 Reactions of N_2H_4 with $ONONO_2$ isomers

The *cis*- and *trans*-$ONONO_2$ isomers of N_2O_4 are known to undergo autoionization to form ionic nitrosonium nitrate ($[ON^+][NO_3^-]$) as has been proven experimentally.[79–83] In this section, the PESs of the $N_2H_4 + ONONO_2$ reactions presented were computed at the G2M (CC3) level by Lai et al.[12] As shown in Fig. 7.6, initial association complex of the N_2H_4 + *trans*-$ONONO_2$ reaction, $IM1_c$, has a 10.1 kcal mol^{-1} binding energy in which the ON—ONO_2 bond is lengthened to 2.286 Å when compared to 1.621 Å in *trans*-ON—ONO_2 isomer and 1.685 Å in *cis*-ON—ONO_2 (C_1), suggesting that N_2H_4 may induce the ionization of

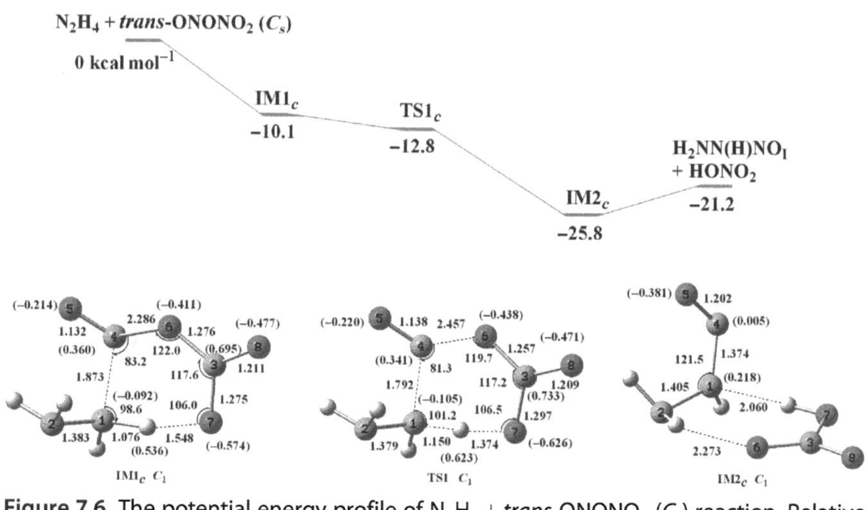

Figure 7.6 The potential energy profile of N_2H_4 + *trans*-$ONONO_2$ (C_s) reaction. Relative energies (kcal mol^{-1}) are calculated with G2M(CC3) based on B3LYP geometries. The geometries include selected bond lengths (Å), angles (degree), and Mulliken charges (atomic unit in parenthesis). (For color version of this figure, the reader is referred to the online version of this chapter.)

ON—ONO$_2$ to [ON$^+$] [NO$_3^-$]. Mulliken analysis also shows that the charges on NO and NO$_3$ are −0.146*e* and +0.768*e*, respectively, where *e* represents the negative charge of an electron. In the IM1$_c$ complex, the NO$_3$ group can readily abstract one of the H atoms in N$_2$H$_4$ via the six-membered ring TS1$_c$ with a negative barrier to produce a H$_2$NN(H)NO + HONO$_2$; the process is exothermic by 21.2 kcal mol^{-1}. The negative value of TS1$_c$ relative to IM1$_c$ is consistent with the rapid transformation to the ionic type TS with a tighter structure. The rate constant for N$_2$H$_4$ + *trans*-ONONO$_2$ → H$_2$NN(H)NO + HONO$_2$ process has a negligible temperature dependence in the range of 250–2500 K and can be expressed as

$$k_{1c} = 4.3 \pm 0.5 \times 10^{-10}\,\mathrm{cm^3\,molecule^{-1}\,s^{-1}}$$

The reaction of N$_2$H$_4$ with *cis*-ONONO$_2$ (C_1) is similar to the *trans*-ONONO$_2$ (C_s) reaction discussed above because both *cis*-ONONO$_2$ (C_s) and *trans*-ONONO$_2$ (C_1) have similar geometries and charge distributions. The computed PES of N$_2$H$_4$ + *cis*-ONONO$_2$ (C_1) reaction is shown in Fig. 7.7. The direct hydrogen abstraction reaction via TS1$_d$ (with a four-membered ring structure) also produces the H$_2$NN(H)NO and HNO$_3$. The TS1$_d$ calculated at the B3LYP level has a small positive barrier of 1.04 kcal mol^{-1} comparing with IM1$_d$. However, at the G2M(CC3) level, the energy of TS1$_d$ has negative activation barrier by 2.8 kcal mol^{-1} due to the ionic transformation. Similar to the *trans*-reaction, the *cis*-reaction also generates the H$_2$NN(H)NO + HONO$_2$ products. The predicted rate constant is very large and temperature independent; it can expressed by

$$k_{1d} = 3.5 \pm 0.5 \times 10^{-10}\,\mathrm{cm^3\,molecule^{-1}\,s^{-1}}$$

Contrary to the reactions of *cis*- and *trans*-isomers with C_1 and C_s symmetries, respectively, discussed above, the reaction of the *cis*-ONONO$_2$ (C_s) isomer with N$_2$H$_4$ occurs with a different mechanism. The reactants have a much weaker pre-reaction complex with a 3.9 kcal mol^{-1} binding energy and a positive 10.6 kcal mol^{-1} energy barrier at TS1$_e$ producing *trans*-HONO and H$_2$NN(H)NO$_2$ instead (see Fig. 7.8). The overall reaction is predicted to be exothermic by 29.5 kcal mol^{-1}, somewhat higher than the two reactions presented above (see Figs. 7.7 and 7.8). Significantly, the predicted high energy barrier, 10.6 kcal mol^{-1}, due to the presence of NO$_2$ complex in TS1$_e$ which has no ionic character as in the N$_2$H$_4$ + *trans*-ONONO$_2$ (C_s) and *cis*-ONONO$_2$ (C_1) reactions.

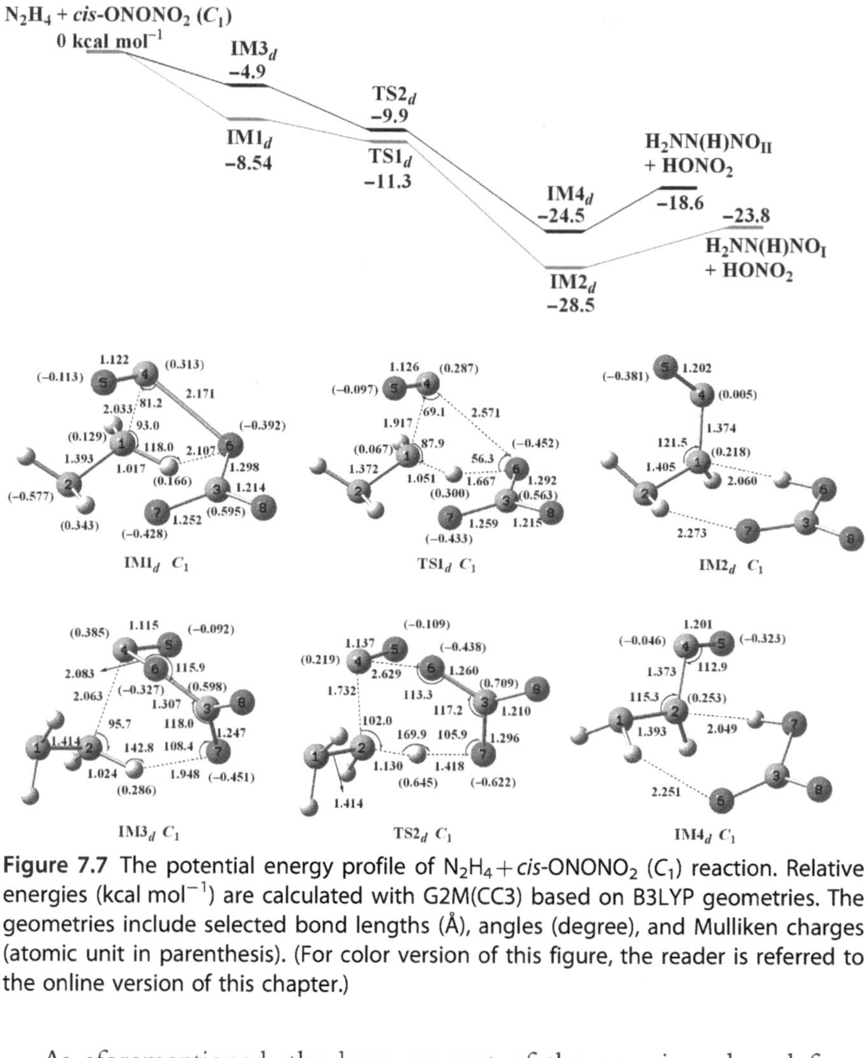

Figure 7.7 The potential energy profile of $N_2H_4 + cis\text{-}ONONO_2$ (C_1) reaction. Relative energies (kcal mol^{-1}) are calculated with G2M(CC3) based on B3LYP geometries. The geometries include selected bond lengths (Å), angles (degree), and Mulliken charges (atomic unit in parenthesis). (For color version of this figure, the reader is referred to the online version of this chapter.)

As aforementioned, the large amount of the energies released from the two very fast, barrieless reactions can be utilized in the dissociation of N_2O_4 and $H_2NN(H)NO$ formed at the liquid interface producing reactive radical species NO_2 and N_2H_3 for chain propagation. In addition, the HNO_3 formed in accompany with $H_2NN(H)NO$ is expected to produce the $[N_2H_5^+][NO_3^-]$ salt instantaneously releasing an even larger amount of energy. For the sublimation of $[NH_4^+][NO_3^-]$ producing $NH_3(g) + HNO_3(g)$, for example, Wight and coworkers[90] reported the value of enthalpy change to be 47 ± 4 kcal mol^{-1}. The value for $[N_2H_5^+][NO_3^-]$ is expected to be similar. This vast amount of acid–base neutralization

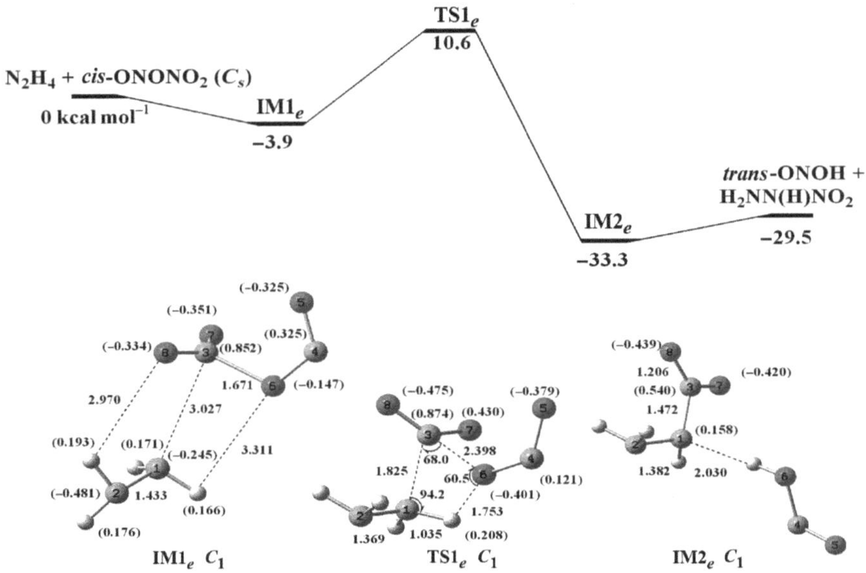

Figure 7.8 The potential energy profile of N_2H_4+*cis*-$ONONO_2$ (C_s) reaction. The geometries include selected bond lengths (Å), angles (degree), and Mulliken charges (atomic unit in parenthesis). (For color version of this figure, the reader is referred to the online version of this chapter.)

energy from the reverse association reaction is expected to have an enormous enhancement in the chain initiation process leading to hypergolic explosion at the droplet-liquid pool interface observed experimentally.[6,7]

3.3.3.3 Unimolecular decomposition of $H_2NN(H)NO$

In the $N_2H_4+ONONO_2$ reaction channel, $H_2NN(H)NO$ is one of the important product which can readily decompose to N_2H_3 and NO radicals without an intrinsic barrier with a dissociation energy of 34.1 kcal mol^{-1} at the G2M level, which may be compared with the value 30.4 kcal mol^{-1} obtained at the CCSD(T)/6-311+G(3df,2p) level of theory. The predicted rate constant for the decomposition reaction, $H_2NN(H)NO \rightarrow N_2H_3+NO$, can be expressed by

$$k_{\mathrm{dis}}=6.24\times 10^{15}T^{-0.15}\exp(-17{,}920/T)s^{-1}$$

over the temperature range of 250–1500 K.

3.4. Reactions of N_2H_3 with NO_2 and N_2O_4

N_2H_3 is one of the most important radicals which are involved in the chain propagation during the explosive interaction of N_2H_4 and NTO. It may be formed by the reactions of N_2H_4 with radicals including NO_2, H, and HO by direct H-atom abstraction as well as by the fragmentation of the excited $H_2NN(H)NO$ product following the reaction of $ONONO_2$ with N_2H_4 as discussed above. The reactions of N_2H_3 with NO_2 and N_2O_4 may produce N_2H_3O which can further fragment to other reactive species for the chain propagation process. To date, the kinetics and mechanisms of these reactions to be presented below are unknown.

3.4.1 Reactions of N_2H_3 with NO_2

According to the potential energy diagram shown in Fig. 7.9, the $N_2H_3 + NO_2$ reaction including isomerization and decomposition processes is quite complex. Optimized geometries of various stable intermediates and transition states are shown in Fig. 7.10. There are two possible entrance pathways for the reaction corresponding to the formation of nitro and nitrite intermediates as shown in Fig. 7.9. In the first pathway, the N_2H_3 radical reacts with the central N atom of the NO_2 forming the nonplanar $NH_2N(H)NO_2$ nitro intermediate through barrierless association, with an association energy of 34.3 kcal mol^{-1} predicted at the CCSD(T)/6-311++G(3df,2p)//B3LYP/6-311++G(3df,2p) level. The intermediate $NH_2N(H)NO_2$ can undergo H-migration from one of H atoms in NH_2 to one of the O atoms through TS1, producing $N_2H_2 +$ *trans*-HONO with a 27.4 kcal mol^{-1} barrier lying 6.9 kcal mol^{-1} below the reactants. Other low-energy isomerization and decomposition pathways of $NH_2N(H)NO_2$ as depicted in Fig. 7.8 include the roaming-type loose transition state RTS2 producing *cis*-$NH_2N(H)ONO$ which can readily further dissociate

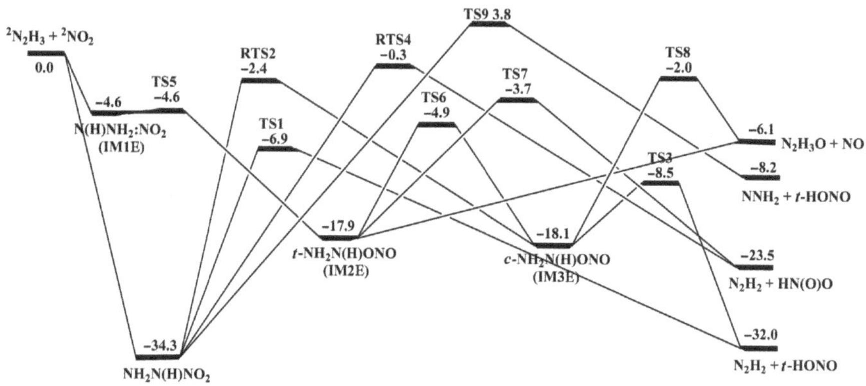

Figure 7.9 The schematic diagram of the potential energy surface for the $N_2H_3 + NO_2$ system computed at the CCSD(T)/6-311++G(3df,2p)//B3LYP/6-311++G(3df,2p) level.

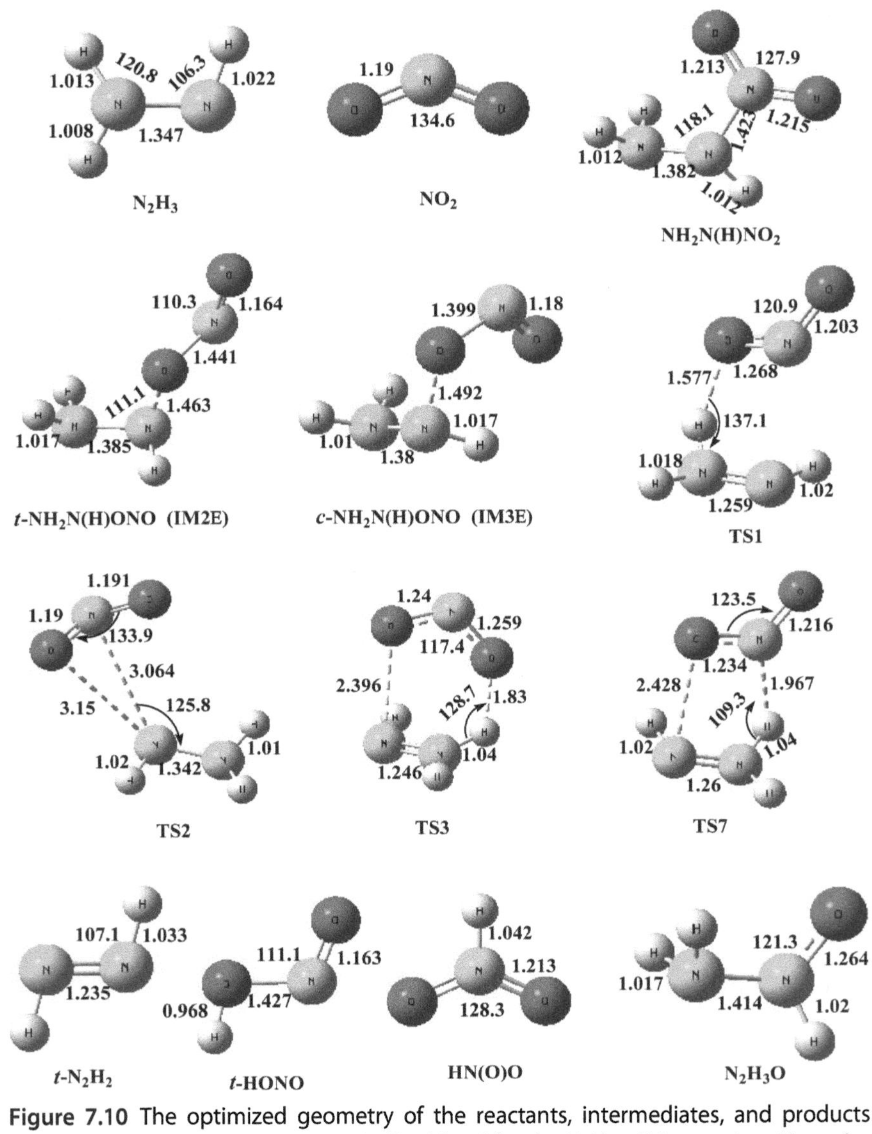

Figure 7.10 The optimized geometry of the reactants, intermediates, and products computed at the B3LYP/6-311++G(3df,2p) level for the $N_2H_3 + NO_2$ reaction. (For color version of this figure, the reader is referred to the online version of this chapter.)

via TS3 with a six-membered ring structure lying 8.5 kcal mol^{-1} below the reactants, giving rise to N_2H_2 + *trans*-HONO. The roaming-type transition state for the nitro-nitrite isomerization has been shown to exist in the CH_3NO_2 system.[91,92] The nitro intermediate $NH_2N(H)NO_2$ can also decompose via another loose roaming transition state RTS4 locating at

0.4 kcal mol^{-1} below the reactants producing $N_2H_2 + HN(O)O$, a metastable isomer of HONO.

The second pathway for the $N_2H_3 + NO_2$ reaction occurs by the direct O—N association producing the *trans*-$NH_2N(H)ONO$ nitrite intermediate via a loose complex ($IM1_e$) and TS5; the process is exothermic by 17.9 kcal mol^{-1}. The nitrite intermediate can dissociate without an intrinsic barrier giving $N_2H_3O + NO$; this is an important process because N_2H_3O is key radical responsible for generation of more reactive chain carriers such as H, OH, and NH_2 as will be discussed below.

The rate constants for the forward reactions of $N_2H_3 + NO_2$ via the low-energy channels have been computed over the temperature range of 300–2000 K and the pressure 760 Torr with the Variflex code,[46] whereas the higher energy production channels are neglected. The mechanisms of these bimolecular reactions giving the most favorable products are described as follows:

$$N_2H_3 + NO_2 \rightarrow NH_2N(H)NO_2^* \rightarrow NH_2N(H)NO_2(+M) \quad (7.4)$$

$$N_2H_3 + NO_2 \rightarrow NH_2N(H)NO_2^* \rightarrow t\text{-}N_2H_2 + t\text{-}HONO \quad (7.5)$$

$$N_2H_3 + NO_2 \rightarrow NH_2N(H)ONO^* \rightarrow N_2H_3O + NO \quad (7.6)$$

$$N_2H_3 + NO_2 \rightarrow NH_2N(H)ONO^* \rightarrow t\text{-}N_2H_2 + HN(O)O \quad (7.7)$$

Based on this energy diagram and the molecular parameters obtained at the B3LYP/6-311++G(3df,2p) level, variational TST and RRKM calculations have been carried out for the bimolecular association reaction rate constants of the $N_2H_3 + NO_2$ system. The energies used in the calculation are plotted in Fig. 7.9, and the molecular parameters were obtained at the B3LYP/6-311++G(3df,2p) level. The Lennard–Jones parameters for collision rate estimates are obtained by using $\sigma = 3.7$ Å and $\varepsilon = 200$ K for $NH_2N(H)—NO_2$, derived from those of N_2H_3 ($\sigma = 3.9$ Å and $\varepsilon = 200$ K) and NO_2 ($\sigma = 3.5$ Å and $\varepsilon = 200$ K)[65] using the approximation $\sigma(NH_2N(H)—NO_2) = [\sigma1 + \sigma2]/2$ and $\varepsilon(NH_2N(H)—NO_2) = [\varepsilon1 * \varepsilon2]^{1/2}$. The *L–J* parameters for N_2H_4, $\sigma = 4.23$ Å and $\varepsilon = 205$ K were taken from the literature.[63]

Due to the absence of an intrinsic transition state for the fragmentation reaction, for example, $NH_2N(H)—NO_2$ dissociation potential function was computed variationally to cover a range of N—N separations from the equilibrium value 1.42–5.5 Å with an interval of 0.1 Å by second-order multireference perturbation theory (CASPT2) based on the CASSCF optimized geometries with eight active electrons and eight active orbitals using the 6-311+G(2d,p) basis set. Other geometric parameters were fully optimized. These calculations were performed with the MOLPRO code.[53] The computed potential energies could be fitted to the Morse function with the parameters of $\beta = 1.96$ Å^{-1}. In the same manner with much computational effort, we also determined the value $\beta = 2.35$ Å^{-1} for the dissociation Morse potential, from $IM1_e$ to $N_2H_3 + NO_2$. These values are used in the rate constant calculations. At 1 atm. in N_2H_4, the individual product rate constants given in units of cm^3 molecule^{-1} s^{-1} can be presented by

$$k_5 = 8.29 \times 10^{22} T^{-11.8} \exp(-3047/T) \text{ for } 300-800\text{K}$$
$$k_5 = 3.98 \times 10^{31} T^{-16.7} \exp(7245/T) \text{ for } 800-3000\text{K}$$
$$k_6 = 1.80 \times 10^{-4} T^{-2.9} \exp(-398.7/T) \text{ for } 300-1000\text{K}$$
$$k_6 = 1.02 \times 10^{-23} T^{2.8} \exp(4455/T) \text{ for } 1000-3000\text{K}$$
$$k_7 = 6.76 \times 10^{-16} T^{0.5} \exp(1205/T) \text{ for } 300-1500\text{K}$$
$$k_7 = 8.50 \times 10^{-17} T^{-0.2} \exp(1377/T) \text{ for } 1500-3000\text{K}$$

Figure 7.11 summarizes the total and individual product rate constants of the $N_2H_3 + NO_2$ reaction at the 760 Torr N_2H_4 pressure, and these results show that *trans*-N_2H_2 + *trans*-HONO is the dominant product pair throughout the entire temperature range.

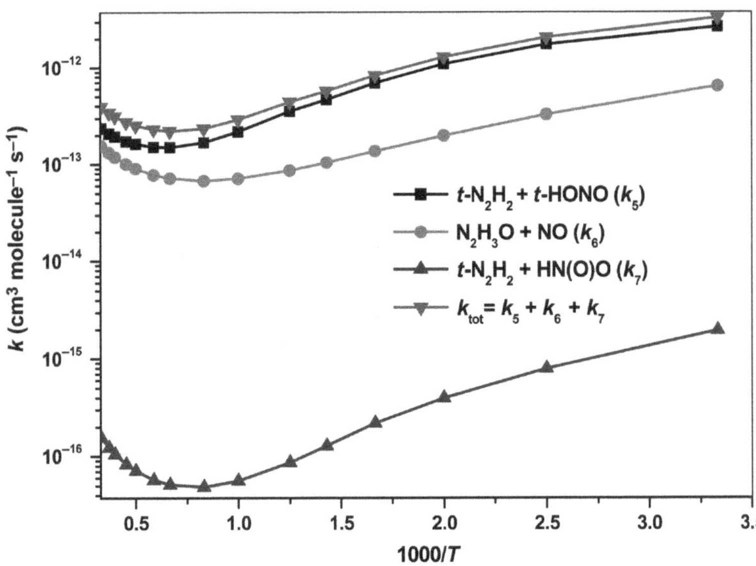

Figure 7.11 The predicted rate constants for all the $N_2H_3 + NO_2$ reaction products calculated at the CCSD(T)/6-311++G(3df,2p)//B3LYP/6-311++G(3df,2p) level. (For color version of this figure, the reader is referred to the online version of this chapter.)

3.4.2 Reactions of N_2H_3 with N_2O_4 (D_{2h})

For the $N_2H_3+N_2O_4$ (D_{2h}) reaction, there are several possible and distinct reaction pathways that lead to the formation of different sets of products, that is.

$$N_2H_3+N_2O_4 \rightarrow NH_2N(H)NO_2+NO_2 \quad (7.8)$$

$$N_2H_3+N_2O_4 \rightarrow \textit{trans}\text{-}N_2H_2+\textit{trans}\text{-}HONO \quad (7.9)$$

$$N_2H_3+N_2O_4 \rightarrow \textit{cis}\text{-}NH_2NHONO+NO_2 \quad (7.10)$$

$$N_2H_3+N_2O_4 \rightarrow N_2H_3O+N_2O_3 \quad (7.11)$$

The potential energy diagram obtained at the CCSD(T)/6-311++G (3df,2p)//B3LYP/6-311++G(3df,2p) level is presented in Fig. 7.12, and the optimized geometries are shown in Fig. 7.13. The association reaction of N_2H_3 and N_2O_4 by a barrierless process forms a loose complex, that is, $IM1_f$, lies 7.5 kcal mol^{-1} below the reactants at the CCSD(T)// B3LYP level.

The lowest energy reaction takes place via the loose complex $IM1_f$ which dissociates to give the reactive radical N_2H_3O and N_2O_3 via TS13 with a small barrier of 6.4 kcal mol^{-1} and a small exothermicity of 1.8 kcal mol^{-1}. The next low-energy reaction channel occurs by direct H-abstraction by one of the O atoms in N_2O_4 from one of the H atoms in the NH_2 group of N_2H_3 via TS11 to produce *trans*-N_2H_2 + *trans*-HONO + NO_2; as shown in Fig. 7.9, TS11 has a 10.4 kcal mol^{-1} barrier with 20.2 kcal mol^{-1} exothermicity. The third low-energy channel occurs by the abstraction of an NO_2 group in N_2O_4 by the N_2H_3 radical via TS10 with 10.5 kcal mol^{-1} barrier leading to the formation of $NH_2N(H)NO_2+NO_2$ with 22.4 kcal mol^{-1} exothermicity. Finally, the fourth product channel taking place via the $IM1_f$ complex by the abstraction of an NO_2 group from N_2O_4 by the N_2H_3 radical with a 13.5 kcal mol^{-1} barrier at TS12 leads to the formation of *cis*-$NH_2NHONO+NO_2$.

Rate constants for all the product channels and branching ratio of the $N_2H_3+N_2O_4$ reaction have been calculated by the TST with Eckart tunneling corrections using the energetics presented in Fig. 7.12 and the vibrational frequencies computed by B3LYP/6-311++G(3df,2p) employing the ChemRate program.[47] For practical applications, the predicted pressure-independent rate constants for these reaction channels listed above have been least-squares fitted for the temperature range 300–3000 K and presented below by the three parameter expressions given in units of cm^3 $molecule^{-1}$ s^{-1}:

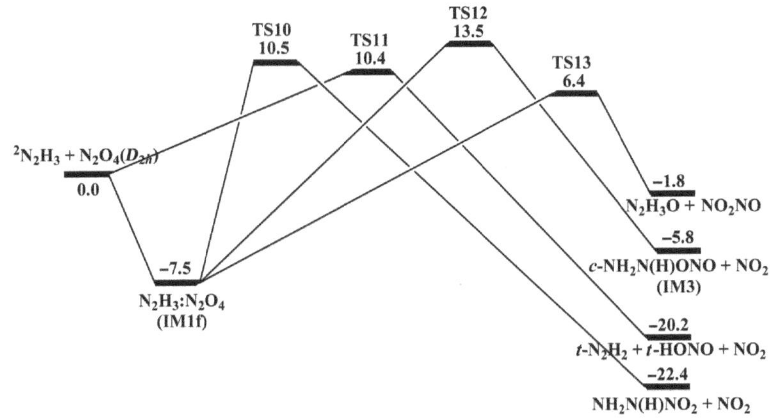

Figure 7.12 The schematic diagram of the potential energy surfaces for the $N_2H_3+N_2O_4$ system computed at the CCSD(T)/6-311++G(3df,2p)//B3LYP/6-311++G (3df,2p) level.

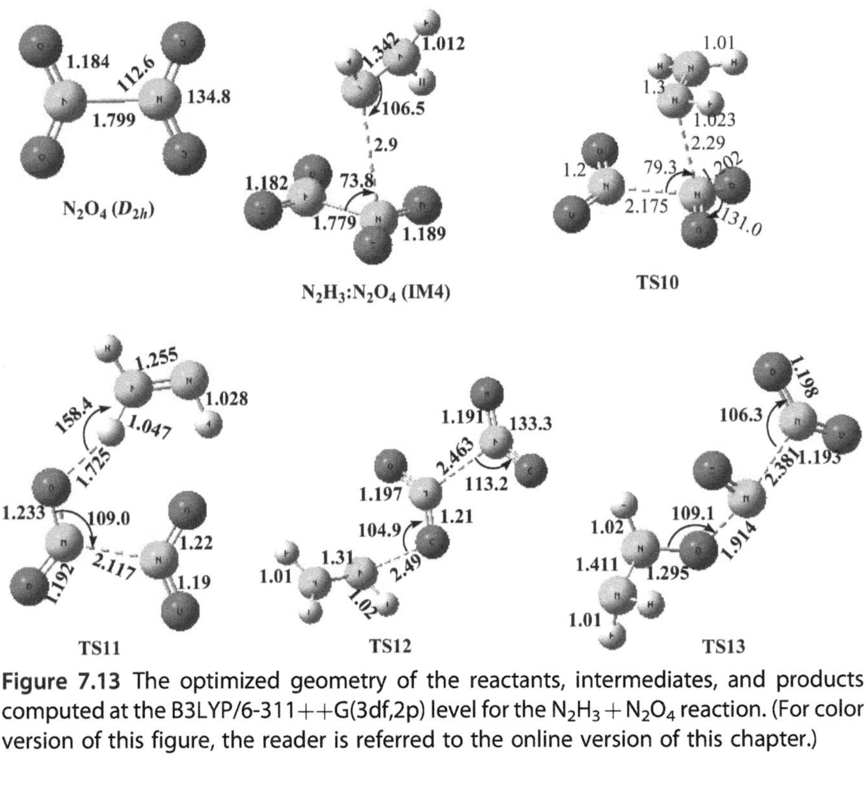

Figure 7.13 The optimized geometry of the reactants, intermediates, and products computed at the B3LYP/6-311++G(3df,2p) level for the $N_2H_3+N_2O_4$ reaction. (For color version of this figure, the reader is referred to the online version of this chapter.)

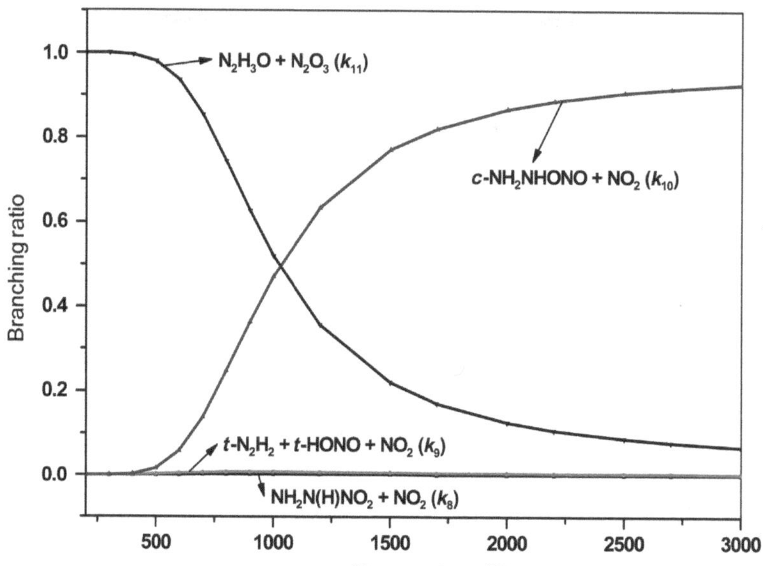

Figure 7.14 Effect of temperature on the product branching ratios of the $N_2H_3+N_2O_4$ reaction covering the range of 300–3000 K. (For color version of this figure, the reader is referred to the online version of this chapter.)

$$k_8=1.83\times10^{-14}T^{0.87}\exp(-5923.7/T)$$
$$k_9=1.42\times10^{-13}T^{0.74}\exp(-5891.1/T)$$
$$k_{10}=7.54\times10^{-11}T^{0.76}\exp(-8031.4/T)$$
$$k_{11}=6.12\times10^{-13}T^{0.87}\exp(-4049.6/T)$$

The branching ratios of these four product channels are shown in Fig. 7.14. Among these channels, the formation of $N_2H_3O+NO_2$—NO (k_{11}) is dominant up to about 1000 K, after which *cis*-$NH_2NHONO+NO_2$ product channel becomes predominant because of its looser TS structure which overcomes its apparent higher energy barrier (13.5 kcal mol^{-1}) at higher temperatures. The result shows that the product channels giving $NH_2N(H)NO_2+NO_2$ (k_8) and *trans*-N_2H_2+*trans*-$HONO+NO_2$ (k_9) are noncompetitive over the entire temperature range due to their tighter TS structures despite the apparent lower energy barriers (10.5 and 10.4 kcal mol^{-1}, respectively).

3.5. Unimolecular decomposition of N_2H_3O

N_2H_3O radical, a key chain propagator in the hydrazine–NTO propellant system, can be produced by the reaction N_2H_3 with NO_2 and N_2O_4 as

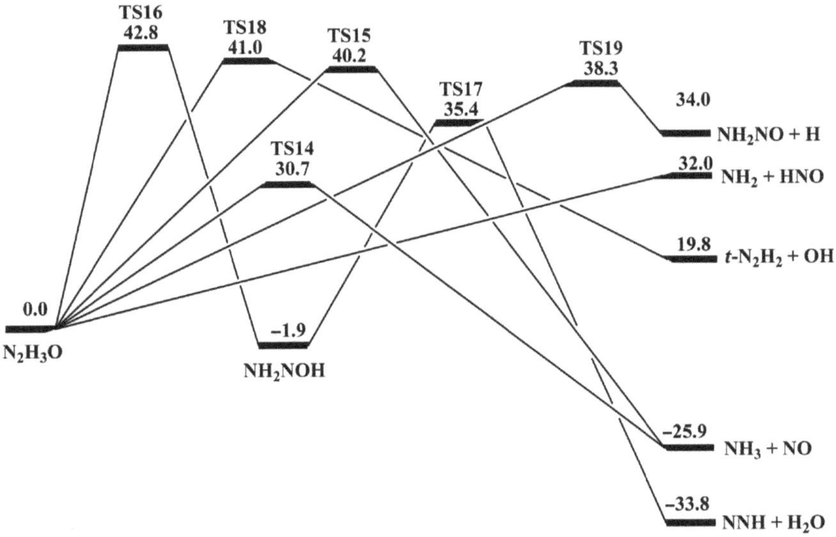

Figure 7.15 Unimolecular decomposition of N_2H_3O system computed at the CCSD(T)/6-311++G(3df,2p)//B3LYP/6-311++G(3df,2p) level.

discussed above. In order to evaluate the role of N_2H_3O radical in the hypergolic ignition process, we investigated its unimolecular dissociation reactions producing potential reactive products such as H, HO, and NH_2. NH_2 is known to be a key reducing agent in combustion process. The reactions of the NH_2 radical with HNO_x ($x=1$–3) are potentially relevant to the thermal NH_3–$deNO_x$ chemistry[93,94] and to the combustion of solid propellant ammonium dinitramide.[95] Xu and Lin[96] have investigated the kinetics and mechanism of NH_2+HNO reaction at the CCSD(T)/6-311+G(3df,2p)//CCSD/6-311++G(d,p) level of theory. However, the decomposition mechanism of N_2H_3O has neither been reported theoretically nor experimentally. In this work, the kinetics and mechanism for the unimolecular decomposition of N_2H_3O giving rise to a variety of radical products have been investigated at the CCSD(T)/6-311++G(3df,2p)//B3LYP/6-311++G(3df,2p) level in conjunction with quantum statistical variational RRKM calculations.

The PES shown in Fig. 7.15 indicates that the unimolecular decomposition of N_2H_3O radical may produce several reactive species:

$$N_2H_3O \rightarrow NNH + H_2O \quad (7.12)$$

$$N_2H_3O \rightarrow NH_3 + NO \quad (7.13)$$

$$N_2H_3O \rightarrow \textit{trans}\text{-}N_2H_2 + OH \quad (7.14)$$

$$N_2H_3O \rightarrow NH_2 + HNO \quad (7.15)$$

$$N_2H_3O \rightarrow NH_2NO + H \quad (7.16)$$

The N_2H_3O radical can dissociate directly to give $NH_3 + NO$, via a roaming-type transition state TS14 with a 30.7 kcal mol^{-1} barrier, at which the NH_2 group nearly dissociates by N—N breaking and roams about the HNO fragment and abstract its H atom during the course of separation. The optimized structures of all the species were calculated at the B3LYP/6-311++G(3df,2p) level and presented in Fig. 7.16. We have also found a tight three-centered transition state TS15 giving the same products requiring to overcome 40.2 kcal mol^{-1} of barrier. Another decomposition reaction producing $NNH + H_2O$ via TS16 (42.8 kcal mol^{-1}) and TS17 (35.4 kcal mol^{-1}) occurs by H migration to O atom followed by elimination of H_2O, respectively. The next low-energy product channel occurs by elimination of an H atom from the NH group to give NH_2NO via TS19 with a 38.3 kcal mol^{-1} barrier. A similar OH-elimination process can take place by

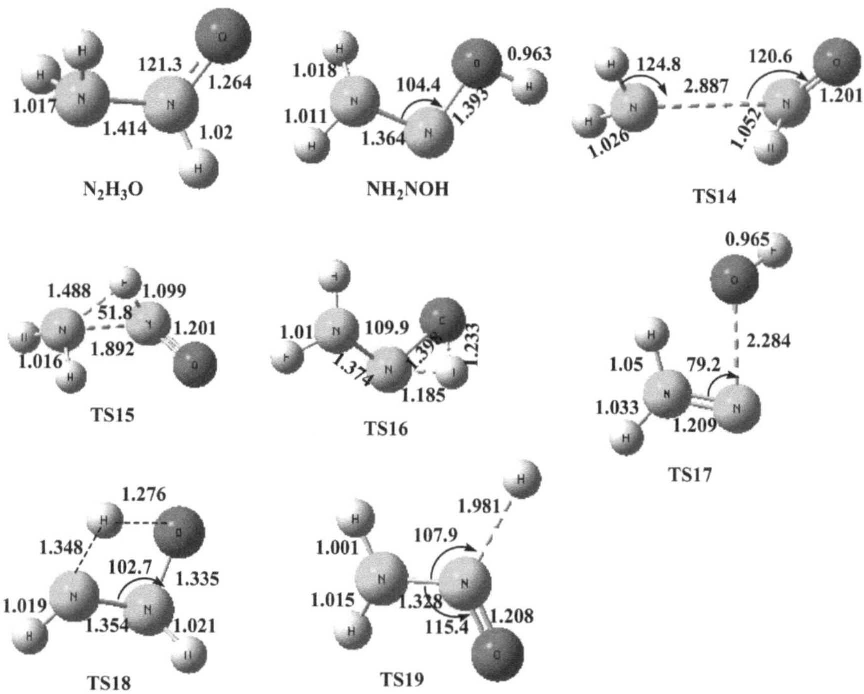

Figure 7.16 The optimized geometry of the reactants, intermediates, and products computed at the B3LYP/6-311++G(3df,2p) level for the N_2H_3O dissociation. (For color version of this figure, the reader is referred to the online version of this chapter.)

the migration of an H atom from the NH_2 group to the O atom via a four-centered TS18 with 41.0 barrier concertedly giving *trans*-N_2H_2+OH. Finally, the most favored path in this PES belongs to the direct cleavages of the N—N bond producing NH_2+HNO without an intrinsic barrier; this reaction is predicted to be endothermic by 32.0 kcal mol^{-1}. This low energy together with the loose variational transition state makes the decomposition channel, which provides the very reactive NH_2 radical, highly important in the present N_2H_4–NTO system.

The thermal decomposition of N_2H_3O under 760 Torr condition process produces predominantly NH_2+HNO because of its zero barrier comparing with those for other products (see Fig. 7.15). In the present reaction, we calculated the lowest energy product channel and their rate constant given below. For barrierless dissociation reaction $N_2H_3O \rightarrow NH_2 + HNO$, we computed variationally the N—N separation from 1.4 to 5.5 Å at an interval of 0.1 Å using the CASPT2(8,8)/6-311+G(2d,p)//CASSCF (8,8)/6-311+G(2d,p) method; the predicted MEP was fitted to the Morse function with $\beta = 2.7$ Å^{-1}. The predicted rate constants over the temperature range of 300–3000 K at 1 atm. N_2H_4 pressure for the three key product channels can be expressed in units of s^{-1} by

$$k_{13} = 2.86 \times 10^{22} T^{-2.8} \exp(-39{,}903/T)$$
$$k_{15} = 9.12 \times 10^{33} T^{-6.68} \exp(-17{,}722/T)$$
$$k_{16} = 1.57 \times 10^{34} T^{-6.63} \exp(-22{,}621/T)$$

3.5.1 Product branching ratios

The branching ratios for all the product channels are shown in Fig. 7.17. The results show that in the N_2H_3O dissociation reaction, the formation of NH_3+NO (k_{13}) is noncompetitive and its branching ratio is near zero throughout the entire temperature range. The predicted branching ratio for NH_2+HNO (k_{15}) formation by direct dissociation without an intrinsic barrier is predominant throughout the entire temperature range. The NH_2NO+H (k_{16}) product channel was also found to contribute negligibly.

3.6. Reactions of N_2H_2 with NO_2, N_2O_4, and OH

Diazene (N_2H_2) is one of the major early products generated in hydrazine (N_2H_4) decomposition; the kinetics for N_2H_2 reactions with various reactive species have been proposed by Konnov et al.[61] in their modeling of N_2H_4 decomposition at high temperatures as aforementioned. The most

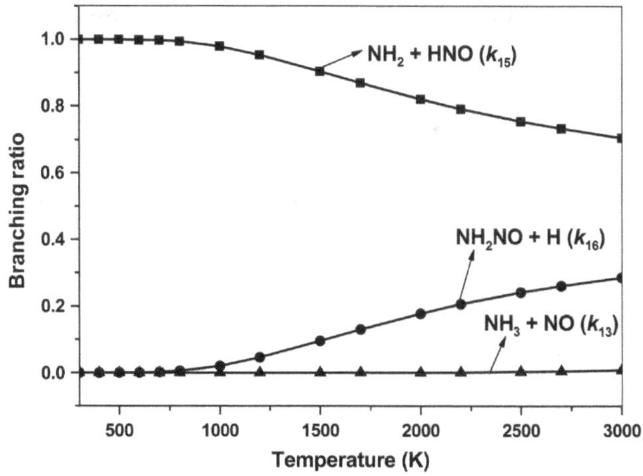

Figure 7.17 Branching ratios of N_2H_3O dissociation products NH_3+NO (k_{13}), NH_2+HNO (k_{15}), and NH_2NO+H (k_{16}) as functions of temperature relative to the total rate constant k_{total} of these three products at pressure 760 Torr N_2H_4 bath gas.

stable structure of N_2H_2 belongs to C_{2h} symmetry with the *trans*-conformation which is 5.1 kcal mol^{-1} more stable than the *cis*-isomer predicted by the CCSD(T)//B3LYP method with the 6-311++G(3df,2p) basis set. This is in exact agreement with the value predicted at the MRCI/aug-cc-pVQZ level, 5.1 kcal mol^{-1}.[97] The heat of formation of *trans*-N_2H_2 was measured by vacuum-ultraviolet spectroscopy above 105 nm which afforded an upper limit at 0 K, $\geq 48.8 \pm 0.5$.[98] N_2H_2 is chemically unstable with a decomposition lifetime of up to a few minutes,[99] which is sufficiently long to allow the molecules to accumulate and inevitably interact with reactive oxidants such as NO_2, N_2O_4, and OH in the reactor. These potential redox reactions are discussed below based on the PESs evaluated in this work.

3.6.1 Reactions of N_2H_2 with NO_2

The PES for the reaction of NO_2 with *cis*- and *trans*-N_2H_2 was optimized by the B3LYP/6-311++G(3df,2p) method. The energies refined by single-point calculations with CCSD(T)/6-311++G(3df,2p) are shown in Fig. 7.18A and B. Figure 7.18A shows that there are two possibilities in the *trans*-N_2H_2 reaction with NO_2: Hydrogen abstraction by NO_2 and O_2N addition to N_2H_2. The lowest energy barrier process is the former

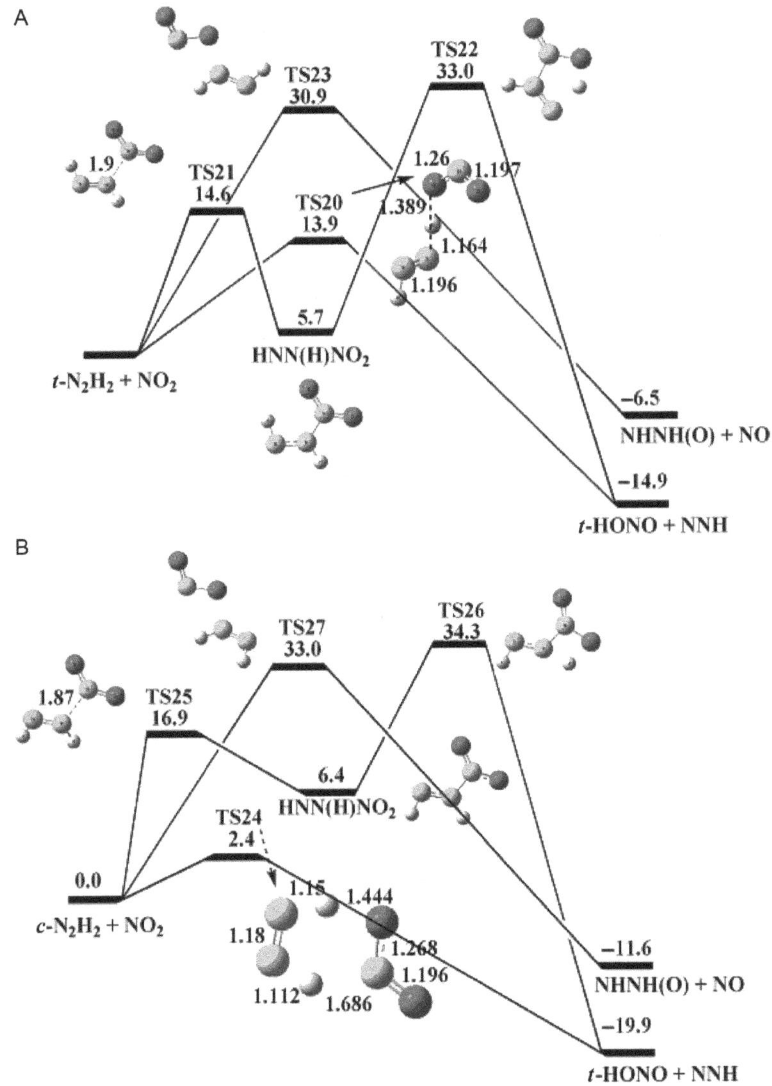

Figure 7.18 The schematic diagram of the potential energy surfaces for (A) t-N_2H_2 + NO_2 and (B) c-N_2H_2 + NO_2 computed at the CCSD(T)/6-311++G(3df,2p)//B3LYP/ 6311++G (3df,2p) level. (For color version of this figure, the reader is referred to the online version of this chapter.)

reaction, where one of the O atoms of NO_2 abstracts one of the H atoms from *trans*-N_2H_2 producing *trans*-HONO+NNH via TS20 with 13.9 kcal mol^{-1} barrier. The exothermicity of the process is predicted to be 14.9 kcal mol^{-1}. Koshi and coworkers[11] have reported a 11.9 kcal mol^{-1} barrier energy for the same reaction by means of the CBS-QB3 method.

As shown in Fig. 7.18A, the same products can be formed by the addition reaction via TS21 giving $HNN(H)NO_2$ followed by its fragmentation via TS22 with the respective barrier energy of 14.6 and 33.0 kcal mol^{-1}. The latter addition/decomposition process obviously cannot compete with the H-abstraction reaction. The *trans*-$N_2H_2+NO_2$ reaction can also occur by direct O abstraction from NO_2 by N_2H_2 to produce $NHN(H)O+NO$ via TS23 with a 30.9 kcal mol^{-1} barrier. This process is also unimportant due to its high barrier.

In the *cis*-N_2H_2 reaction with NO_2, similar to the *trans*-N_2H_2 case, two possible product channels were identified as shown in Fig. 7.18B. Significantly in the *cis*-reaction case, the H-abstraction channel was found to have a much lower energy barrier, 2.4 kcal mol^{-1}, comparing with that of the *trans*-reaction, 13.9 kcal mol^{-1}, attributable to the double hydrogen bondings as revealed by the TS structure. For the other two types of reactions by O_2N addition/decomposition and by direct O-transfer producing $HONO+N_2H$ and HNN(H)O, respectively, the energy barriers involved are very similar in both *cis*- and *trans*-reactions.

The rate constants for the lowest energy product channels of NO_2 reactions with both isomers of N_2H_2 via TS20 and TS24, respectively, producing similar products have been calculated by the TST with Eckart tunneling corrections employing the ChemRate program:

$$trans\text{-}N_2H_2+NO_2 \rightarrow trans\text{-}HONO+NNH \quad (7.17)$$

$$cis\text{-}N_2H_2+NO_2 \rightarrow trans\text{-}HONO+NNH \quad (7.18)$$

For the rate constant prediction, the CCSD(T)//B3LYP barriers and the B3LYP/6-311++G(3df,2p) frequencies were used. In the calculation, the lowest vibrational modes were treated as one-dimensional free rotors. The calculated rate constant expressions for reactions (7.17) and (7.18), obtained by three-parameter fitting for the 300–2500 K temperature range, are given in units of cm^{3}molecule^{-1} s^{-1} below:

$$k_{17}=2.08\times10^{-24}T^{3.8}\exp(-5238.3/T)$$
$$k_{18}=3.87\times10^{-25}T^{3.5}\exp(64.7/T)$$

3.6.2 *Reactions N_2H_2 with N_2O_4 (D_{2h})*

Similar to the *cis-/trans*-$N_2H_2+NO_2$ reactions discussed above, we have carried out an extensive search for the potential reaction paths of the reactions of the two isomers of N_2H_2 with N_2O_4 (D_{2h}), primarily by

H-abstraction reactions, at the B3LYP/6-311++G(3df,2p) level. Interestingly, as shown in Fig. 7.19 for the *trans*-isomer, the reactants form a loose complex with N_2O_4, $IM1_g$, with 2.1 kcal mol^{-1} binding energy. The major low-energy path produces *trans*-HONO + NO_2 + NNH by the direct transfer of an H atom from *trans*-N_2H_2 to one of the NO_2 groups in N_2O_4 via TS28 with 23.7 kcal mol^{-1} barrier. In addition, the same products can derive from the same complex $IM1_g$ via TS29 with a higher barrier energy of 33.1 kcal mol^{-1} as shown in Fig. 7.19A. A similar H-transfer reaction via TS30 with a much higher energy barrier of 45.3 kcal mol^{-1} produces *c*-HONO + NO_2 + NNH. These high barrier processes are not expected to compete with the very fast fragmentation reaction of N_2O_4 producing the more reactive monomers, NO_2.

Significantly, similar to the *trans*-N_2H_2 + NO_2 reaction, we found that the *cis*-isomer reaction with N_2O_4 occurs with a much lower barrier by the transfer of one H atoms from N_2H_2 to one of the NO_2 groups and another H atom concurrently transfers to another NO_2 group via TS31 with 15.7 kcal mol^{-1} barrier, producing *trans*-HONO + HNO(O) + N_2. The exothermicity of the process is predicted to be as much as 77.2 kcal mol^{-1}. Similar to the *trans*-N_2H_2 reaction, the *cis*-N_2H_2 also forms an association complex, $IM1_h$, with 2.2 kcal mol^{-1} binding energy, to be followed by H abstraction by one of the NO_2 groups with 32.1 kcal mol^{-1} barrier via TS32 to yield the *trans*-HONO + NO_2 + NNH products.

The rate constant for following low-energy product channels of *cis*-and *trans* isomers of N_2H_2 with N_2O_4 reactions calculated in the range of temperature 300–2500 K and can be expressed as

$$trans\text{-}N_2H_2 + N_2O_4(D_{2h}) \rightarrow trans\text{-}HONO + NO_2 + NNH \quad (7.19)$$

$$k_{19} = 1.46 \times 10^{-23} T^{3.1} \exp(-14{,}486/T)\,\text{cm}^3\,\text{molecule}^{-1}\,\text{s}^{-1}$$

$$cis\text{-}N_2H_2 + N_2O_4(D_{2h}) \rightarrow trans\text{-}HONO + HNO(O) + N_2 \quad (7.20)$$

$$k_{20} = 3.96 \times 10^{-26} T^{3.9} \exp(-6723/T)\,\text{cm}^3\,\text{molecule}^{-1}\,\text{s}^{-1}$$

3.6.3 Reactions N_2H_2 with OH

The reaction of OH, the most important chain carrier, with the metastable N_2H_2 accumulated in the chain reactions, is of considerable importance in the present system. In this section, we describe in detail the mechanism for the OH reaction with *cis*- and *trans*-conformations of N_2H_2 based on the

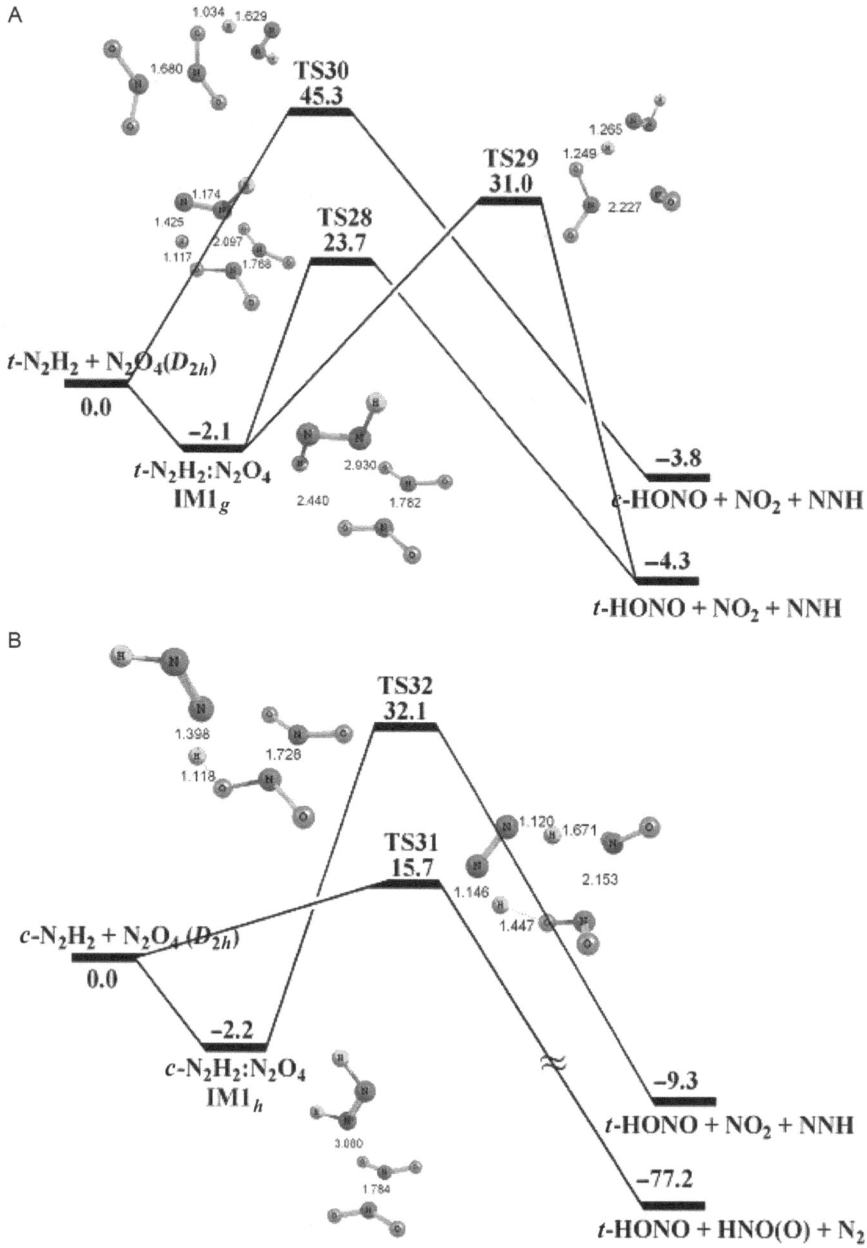

Figure 7.19 The schematic diagram of the potential energy surfaces for (A) t-N_2H_2 + N_2O_4 (D_{2h}) and (B) c-N_2H_2 + N_2O_4 (D_{2h}) computed at the CCSD(T)/6-311++G(3df,2p)// B3LYP/ 6311++G (3df,2p) level. (For color version of this figure, the reader is referred to the online version of this chapter.)

predicted PESs and try to present the similarity and difference between the reactions of the two conformers. The energetics and structures of all the species involved were calculated at the MP2/6-311++G(2df,p) level of theory. The reactions of both N_2H_2 conformers can take place by direct abstraction or by association processes. A schematic potential energy diagram for the *trans*-N_2H_2+OH reaction is shown in Fig. 7.20A. The first mechanism of the reaction is the direct H abstraction occurring by the attack of OH at one of the hydrogen atoms of *trans*-N_2H_2 via the pre-reaction complex $IM1_i$, OH...N(H)NH, with a binding energy of 3.8 kcal mol^{-1} as depicted in Fig. 7.21. There are two distinct reaction configurations with *cis*- and *trans*-type structures via TS33 and TS34 with 2.2 and 3.1 kcal mol^{-1} barriers, respectively, as shown in Fig. 7.20A. The predicted barrier heights are in good agreement with a previous study on the same reaction, 1.6 and 2.7 kcal mol^{-1} computed at the MRCI33/m-cc-pVTZ level of theory.[100] The exothermicity of the process producing NNH+H_2O is predicted to be 51.7 kcal mol^{-1}. In the second mechanism, the reaction takes place by the attack of the OH at one of the nitrogen atoms of N_2H_2 via TS35 to give the *trans*-HNN(H)OH adduct with 14.7 kcal mol^{-1} barrier and 11.2 kcal mol^{-1} exothermicity.

The PES for the analogous *cis*-N_2H_2+OH reaction is presented in Fig. 7.20B. The H-abstraction reaction can take place via a van der Waals complex, $IM1_j$, with C_s symmetry; it can fragment readily via TS36 with 1.6 kcal mol^{-1} barrier to give NNH and H_2O with an overall exothermicity of 57.6 kcal mol^{-1}. The complex has 2.6 kcal mol^{-1} binding energy with 2O...H hydrogen bond length of 2.288 Å predicted at the MP2/6-311++G(2df,p) level. Significantly, the structures of the N_2H_2...OH complexes shown in Fig. 7.21 are very different from the isoelectronic structure of the C_2H_4...OH complex with C_{2v} symmetry studied previously.[101] The binding energy of this π complex is 1.9 kcal mol^{-1} calculated at PMP2/aug-cc-PVQZ//MP2/cc-PVTZ level by Zhu et al.[101] In the C_2H_4:OH complex, the electron-deficient H atom of OH was shown to point perpendicularly toward the C=C π bond. The hydrogen bond length of OH...C predicted at the MP2/cc-PVTZ level is 2.455 Å (see Fig. 7.21), which is similar to the value we computed at the MP2/6-311++G(2df,p) level. Finally, similar to the *trans*-N_2H_2 reaction, the formation of the *cis*-HNN(H)OH intermediate in the *c*-N_2H_2 reaction with OH by association requires 15.3 kcal mol^{-1} barrier energy (TS37) which is significantly higher than that of the abstraction process. For the analogous, isoelectronic

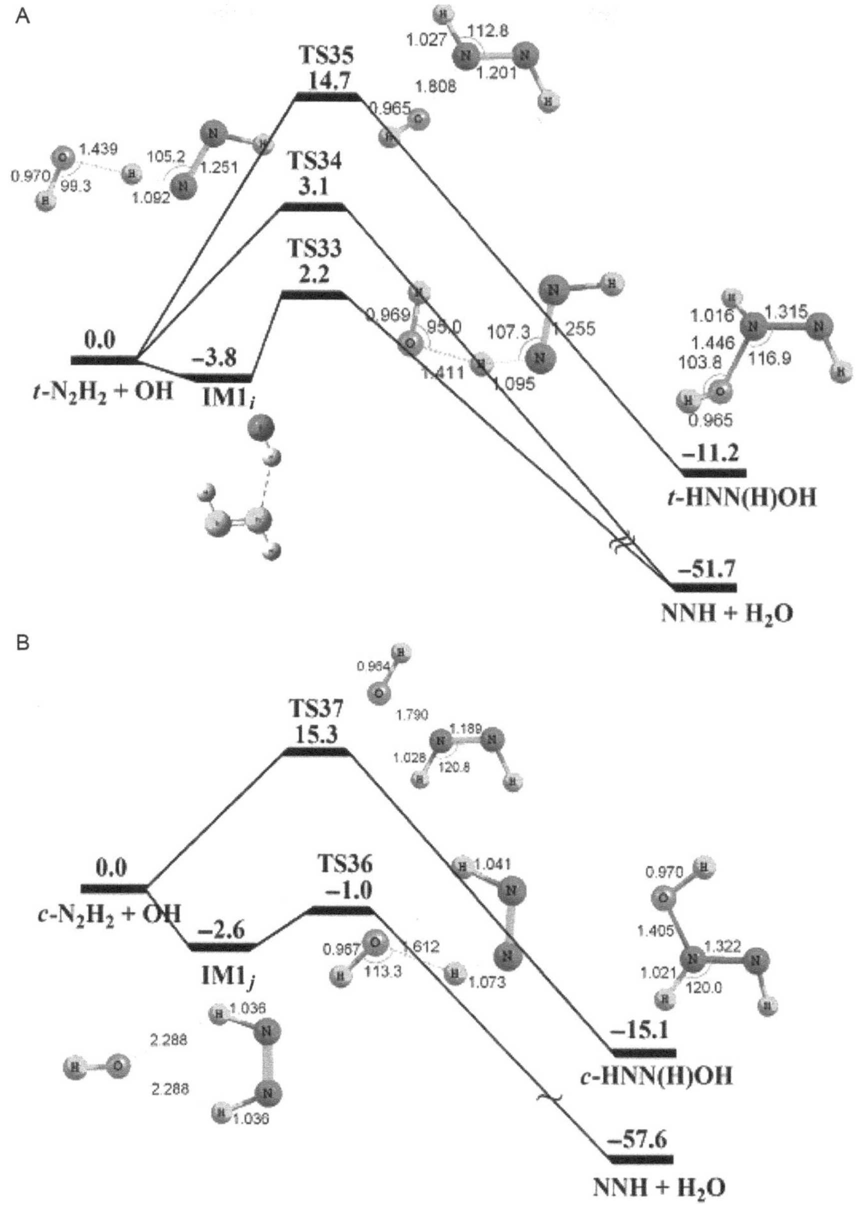

Figure 7.20 The schematic diagram of the potential energy surfaces for (A) t-N_2H_2 + OH and (B) c-N_2H_2 + OH computed at the CCSD(T)/6-311++G(3df,2p)//B3LYP/ 6311++G (3df,2p) level. (For color version of this figure, the reader is referred to the online version of this chapter.)

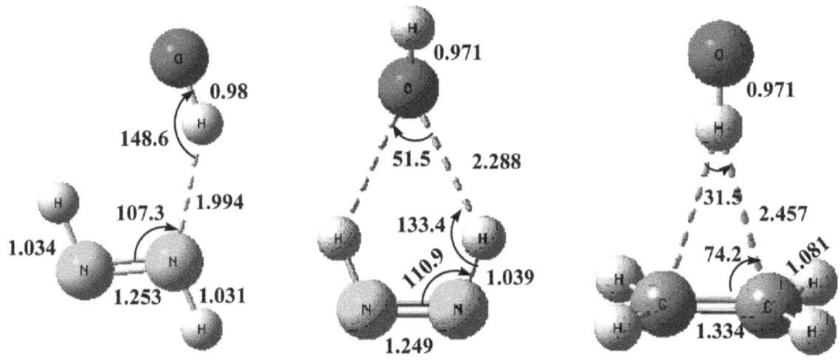

Figure 7.21 Optimized geometries of OH complexing with N_2H_2 and C_2H_4 computed at the MP2/6-311++g(2df,p) level. (For color version of this figure, the reader is referred to the online version of this chapter.)

C_2H_4+OH reaction, it take place almost exclusively by addition without thermal activation, instead of H abstraction.[101–103]

The rate constants for the both hydrogen abstraction product channels were computed with the transition state with Eckart tunneling corrections for *cis*- and *trans*-N_2H_2+OH reactions:

$$\textit{trans}\text{-}N_2H_2+OH\rightarrow NNH+H_2O, \tag{7.21}$$

$$\textit{cis}\text{-}N_2H_2+OH\rightarrow NNH+H_2O. \tag{7.22}$$

The predicted rate constant for the *trans*-N_2H_2+OH reaction channel producing the low-energy product $NNH+H_2O$ via TS33 and TS34 is $1.53\times10^{-20}\ T^{2.51}\ \exp(453/T)$ and $1.72\times10^{-20}\ T^{2.7}\ \exp(12/T)\,\text{cm}^3$ molecule^{-1} s^{-1} in the temperature range 300–2500 K, respectively. The total rate constant for the two production channels can be given by

$$k_{21}=1.12\times10^{-20}T^{2.8}\exp(255/T)\,\text{cm}^3\,\text{molecule}^{-1}\text{s}^{-1}.$$

The absolute value of our rate constant agrees very well with the theoretical result $k_{18}=9.84\times10^{-23}\ T^{3.40}\ \exp\ (686.3/T)\,\text{cm}^3$ molecule^{-1} s^{-1} predicted for the temperature range of 300–3000 K by Linder et al. based on the variational TST calculation using the MRCI method.[100]

In the *cis*-reaction, because of the existence of the van der Waals complex and the low-exist barrier, the effect of the Hirschfelder–Wigner–Miller multiple reflections[104] has been included. The predicted rate constants for *cis*-N_2H_2+OH reaction channel 19 has been represented by the following three parameter expressions given in units of cm^3 $molecule^{-1}$ s^{-1} for the temperature range 250–2500 K:

$$k_{22} = 1.27 \times 10^{-19} T^{2.25} \exp(1183/T).$$

3.7. Thermochemistry

The predicted heats of formation of all the important species involved in the hypergolic reaction of N_2H_4 and N_2O_4 are presented in Table 7.1 on the basis of the energies computed with different *ab initio* molecular orbital methods. These values were determined by combining the computed heats of reaction ($\Delta_r H_0^\circ$) and experimental heats of formation ($\Delta_f H_0^\circ$) of better known species in the reactions at 0 K. All the methods used in the reactions are presented in the footnote of Table 7.1. The experimental heats of formation of all the available species used in the reactions at 0 K are taken from NIST JANAF tables and the relevant literature.[58,98,105,106] We have given these reference values in the footnote of the Table 7.1. The heat of formation is calculated using the general formula given below using the $N_2H_4 + NO_2 \rightarrow N_2H_3 + c$-HONO reaction as an example,

$$\Delta_f H_0^\circ(N_2H_4) = \Delta_f H_0^\circ(N_2H_3) + \Delta_f H_0^\circ(c-\mathrm{HONO}) - \Delta_f H_0^\circ(NO_2) \\ - \Delta_r H_0^\circ$$

Our predicted heats of formation of the species listed in Table 7.1 are in good agreement with the values derived from available experimental and theoretical data within ±1 kcal mol^{-1}.[58,98,105,106]

4. CONCLUDING REMARKS

The mechanism for the hypergolic ignition of hydrazines (N_2H_4, CH_3NHNH_2, and $(CH_3)_2NNH_2$) with NTO (N_2O_4) at their liquid interface and in their gaseous state upon mixing at low temperatures is still unknown. There has been no realistic and meaningful simulation on how a falling droplet of N_2H_4 on a pool of liquid N_2O_4, and vice versa, observed experimentally is hypergolically ignited and exploded. Although the direct comparison of the two types of tests using the droplet of N_2H_4 falling on a pool of N_2O_4 versus the droplet of N_2O_4 falling on N_2H_4 indicates that they are distinctly different in the apparent splashing of the liquids upon contact, the probability of occurrences and the strength of explosions appeared to be similar. This important observation, from so far the most well-executed

Table 7.1 Heats of formation ($\Delta_f H_0^\circ$) of species at 0 K predicted at the various levels of theory given in kcal mol^{-1} with the listed reference reaction(s)

Species	Reaction(s)[a]	Heat of formation $\Delta_f H_0^\circ$ Calculated[b]	Literature
N_2H_4	$N_2H_4 \rightarrow NH_2 + NH_2$[c]	26.8 ± 0.2	26.2[58]
	$N_2H_4 + NO_2 \rightarrow N_2H_3 + c\text{-}HONO$[d]	27.3 ± 0.5	
N_2H_3	$N_2H_4 \rightarrow N_2H_3 + H$[c]	55.9	56.2[105]
	$N_2H_4 + NO_2 \rightarrow N_2H_3 + c\text{-}HONO$[d]	55.1 ± 0.5	
$t\text{-}N_2H_2$	$N_2H_3 \rightarrow t\text{-}N_2H_2 + H$[c]	50.0	≥48.8 ± 0.5[98]
	$N_2H_3 + NO_2 \rightarrow t\text{-}N_2H_2 + t\text{-}HONO$[e]	50.2	
$c\text{-}N_2H_2$	$N_2H_3 \rightarrow c\text{-}N_2H_2 + H$[c]	54.9	54.9[58]
	$N_2H_3 + NO_2 \rightarrow c\text{-}N_2H_2 + t\text{-}HONO$[e]	55.2	
NNH_2	$N_2H_3 + NO_2 \rightarrow NNH_2 + t\text{-}HONO$[e]	74.0	–
NNH	$t\text{-}N_2H_2 + NO_2 \rightarrow t\text{-}HONO + NNH$[e]	59.9 ± 1.7	–
$H_2NN(H)\text{-}NO_2$	$N_2H_4 + N_2O_4\ (D_{2h}) \rightarrow t\text{-}HONO + H_2NN(H)NO_2$[d]	28.4	–
N_2H_3O	$N_2H_3 + NO_2 \rightarrow N_2H_3O + NO$[e]	37.2	–
	$N_2H_3O \rightarrow NH_2 + HNO$[e]	37.7 ± 0.2	
	$N_2H_3 + N_2O_4 \rightarrow N_2H_3O + N_2O_3$[e]	37.5	
NH_2NO	$N_2H_3O \rightarrow NH_2NO + H$[e]	20.0	–
$H_2NN(H)\text{-}NO$	$N_2H_4 + t\text{-}ONONO_2(C_s) \rightarrow HONO_2 + H_2NN(H)NO$[d]	46.3	–
	$H_2NN(H)NO \rightarrow N_2O_3 + NO$[e]	47.3	–
$NH_2N(H)\text{-}NO_2$	$N_2H_3 + NO_2 \rightarrow NH_2N(H)NO_2$[e]	30.5	–
HNNH(O)	$t\text{-}N_2H_2 + NO_2 \rightarrow HNNH(O) + NO$[e]	29.4 ± 1.5	–
HNO(O)	$N_2H_3 + NO_2 \rightarrow t\text{-}N_2H_2 + HNO(O)$[e]	−7.5	–
$N_2O_4\ (D_{2h})$	$NO_2 + NO_2 \rightarrow N_2O_4\ (D_{2h})$[e]	5.1	4.5 ± 0.4[58]
$t\text{-}ONONO_2$	$NO_2 + NO_2 \rightarrow t\text{-}ONONO_2$[e,(c)]	11.5, (9.1)	–
$c\text{-}ONONO_2$	$NO_2 + NO_2 \rightarrow c\text{-}ONONO_2$[e,(c)]	14.1, (11.5)	–
N_2O_3	$N_2H_3 + N_2O_4 \rightarrow N_2H_3O + N_2O_3$[e]	21.2	21.4[58]

[a]Heats of reaction of the reaction(s) used for the heat of formation prediction using the methods as indicated.
[b]Experimental values used for calculations of the heats of formation at 0 K are $N_2O_4 = 4.5 \pm 0.4$ kcal mol^{-1}; $N_2H_4 = 26.2$ kcal mol^{-1}; $NO_2 = 8.6 \pm 0.2$ kcal mol^{-1}; $NO = 21.5 \pm 0.1$ kcal mol^{-1}; t-HONO = −17.4 ± 0.3 kcal mol^{-1}; c-HONO = −16.9 ± 0.3 kcal mol^{-1}; $N_2O_3 = 21.4$ kcal mol^{-1}; $HONO_2 = -29.8 \pm 0.1$ kcal mol^{-1}; HNO = 24.5 kcal mol^{-1}; $H = 51.7$ kcal mol^{-1} (Ref. 58); $N_2H_3 = 56.2$ kcal mol^{-1} (Ref. 105); $t\text{-}N_2H_2 = \geq 48.8 \pm 1.2$ kcal mol^{-1} (Ref. 98); $c\text{-}N_2H_2 = 54.9$ kcal mol^{-1} (Ref. 105); $NH_2 = 45.2 \pm 0.24$ kcal mol^{-1} (Ref. 106).
[c]CBS//B3LYP/6-311++G(3df,2p).
[d]G2M(CC3)//B3LYP/6-311++G(3df,2p).
[e]CCSD(T)/6-311++G(3df,2p)//B3LYP/6-311++G(3df,2p).

experiment on the hydrazine–NTO system, suggests that the initiation reaction involved in the two types of experiments by Kimura and coworkers[7] is essentially the same. This low-temperature liquid-phase initiation has been attributed to the nonthermally activated $N_2H_4 + ONONO_2$ reaction which produces $H_2NN(H)NO + HNO_2$ with 21 kcal mol^{-1} of exothermicity.[12] The HNO_3 formed in the reaction can instantaneously react with N_2H_4 giving rise to $[N_2H_5^+][NO_3^-]$ salt by acid–base neutralization process releasing twice as much energy as the initiation reaction. The energies released from the two nonthermally activated processes plus the mechanical energy of the falling droplet should set off the explosion as observed experimentally.

In order to assist future computer modeling of this important propellant system, we have further carried out a systematic study of potential chain initiating and propagating reactions by high-level quantum chemical and statistical theory calculations to elucidate the mechanisms involved and to provide their rate constants over a wide range of temperature and pressure. The elementary reactions investigated include the unimolecular decomposition reactions of N_2H_x ($x=3$, 4), N_2H_3NO, and N_2H_3O, and the bimolecular reactions of N_2H_3 radicals with NO_2 and N_2O_4, and N_2H_2 isomers with OH, NO_2, and N_2O_4. These reactions are believed to be important to the chain initiation and propagation in the early stages of explosion. The inclusion of these reactions with the detailed N_2H_4 decomposition mechanism established by Konnov and Ruyck[61] and the HNO_x system established by Park et al. for $NH_x + NO_x$ reaction[107] should allow us to more realistically model the complex chemistry of the hydrazine–NTO combustion process.

A preliminary test for the ignition of a mixture containing atmospheric pressure of both compounds indicated that at room temperature, the temperature and pressure of the system sharply increase within 0.1 μs upon mixing. The result clearly indicates the feasibility of understanding the hypergolic explosion in aforementioned experiments through a 3D-CFD simulation, for example.

ACKNOWLEDGMENTS

The authors deeply appreciate the support by Taiwan's National Science Council (NSC) under contract No. NSC100-2113-M-009-013 and by the Ministry of Education's ATU program. M. C. L. also acknowledges the support from the NSC for the distinguished visiting professorship at National Chiao Tung University in Hsinchu, Taiwan. We are also grateful to the National Center for High-Performance Computing for computer time and the use of its facilities. The authors would also like to acknowledge the help of Dr. Y. M. Lee for the crossing point search in the decomposition of N_2H_4 to triplet $NH + NH_3$ and BiRen Gu for preliminary ignition modeling.

REFERENCES

1. Friedman, R.; Barnes, W. P.; Markels, M. A., Jr. "Study of Explosions Induced by Contact of Hydrazine-type Fuels with Nitrogen Tetroxide", *Technical Documentary Report ASD-TDR-62-685*, 1962.
2. Weiss, H. G. Chemical Factors in N_2O_4-N_2H_4 Liquid Reactions. *NASA Contractor Report 64338*, 1965.
3. Burrows, M. C. Mixing and Reaction Studies of Hydrazine and Nitrogen Tetroxide using Photographic and Spectral Techniques. *NASA Contractor Report TND-4467*, 1968; p 22.
4. Hersch, M. Performance and Stability Characteristics of Nitrogen Tetroxide Hydrazine Combustors. *NASA Technical Note- D-4776*, 1968, 20.
5. Lawver, B. P.; Breen, B. P. Hypergolic Stream Impingement Phenomena-Nitrogen Tetroxide/Hydrazine. *NASA Contractor Report* 72444, 1968; p 25.
6. Lawver, B. R. Some Observations on the Combustion of N_2H_4 Droplets. *AIAA J.* **1966**, *4*, 659.
7. Daimon, W.; Tanaka, M.; Kimura, I. The Mechanism of Explosions Induced by Contact of Hypergolic Liquid Propellants, Hydrazine and Nitrogen Tetroxide. *Proc. Combust. Inst.* **1984**, *20*, 2065–2071; Tanaka, M.; Daimon, W.; Kimura, I. Explosion Phenomenon from Contact of Hypergolic Liquids. *J. Propul. Power* **1984**, *1*, 314–316; Daimon, W.; Goto, Y.; Kimura, I. Mechanism of Explosion Induced by Contact of Hypergolic Liquids. *J. Propul. Power* **1991**, 7, 946–952.
8. Sawyer, R. F.; Glassman, I. Gas-Phase Reactions of Hydrazine with Nitrogen Dioxide and Oxygen. *Proc. Combust. Inst.* **1967**, *11*, 861–869.
9. Miyajima, H.; Sakamoto, H. Gas Phase Ignition of Hydrazine with Nitrogen Dioxide. *Combust. Sci. Technol.* **1973**, *8*, 199.
10. Frank-Kamenetskii, D. A. *Diffusion and Heat Transfer in Chemical Kinetics*. Plenum Press: New York, 1969: p 353.
11. Daimon, Y.; Terashima, H.; Koshi, M. Evaluation of Rate Constants Relavant to the hypergolic Reaction of Hydrazine with Nitrogen Dioxide. In: *7th Mediterranean Combustion Symposium, Chia Laguna, Cagliari, Sardinia, Italy, September 11–15*, 2011.
12. Lai, K.-Y.; Zhu, R. S.; Lin, M. C. Why Mixtures of Hydrazine and Dinitrogen Tetroxide Are Hypergolic? *Chem. Phys. Lett.* **2012**, *537*, 33–37.
13. Pople, J. A.; Gordon, M. H.; Fox, D. J.; Raghavachari, K.; Curtiss, L. A. Gaussian-1 Theory: A General Procedure for Prediction of Molecular Energies. *J. Chem. Phys.* **1989**, *90*, 5622–5629.
14. Curtiss, L. A.; Raghavachari, K.; Pople, J. A. Gaussian-2 Theory : Use of Higher Level Correlation Methods, Quadratic Configuration Interaction Geometries, and Second-Order Moller-Plesset Zero-Point Energies. *J. Chem. Phys.* **1995**, *103*, 4192–4200.
15. Curtiss, L. A.; Raghavachari, K.; Redfern, P. C.; Rassolov, V.; Pople, J. A. Gaussian-3 (G3) Theory for Molecules Containing First and Second-Row Atoms. *J. Chem. Phys.* **1998**, *109*, 7764–7776.
16. Curtiss, L. A.; Redfern, P. C.; Raghavachari, K.; Pople, J. A. Gaussian-3X (G3X) Theory: Use of Improved Geometries, Zero-Point Energies, and Hartree-Fock Basis Sets. *J. Chem. Phys.* **2001**, *114*, 108–117.
17. Mebel, A. M.; Morokuma, K.; Lin, M. C. Modification of the Gaussian2 Theoretical Model: The Use of Coupled Cluster Energies, Density Functional Geometries, and Frequencies. *J. Chem. Phys.* **1995**, *103*, 7414–7421.
18. Becke, A. D. Density Functional Thermochemistry. I. The Effect of the Exchange Only Gradient Correction. *J. Chem. Phys.* **1992**, *96*, 2155–2160.
19. Becke, A. D. Density Functional Thermochemistry. III. The Role of Exact Exchange. *J. Chem. Phys.* **1993**, *98*, 5648–5652.

20. Lee, C.; Yang, W.; Parr, R. G. Development of the Colle-Salvetti Correlation-Energy Formula into a Functional of the Electron Density. *Phys. Rev. B* **1988**, *37*, 785–789.
21. Park, J.; Xu, Z. F.; Lin, M. C. Kinetics Study of the $C_{10}H_7+O_2$ Reaction. *J. Phys. Chem. A* **2009**, *113*, 5348–5354.
22. Park, J.; Lin, M. C. Temperature and Pressure Effects on Formation and Decomposition of Phenylvinylperoxy Radicals in the $C_6H_5C_2H_2+O_2$ Reaction. *Proc. Combust. Inst.* **2009**, *32*, 305–310.
23. Nam, G.; Tokmakov, I. V.; Park, J.; Lin, M. C. Kinetics for the Reaction of Phenyl radical with Phenylacetylene and Styrene. *Proc. Combust. Inst.* **2007**, *31*, 249–256.
24. Xu, Z. F.; Lin, M. C. Computational Study on the Mechanism and Rate Constant for the $C_6H_5+C_6H_5NO$ Reaction. *J. Phys. Chem. A* **2005**, *109*, 9054–9060.
25. Choi, Y. M.; Park, J.; Wang, L.; Lin, M. C. Formation and Decomposition of Phenylvinylperoxy Radicals in the Reaction: $C_6H_5C_2H_2+O_2$. *ChemPhysChem* **2004**, *5*, 1231–1234.
26. Zhu, R. S.; Lin, M. C. Ab Initio Study of Ammonium Perchlorate Combustion Initiation Processes: Unimolecular Decomposition of Perchloric Acid and the Related $OH+ClO_3$ Reaction. *PhysChemComm* **2001**, *4*, 127–132.
27. Zhu, R. S.; Lin, M. C. Ab Initio Studies of ClO_x Reactions. 2. Unimolecular Decomposition of s-ClO_3 and the Bimolecular O+OClO Reaction. *J. Phys. Chem. A* **2002**, *106*, 8386–8390.
28. Zhu, R. S.; Lin, M. C. Ab Initio Studies of ClO_x Reactions. VII. Isomers of Cl_2O_3 and Their Roles in the ClO + OClO Reaction. *J. Chem. Phys.* **2003**, *118*, 8645–8655.
29. Xu, Z. F.; Lin, M. C. Ab Initio Studies of ClO_x Reactions. IX. Combination and Disproportionation Reactions of ClO and s-ClO_3 Radicals. *J. Chem. Phys.* **2003**, *119*, 8897–8904.
30. Zhu, R. S.; Xu, Z. F.; Lin, M. C. Ab Initio Studies of ClO_x Reactions. I. Kinetics and Mechanism for the OH+ClO Reaction. *J. Chem. Phys.* **2002**, *116*, 7452–7460.
31. Xu, Z. F.; Zhu, R. S.; Lin, M. C. Ab Initio Studies of ClO_x Reactions. 3. Kinetics and Mechanism for the OH+OClO Reaction. *J. Phys. Chem. A* **2003**, *107*, 1040–1049.
32. Moskaleva, L. V.; Lin, M. C. Computational Study of the Kinetics and Mechanisms for the Reaction of H Atoms with c-C_5H_6. *Proc. Combust. Inst.* **2002**, *29*, 1319–1327.
33. Zhu, R. S.; Lin, M. C. Ab Initio Studies of ClO_x Radical Reactions: V. Evidence for a New Path in the Cl+ClOOCl Reaction. *J. Phys. Chem. A* **2003**, *107*, 3836–3840.
34. Xu, Z. F.; Zhu, R. S.; Lin, M. C. Ab Initio Studies of ClO_x Reactions: VI. Theoretical Prediction of Total Rate Constant and Product Branching Probabilities for the HO_2+ClO Reaction. *J. Phys. Chem. A* **2003**, *107*, 3841–3850.
35. Zhu, R. S.; Lin, M. C. Ab Initio Studies of ClO_x Reactions. VIII. Isomerization and Decomposition of ClO_2 Radicals and Related Bimolecular Processes. *J. Chem. Phys.* **2003**, *119*, 2075–2082.
36. Zhu, R. S.; Lin, M. C. Ab Initio Studies of ClO_x Reactions: Prediction of the Rate Constants of ClO+NO for the Forward and Reverse Processes. *ChemPhysChem* **2004**, *5*, 1864–1870.
37. Zhu, R. S.; Lin, M. C. Ab Initio Studies of ClO_x Reactions: Prediction of the Rate Constants of $ClO+NO_2$ for the Forward and Reverse Processes. *ChemPhysChem* **2005**, *6*, 1514–1521.
38. Woon, D. E., Jr.; Dunning, T. H. Gaussian Basis Sets for Use in Correlated Molecular Calculations. V. Core-valence Basis Sets for Boron Through Neon. *J. Chem. Phys.* **1995**, *103*, 4572–4585.
39. Peterson, K. A.; Woon, D. E.; Dunning, T. H., Jr. Benchmark Calculations with Correlated Molecular Wave Functions. IV. The Classical Barrier Height of the $H+H_2 \rightarrow H_2+H$ Reaction. *J. Chem. Phys.* **1994**, *100*, 7410–7415.

40. Johnson, B. G.; Frisch, M. J. An Implementation of Analytic Second Derivatives of the Gradient-Corrected Density Functional Energy. *J. Chem. Phys.* **1994**, *100*, 7429–7442; *Chem. Phys. Lett.* **1993**, *216*, 133–140.
41. Halpern, A. M.; Glendening, E. D. Ab Initio Study of the Torsional Potential Energy Surfaces of N_2O_3 and N_2O_4: Origin of the Torsional Barriers. *J. Chem. Phys.* **2007**, *126*, 154305–154308.
42. Gonzalez, C.; Schlegel, H. B. An Improved Algorithm for Reaction Path Following. *J. Chem. Phys.* **1989**, *90*, 2154–2161.
43. Pople, J. A.; Head-Gordon, M.; Raghavachari, K. Quadratic Configuration Interaction. A General Technique for Determining Electron Correlation Energies. *J. Chem. Phys.* **1987**, *87*, 5968–5975.
44. Frisch, M. J.; et al. *Gaussian 03*. Gaussian Inc.: Wallingford, CT, 2004.
45. Marcus, R. A. Unimolecular Dissociations and Free Radical Recombination Reactions. *J. Chem. Phys.* **1952**, *20*, 359–364.
46. Klippenstein, S. J.; Wagner, A. F.; Dunbar, R. C.; Wardlaw, D. M.; Robertson, S. H. VARIFLEX: Version 1.00, 1999.
47. Mokrushin, V.; Bedanov, V.; Tsang, W.; Zachariah, M.; Knyazev, V. *ChemRate*, version 1.5.8; NIST: Gaithersburg, MD, 2009.
48. Wardlaw, D. M.; Marcus, R. A. RRKM Reaction Rate Theory for Transition States of Any Looseness. *Chem. Phys. Lett.* **1984**, *110*, 230–234; Wardlaw, D. M.; Marcus, R. A. Unimolecular Reaction Rate Theory for Transition States of Partial Looseness. II. Implementation and Analysis with Applications to NO_2 and C_2H_6 Dissociations. *J. Chem. Phys.* **1985**, *83*, 3462–3480.
49. Klippenstein, S. J. An Efficient Procedure for Evaluating the Number of Available States within a Variably Defined Reaction Coordinate Framework. *J. Phys. Chem.* **1994**, *98*, 11459–11464; Klippenstein, S. J. A Bond Length Reaction Coordinate for Unimolecular Reactions. II. Microcanonical and Canonical Implementations with Application to the Dissociation of NCNO. *J. Chem. Phys.* **1991**, *94*, 6469–6482.
50. Gilbert, R. G.; Smith, S. C. *Theory of Unimolecular and Recombination Reactions*; Blackwell Scientific: Carlton, Australia, 1990.
51. Holbrook, K. A.; Pilling, M. J.; Robertson, S. H. *Unimolecular Reaction*; Wiley: New York, 1996.
52. Truhlar, D. G.; Garrtett, B. C.; Klippenstein, S. J. Current Status of Transition-State Theory. *J. Phys. Chem.* **1996**, *100*, 12771–12800.
53. MOLPRO Version 2009.1 is a package of *ab initio* programs written by Werner, H.-J.; Knowles, P. J. with contributions from Almlof, J.; Amos, R. D.; Berning, A. et al.
54. Truhlar, D. G. *The Reaction Path in Chemistry: Current Approaches and Perspectives*; Kluwar Academic Publishers: Dordrecht, 1995.
55. Hsu, C. C.; Mebel, A. M.; Lin, M. C. Ab Initio Molecular Orbital Study of the $HCO + O_2$ Reaction: Direct Versus Indirect Abstraction Channels. *J. Chem. Phys.* **1996**, *105*, 2346–2352.
56. Chakraborthy, D.; Hsu, C. C.; Lin, M. C. Theoretical Studies of Nitroamino Radical reactions: Rate Constants for the Unimolecular Decomposition of $HNNO_2$ and Related Bimolecular Processes. *J. Chem. Phys.* **1998**, *109*, 8887–8896.
57. Troe, J. Theory of Thermal Unimolecular Reactions at Low Pressures. I. Solutions of the Master Equation. *J. Chem. Phys.* **1977**, *66*, 4745–4757; Troe, J. Theory of Thermal Unimolecular Reactions at Low Pressures. II. Strong Collision Rate Constants. Applications. *J. Chem. Phys.* **1977**, *66*, 4758–4775.
58. Chase, M. W. J. NIST-JANAF Thermochemical. Tables, 4th ed.; *J. Phys. Chem. Ref. Data Monograph* No. 9 (Parts I and II), 1998.

59. Ruscic, B.; Berkowitz, J. Photoionization Mass Spectrometric Study of N_2H_2 and N_2H_3:N-H, N=N Bond Energies and Proton Affinity of N_2. *J. Chem. Phys.* **1991**, *95*, 4378–4384.
60. Rubick, A.; Joseph, W. B.; Gabriel, S.; Saartje, S.; Nguyen, M. T. Formation and Decomposition of Chemically Activated and Stabilized Hydrazine. *J. Phys. Chem. A* **2010**, *114*, 6235–6249.
61. Konnov, A. A.; Ruyck, J. D. Kinetic Modeling of the Decomposition and Flames Hydrazine. *Combust. Flame* **2001**, *124*, 106–126.
62. Chang, J. G.; Chen, H. T.; Xu, S.; Lin, M. C. Computational Study on the Kinetics and Mechanisms for the Unimolecular Decomposition of Formic and Oxalic Acids. *J. Phys. Chem. A* **2007**, *111*, 6789–6797.
63. Carl, L. Y. 1st ed.; *Handbook of Thermodynamic Diagrams;* William Andrew Publishing: USA, 1996.
64. Meyer, E.; Oslchewski, H. A.; Troe, J.; Wagner, H. G. Investigation of N_2H_4 and H_2O_2 Decomposition in Low and High Pressure Shock Waves. *Symp. Int. Combust. Proc.* **1969**, *12*, 345–355.
65. McHale, E. T.; Knox, B. E.; Palmer, H. B. Determination of the Decomposition Kinetics of Hydrazine Using a Single-Pulse Shock Tube. *Symp. Int. Combust. Proc.* **1965**, *10*, 341–351.
66. Eberstein, I. J.; Glassman, I. The Gas-Phase Decompisition of Hydrazine and Its Methyl Derivatives. *Symp. Int. Combust. Proc.* **1965**, *10*, 365–374.
67. Moberly, W. H. Shock Tube Study of Hydrazine Decomposition. *J. Phys. Chem.* **1962**, *66*, 366–368.
68. Kerr, J. A.; Sekhar, R. C.; Trotman-Dickenson, A. F. The Pyrolyses of Hydrazine and Benzylamines. C-C and N-N Bond Dissociation Energies. *J. Chem. Soc.* **1963**, 3217–3225.
69. Genich, A. P.; Zhirnov, A. A.; Maneils, G. B. Decomposition of Hydrazine Behind Reflected Shock Waves at High Pressures. *Russ. J. Phys. Chem.* **1974**, *48*, 728–729.
70. Szwarc, M. The Dissociation Energy of N-N Bond in Hydrazine. *Proc. R. Soc. Lond.* **1949**, *198*, 267–284.
71. Liu, W. G.; Goddard, W. A., III First-Principle Study of the Role of Interconversion Between NO_2, N_2O_4, cis-ONO-NO_2, and trans-ONO-NO_2 in Chemical Processes. *J. Am. Chem. Soc.* **2012**, *134*, 12970–12978.
72. Zhu, R. S.; Lai, K. Y.; Lin, M. C. Ab Initio Chemical Kinetics for the Hydrolysis of N_2O_4 Isomers in the Gas Phase. *J. Phys. Chem. A* **2012**, *116*, 4466–4472.
73. Pimentel, A. S.; Lima, F. A. A.; da Silva, A. B. F. The Isomerization of Dinitrogen Tetroxide: O_2N-$NO_2 \rightarrow$ ONO-NO_2. *J. Phys. Chem. A* **2007**, *111*, 2913–2920.
74. Pinnick, D. A.; Agnew, S. F.; Swanson, B. I. Fluid Dinitrogen Tetroxide at Very High Pressure and High Temperature: Observation of the Nitrite Isomer. *J. Phys. Chem.* **1992**, *96*, 7092–7096.
75. Fateley, W. G.; Bent, H. A.; Crawford, B. J. Infrared Spectra of the Frozen Oxides of Nitrogen. *J. Chem. Phys.* **1959**, *31*, 204–217.
76. Hisatsune, I. C.; Devlin, J. P.; Wada, Y. J. Infrared Spectra of Some Unstable Isomers of N_2O_4 and N_2O_3. *J. Chem. Phys.* **1960**, *33*, 714–719.
77. Louis, R. V. S.; Crawford, B., Jr. Infrared Spectrum of Matrix-Isolated NO_2. *J. Chem. Phys.* **1965**, *42*, 857–864.
78. Varetti, E. L.; Pimental, G. C. Isomeric Forms of Dinitrogen Trioxide in a Nitrogen Matrix. *J. Chem. Phys.* **1971**, *55*, 3813–3821.
79. Parts, L.; Miller, J. T. Nitrosonium Nitrate. Isolation at 79-250 K and Infrared Spectra of the Polymorphic Compound. *J. Chem. Phys.* **1965**, *43*, 136–139.
80. Bolduan, F.; Jodl, H. J. Raman Spectroscopy on Matrix-Isolated NO^+, NO_3^- and N_2O_4 in Ne. *Chem. Phys. Lett.* **1982**, *85*, 283–286.

81. Bolduan, F.; Jodl, H.; Loewenschuss, A. Raman Study of Solid N_2O_4: Temperature Induced Autoionization. *J. Chem. Phys.* **1984**, *80*, 1739–1743.
82. Agnew, S. F.; Swanson, B. I.; Jones, L. H.; Mills, R. L.; Schiferl, D. Chemistry of Nitrogen Oxide (N_2O_4) at High Pressure : Observation of a Reversible Transformation Between Molecular and Ionic Crystalline Forms. *J. Phys. Chem.* **1983**, *87*, 5065–5068.
83. Jones, L. H.; Swanson, B. I.; Agnew, S. F. Infrared Studies of Autoionization of Thin Films of Dinitrogen Tetroxide. *J. Chem. Phys.* **1985**, *82*, 4389–4390.
84. Givan, A.; Loewenschuss, A. Fourier Transform Infrared and Raman Studies on Solid Nitrogen Dioxide: Temperature Cycling of Ordered, Disordered, and Multicompotent Layers. *J. Chem. Phys.* **1989**, *90*, 6135–6142; Givan, A.; Loewenschuss, A. On the Intermolecularity or Intramolecularity of Nitrosonium Nitrate Formation in Thin Films of Nitrogen Dioxide: A Fourier Transform Infrared Study. *J. Chem. Phys.* **1989**, *91*, 5126–5127.
85. Stirling, A.; Papai, I.; Mink, J.; Salahub, D. R. Density Functional Study of Nitrogen Oxides. *J. Chem. Phys.* **1994**, *100*, 2910–2923.
86. McKee, M. L. Ab Initio Study of the N_2O_4 Potential Energy Surface. Computational Evidence for a New N_2O_4 Isomer. *J. Am. Chem. Soc.* **1995**, *117*, 1629–1637.
87. Zakharov, I. I.; Kolbasin, A. I.; Zakharova, O. I.; Kravchenko, I. V.; Dyshlovoi, V. I. Quantum Chemical Evidence for the Possible Existence of a New Isomer of Dinitrogen Tetroxide. *Theor. Exp. Chem.* **2008**, *44*, 26–31.
88. Wang, X.; Qin, Q.-Z.; Fan, K. Ab Initio Study on the Characterization of N_2O_4 Isomers. *J. Mol. Struct. (Theochem)* **1998**, *432*, 55–62.
89. Wille, U.; Dreessen, T. Mechanistic Insights into NO_3 Induced Self-Terminating Radical Oxygenations, Part1: A Computational Study on $NO_3$and Its Addition to Alkynes. *J. Phys. Chem. A* **2006**, *110*, 2195–2203.
90. Vyazovkin, S.; Clawson, J. S.; Wight, C. A. Thermal Dissociation Kinetics of Solid and Liquid Ammonium Nitrate. *Chem. Mater.* **2001**, *13*, 960–966.
91. Zhu, R. S.; Lin, M. C. CH_3NO_2 Decomposition/Isomerization Mechanism and Product Branching Ratios: An Ab Initio Chemical Kinetics Study. *Chem. Phys. Lett.* **2009**, *478*, 11–16.
92. Zhu, R. S.; Raghunath, P.; Lin, M. C. Effect of Roaming Transition States upon Product Branching in the Thermal Decomposition of CH_3NO_2. *J. Phys. Chem. A* **2013**, *117*, 7308–7313.
93. Glarborg, P.; Dam-Johansen, K.; Miller, J. A.; Kee, R. J.; Coltrin, M. E. Modeling the Thermal $DENO_x$ Process in Flow Reactors. Surface Effects and Nitrous Oxide Formation. *Int. J. Chem. Kinet.* **1994**, *26*, 421–453.
94. Miller, J. A.; Bowman, C. T. Mechanism and Modeling of Nitrogen Chemistry in Combustion. *Prog. Energy Combust. Sci.* **1989**, *15*, 287–338.
95. Brill, T. B.; Brush, P. J.; Patil, D. G. Thermal Decomposition of Energetic Materials 58. Chemistry of Ammonium Nitrate and Ammonium Dinitramide Near the Burning Surface Temperature. *Combust. Flame* **1993**, *92*, 178–186.
96. Xu, S.; Lin, M. C. Ab Initio Chemical Kinetics for the NH_2+HNO_x Reactions. Part I: Kinetics and Mechanism for NH_2+HNO. *Int. J. Chem. Kinet.* **2009**, *41*, 667–677.
97. Biczysko, M.; Poveda, L. A.; Varandas, A. J. C. Accurate MRCI Study of Ground State N_2H_2 Potential Energy Surface. *Chem. Phys. Lett.* **2006**, *424*, 46–53.
98. Biehl, H.; Stuhl, F. Vacuum Ultraviolet Photolysis of N_2H_2: Generation of NH Fragments. *J. Chem. Phys.* **1994**, *100*, 141–145.
99. Willis, C.; Back, R. A. Di-imide: Some Physical and Chemical Properties, and the Kinetics and Stoichiometry of the Gas-Phase Decomposition. *Can. J. Chem.* **1973**, *51*, 3605–3619.

100. Linder, D. P.; Duan, X.; Page, M. Thermal Rate Constants for $R + N_2H_2 \rightarrow RH + N_2H$ ($R = H$, OH, NH_2) Determined from Multireference Configuration Interaction and Variational Transition State Theory Calculations. *J. Chem. Phys.* **1996**, *104*, 6298–6307.
101. Zhu, R. S.; Park, J.; Lin, M. C. Ab Initio Kinetic Study on the Low-Energy Paths of the $HO + C_2H_4$ Reaction. *Chem. Phys. Lett.* **2005**, *408*, 25–30.
102. Fulle, D.; Hamann, H. F.; Hippler, H.; Jansch, C. P. The High Pressure Range of the Addition of OH to C_2H_2 and C_2H_4. *Ber. Bunsen. Phys. Chem.* **1997**, *101*, 1433–1442.
103. Diau, E. W. G.; Lee, Y. P. Detailed Rate Coefficients and the Enthalpy Change of the Equilibrium Reaction $OH + C_2H_4 = HOC_2H_4$ over the Temperature Range 544-673 K. *J. Chem. Phys.* **1992**, *96*, 377–386.
104. Miller, W. H. Unified Statistical Model for "Complex" and "Direct" Reaction Mechanisms. *J. Chem. Phys.* **1976**, *65*, 2216–2223.
105. Matus, M. H.; Arduengo, A. J., III; Dixon, D. A. The Heats of Formation of Diazene, Hydrazine, $N_2H_3^+$, $N_2H_5^+$, N_2H, and N_2H_3 and the Methyl Derivatives CH_3NNH, CH_3NNCH_3, and $CH_3HNNHCH_3$. *J. Phys. Chem. A* **2006**, *110*, 10116–10121.
106. Ruscic, B.; et al. IUPAC Critical Evaluation of Thermochemical Properties of Selected Radicals. Part I. *J. Phys. Chem. Ref. Data* **2005**, *34*, 573–656.
107. Park, J.; Gates, S.; Lin, M. C. Photolytically and Thermally Initiated Reactions of NH_3 with NO_x ($x = 1,2$). *Combust. Sci. Technol.* **2010**, *182*, 365–379.

CHAPTER EIGHT

Material Dependence of Water Interactions with Metal Oxide Nanoparticles: TiO_2, SiO_2, GeO_2, and SnO_2

Marta Gałyńska, Petter Persson[1]
Theoretical Chemistry Division, Chemistry Department, Lund University, Box 124, Lund, Sweden
[1]Corresponding author: e-mail address: petter.persson@teokem.lu.se

Contents

Abstract

Surface interactions of water with nano-TiO_2 and structurally related group IV metal dioxide nanoparticles have been investigated using density functional theory (DFT) and time-dependent DFT (TD-DFT) calculations. The investigated clusters include 46 and 92 unit models of TiO_2 (i.e., $(TiO_2)_{46}$ and $(TiO_2)_{92}$) of the three low-energy polymorphs of TiO_2: anatase, rutile, and brookite. The investigation also includes studies of corresponding rutile-type clusters of three structurally related forms of group IV oxides: SnO_2, GeO_2, and SiO_2. The calculations demonstrate the strong influence of water surface adsorption on several important nanoparticle properties including structure, charge distribution, and electronic structure for all investigated materials. Significant differences between the various materials are also revealed by the calculations, including an unusual stability of the electronic structure of TiO_2 in the environment of particular relevance for its photoelectrochemical and photocatalytic applications.

Advances in Quantum Chemistry, Volume 69
ISSN 0065-3276
http://dx.doi.org/10.1016/B978-0-12-800345-9.00008-8

1. INTRODUCTION

The fast development of several modern photoelectrochemical and photocatalytic applications such as dye-sensitized solar cells,[1–4] water splitting, and solid gas sensors[5–8] has made TiO_2 and related metal oxide materials highly relevant for current research. Moreover, the advanced functionality of these materials is often based on the unique properties of nanocrystal materials in general, and their nanoscale surface properties in particular. This creates a great demand for the careful characterization of nanoparticle materials beyond the scope of traditional material investigations of bulk and solid surface properties. Furthermore, since many important processes take place in solution, it is often not enough to consider only bare cluster models of nanoparticles in vacuum. The interface between water and nanoparticle surfaces is particularly important with respect to understanding the behavior of nanoparticles in solution, but the inclusion of surface adsorbate layers can also be crucial for understanding a wide range of chemical conversion processes. Therefore, it constitutes a research area where there are both significant opportunities and challenges for first principles modeling of nanoscale material properties and processes.[9]

There have been many computational studies dedicated to model interactions of metal oxide surfaces with water molecules. Significant attention has been directed, in particular, toward the investigation of water interactions with different polymorphs of titanium dioxide[10] using quantum chemical periodic[11–14] or cluster model[15–18] calculations. Dynamical simulations have also been carried out recently using first principles molecular dynamics[19] and classical molecular dynamics simulations[11,20,21] including, for example, ReaxFF reactive force field simulations.[22] Several of these studies address complicated questions about the type of water adsorption (associative, dissociative, or mixed associative/dissociative) on different TiO_2 surfaces. Other studies investigate the influence of saturation on the structural and electronic properties of TiO_2 nanoparticles.

Theoretical studies of nanostructured TiO_2 have also emerged as an important focal area of broader significance as the application of nano-TiO_2 and related metal oxide materials to solar energy conversion in dye-sensitized solar cells and related photoelectrochemical and photocatalytic processes have emerged as a premier showcase of modern nanoscience at the boarder of chemistry, physics, and material science.[9,23–27] Key intrinsic material properties for these applications include, for example, the electronic structure and optical properties of nanostructured metal oxides.[27,28] There is also significant interest in understanding surface sensitization properties, in particular the photoinduced interfacial electron transfer processes between molecular adsorbates and wide band gap semiconducting metal oxide nanocrystals.[27] This includes, for example, interfacial charge-transfer excitation mechanisms,[29] ultrafast electron injection from champion dyes to nano-TiO_2,[30] and anchor/spacer control of interfacial electron transfer.[31] In this context, it is important to include surface interactions of the nanomaterials with the environment in order to facilitate progress toward a more realistic modeling of such interfaces.[27]

For purposes of comparison, which allows more general material conclusions to be drawn, it is also interesting to compare nano-TiO_2 properties with those of structurally related metal oxides, notably, some group IV MO_2 materials. Several studies have also been carried out of SnO_2 saturated by water as another example of a currently important material.[32] Also, mechanistic differences in the water adsorption between rutile-type surfaces of SnO_2 and TiO_2 have been investigated for selected surfaces.[33–35] SiO_2 is another group IV material that can, under special conditions, also be found in a crystal structure belonging to the rutile group, that is, in the so-called stishovite form. Stishovite is a high-pressure SiO_2 polymorph with an unusual 6:3 coordination of the ions, which has been found to occur naturally in meteorite impact craters. Similarly, the argutite form of GeO_2 belongs to the rutile group, and including GeO_2 in our study provides for a systematic comparison of group IV MO_2 model nanocrystals.

Here, we present results from quantum chemical cluster calculations of three low-energy polymorphs of titanium dioxide: anatase, brookite, and rutile, as well as rutile-type forms of tin dioxide (SnO_2), germanium dioxide (GeO_2), and silicon dioxide (SiO_2) saturated by dissociative water molecules. Water interactions with nanoparticle surfaces have thus been investigated for $(MO_2)_n$ with M = Ti, Sn, Ge, and Si, and $n = 46$, 92 using cluster models of anatase, brookite, and rutile. The clusters have been constructed using a procedure based on stoichiometry, high coordination of all ions, and balanced charge distribution that has previously been employed and validated for cluster calculations of nanostructured TiO_2.[28,36,37] Results related to the geometry, stability, and electronic structure of the bare TiO_2 clusters corresponding to those that are water saturated were recently reported.[28]

In this study, under-coordinated surface titanium and oxygen atoms have been saturated with hydroxyl groups and hydrogen atoms in a 1:1 proportion to get fully coordinated ions, corresponding to the addition of water molecules to form $(MO_2)_n(H_2O)_m$ clusters, where both n and m are integers. This surface saturation model keeps the clusters neutral, in addition to avoiding surface defects that are created because of under-coordinated surface ions. A computational advantage of this procedure is that because the bare stoichiometric oxide clusters, $(MO_2)_n$, are already electroneutral, and an integral number of water molecules, $(H_2O)_m$, is added—which can be regarded as adding either hydroxyl and hydrogen atoms ($^{\bullet}OH/^{\bullet}H$) or hydroxide ions and protons ($OH^{-\bullet}/H^{+}$) in equal amounts—the charges of all the combined $(MO_2)_m(H_2O)_n$ systems can unambiguously be taken to be zero regardless of the (partial) ionicity of the substrate and adsorbate components. For simplicity of notation, the adsorbed molecular fragments are, when not noted otherwise, typically referred to as surface ions, that is, hydroxide ions and protons, in the following sections without taking the actual ionicity of the molecular fragments into consideration.

This study systematically investigates changes in geometry, charge distribution, and electronic structure between the bare and water-saturated clusters for several metal oxide nanocrystal materials in a consistent manner. This approach provides significant opportunities to provide a comparative analysis of the material dependence of surface interactions of nanostructured MO_2 materials. As mentioned above, it should, however, be noted that water adsorption on metal oxide surfaces is in general a complex problem that often involves mixed associative/dissociative adsorption, and the present investigation of clusters embedded by dissociative water is in this sense a stepping stone to a more general understanding and realistic modeling of water at the interface of nanocrystal surfaces in the solution. The question of water adsorption type and mechanism of water dissociation on different oxide nanocrystal surfaces thus remains open for further research.

2. COMPUTATIONAL METHODS

Density functional theory (DFT) and time-dependent density functional theory (TD-DFT) with the standard B3LYP hybrid functional[38] have been used for all calculations. Water-saturated TiO_2 clusters have been fully relaxed with an ECP basis set introduced by Bouteiller et al.[39] This basis set is referred to as VDZ and has been used successfully in several previous investigations of TiO_2 materials.[37,40] The electronic structure and the first excitation energies have been computed using the all electron triple-zeta quality 6-311G basis set.[41–43] The water-passivated rutile-type SnO_2, GeO_2, and SiO_2 clusters have been fully investigated using the LANL2DZ basis set.[44–46] In addition to the explicit adsorbed water molecules, a polarization continuum model (PCM) has been used to model the effect of an implicit solvent. All calculations have been made using the Gaussian 09 package.[47]

3. RESULTS

3.1. TiO_2 clusters

Changes in geometry, charge distribution, and electronic structure between bare and embedded clusters of three polymorphic forms of titanium dioxide: anatase, brookite, and rutile have been studied. The clusters are labeled as x-$(TiO_2)_n(H_2O)_m$, where x is the first letter of polymorphs (r for rutile, a for anatase, and b for brookite), n is the number of TiO_2 units, and m is the number of dissociative water molecules in the saturated clusters. Different numbers of water molecules need to be added to the different clusters to fully saturate all under-coordinated surface Ti and O atoms. The anatase, brookite, and rutile $(TiO_2)_{46}$ clusters were passivated by 48, 52, and 54 water molecules, respectively. Similarly, 72, 80, and 84 water molecules were used to saturate the anatase, brookite, and rutile $(TiO_2)_{92}$ clusters, respectively.

3.1.1 Geometries

The bare and saturated titanium oxide clusters were relaxed using the B3LYP/VDZ level of theory, and the resulting optimized structures are shown in Fig. 8.1. The optimized geometries have also been compared in terms of Ti—O bond distances and Ti—O—Ti angle distributions, as recently done for the bare TiO_2 clusters.[28] These results are shown for all investigated TiO_2 clusters in Fig. 8.2.

The Ti—O bond distances for the bare and saturated TiO_2 clusters have been compared in Fig. 8.2. Each point in the diagrams represents the number of Ti—O bonds within a 0.3 Å length interval. The Ti—OH bonds created by the surface hydroxide groups have been included separately. The comparison shows that the Ti—O bond distances within the cluster become longer in the water-passivated cluster regardless of the TiO_2 polymorph type. The average bond lengths are furthermore similar for all TiO_2 clusters according to the results listed in Table 8.1. A sharp maximum and smaller mean deviation of the calculated average bond lengths, especially in the case of the anatase and rutile clusters, indicate a more uniform bond character in the saturated clusters. The lack of very short bonds in the water-saturated clusters is, furthermore, indicative of significantly less reconstructed surfaces and bulk compared to the bare clusters as seen also in Fig. 8.1. The Ti—OH

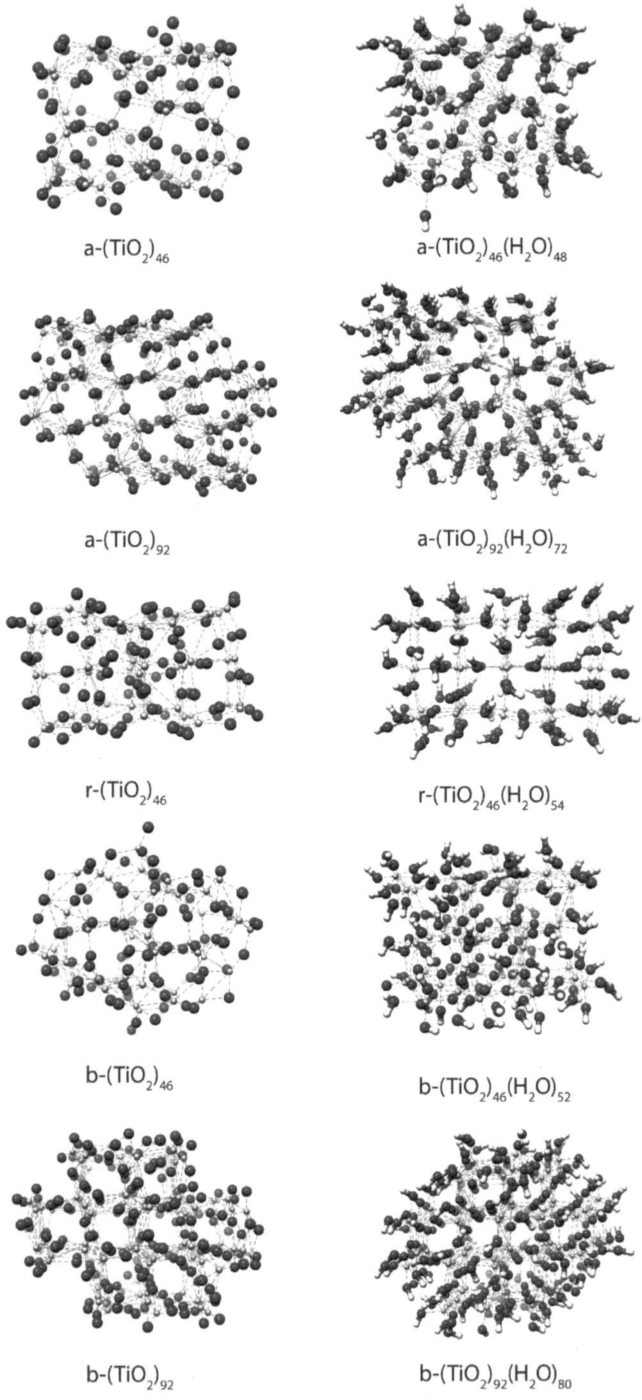

Figure 8.1 B3LYP/VDZ relaxed structures of the bare (first column) and water-saturated (second column) TiO_2 clusters. (For color version of this figure, the reader is referred to the online version of this chapter.)

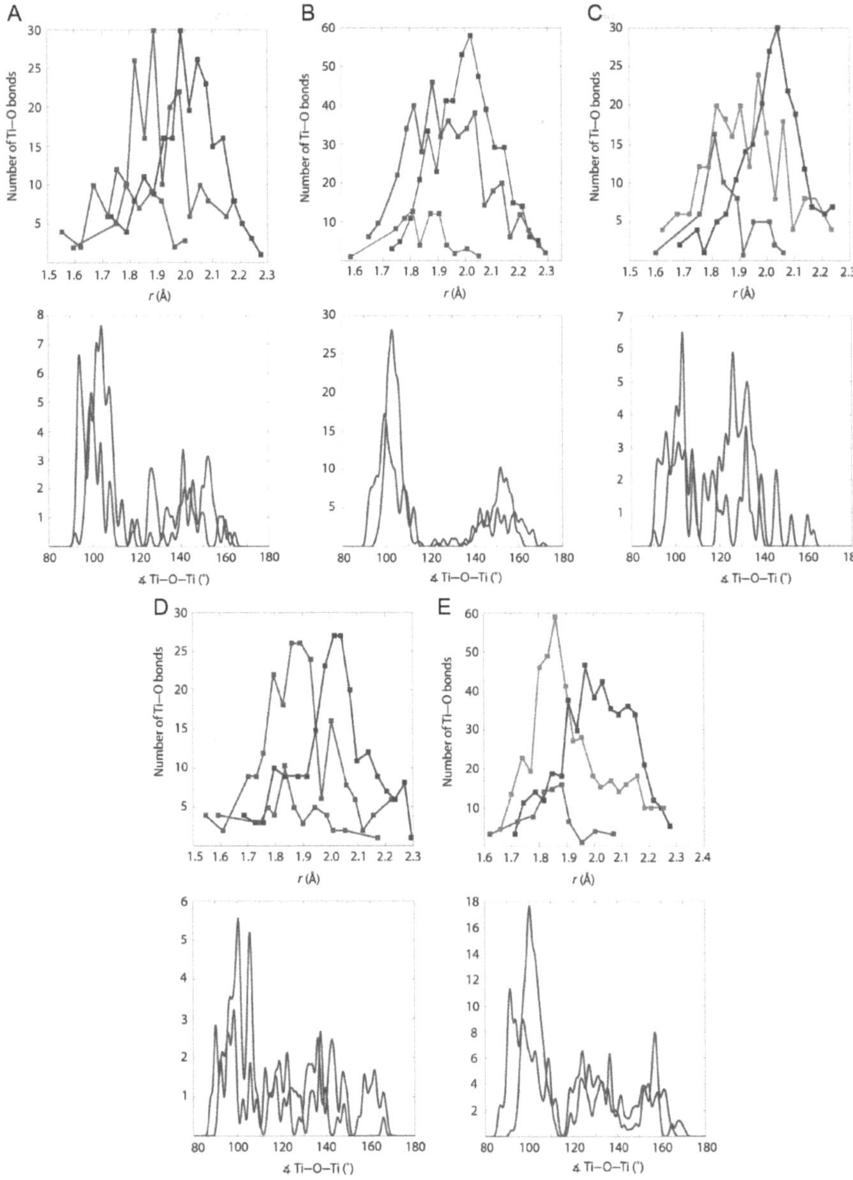

Figure 8.2 Ti—O bond distances (top) and Ti—O—Ti angle distribution (bottom) of the bare (gray line) and embedded (red line) (A) a-$(TiO_2)_{46}$, (B) a-$(TiO_2)_{92}$, (C) r-$(TiO_2)_{46}$, (D) b-$(TiO_2)_{46}$, and (E) b-$(TiO_2)_{92}$ clusters. The Ti—OH bonds are marked in blue. (For interpretation of the references to color in this figure legend, the reader is referred to the online version of this chapter.)

Table 8.1 Average Ti—O and Ti—OH bond lengths with mean deviation (in Å) for the bare and embedded TiO_2 clusters

	$Ti—O_{bare}$	$Ti—O_{embedded}$	Ti—OH
a-$(TiO_2)_{46}$	1.92 ± 0.14	2.02 ± 0.11	1.90 ± 0.13
a-$(TiO_2)_{92}$	1.93 ± 0.14	2.01 ± 0.11	1.87 ± 0.08
r-$(TiO_2)_{46}$	1.94 ± 0.14	2.03 ± 0.12	1.90 ± 0.10
b-$(TiO_2)_{46}$	1.90 ± 0.15	2.02 ± 0.13	1.90 ± 0.15
b-$(TiO_2)_{92}$	1.92 ± 0.14	2.02 ± 0.13	1.88 ± 0.10

bonds are significantly shorter compared to the intracluster Ti—O bonds of the saturated clusters, with average Ti—OH bond lengths comparable to average Ti—O bond values in the bare clusters.

The titanium dioxide structure can be considered as a system of octahedral TiO_6 units arranged differently in the different TiO_2 polymorphs. Alternatively, the TiO_2 structure can be viewed as a system created by flat, triangular OTi_3 units with an oxygen atom in the middle and three titanium atoms on the corners, which are unique for every TiO_2 polymorph. This way of analyzing the TiO_2 structure has been used in a few theoretical bulk[48,49] and nanocrystal[28] studies. The anatase-like triangles have one large (155.5°) and two small (102.3°) angles,[50] in contrast to rutile-like triangular units, which contain one small (99.2°) and two large (130.5°) angles.[51] The brookite structure consists of a mixture of anatase-like (159.2°) and rutile-like (100.7°) units in a 1:1 ratio.[52] The Ti—O—Ti angle distribution of the bare and embedded a-$(TiO_2)_{92}$ clusters, shown in Fig. 8.2B, clearly exposes the anatase-like character of the a-$(TiO_2)_{92}$ cluster, with two clear maxima around 105° and 155° angles. Although the anatase structure can be distinguished from the angle profile in both cases, the distribution is more pronounced for the saturated cluster, since the two maxima are sharper. This indicates a significant reduction in bulk and surface structural reorganization for the water-saturated clusters. In the case of the 46 unit clusters, the effect of saturation is even more visible. Because of the large relaxation of the surfaces, the characteristic angle pattern can barely be seen in the case of the bare 46 clusters. When the water molecules have been added, the diffuse two maxima can be clearly distinguished for the anatase and rutile clusters. In contrast, the angle pattern is not yet well developed for the b-$(TiO_2)_{46}$, clusters even when the water saturation has been applied. The emerging brookite angle pattern is instead only seen with b-$(TiO_2)_{92}(H_2O)_{80}$ cluster.

At the surfaces, surface hydroxide groups, surface protons, and surface oxygen ions interact with each other by a system of hydrogen bonds. In some cases, the interaction between neighboring OH^- groups with each other, or with neighboring H^+ is strong enough to spontaneously form a molecular water molecule on the cluster surface during the structure optimization. It is also found that such formation of molecular water often leads to the concomitant creation of unsaturated Ti=O double bonds on the cluster surface. For the small $(TiO_2)_{46}$ clusters, after the full system relaxation, molecular water is spontaneously recovered for 1.85%, 4.2%, and 9.6% of all adsorbed water molecules in the rutile, anatase, and brookite clusters, respectively. This percentage decreases to 1.4% for anatase and 2.5% for brookite with the larger $(TiO_2)_{92}$ clusters. This behavior is indicative of the well-known phenomenon of water molecules coexisting in dissociative as well as molecular forms on titanium dioxide surfaces. Computational studies directed toward a more complete understanding of mixed molecular/dissociative water binding in explicitly nanostructured oxides will be interesting for further studies but are beyond the scope of the present investigation.

3.1.2 Charge distributions

Mulliken charge distributions of bare and saturated a-$(TiO_2)_{92}$ clusters have been computed at the B3LYP/6-311G//B3LYP/VDZ level of theory, and the calculated charges are plotted as a function of increasing distance from the center of the clusters in Fig. 8.3. The charge distribution analysis has been extended with an implicit solvent for the a-$(TiO_2)_{92}(H_2O)_{72}$ system.

The charge distribution for the titanium ions is shown in Fig. 8.3A. In the a-$(TiO_2)_{92}$ cluster, the most positive charges are found to belong to sixfold coordinated Ti atoms, which are located mostly in the middle of the cluster. The charge decreases with increasing distance from the center, and the least positive titanium ions are localized on the edges of the cluster, mainly with a fourfold coordination. When the water saturation is added to the cluster, the Ti charges generally become more ionic with the average Ti charge increasing from 1.76 to 1.90 as seen in Fig. 8.3A. It is also noticeable that the charge profile becomes more uniform, with no significant difference remaining between charges in the middle of the clusters and on the surfaces. As also seen in Fig. 8.3A, the Ti charge distribution did not change any further after addition of an implicit PCM solvent to the embedded cluster.

The corresponding Mulliken charge distribution for the oxygen atoms is shown in Fig. 8.3B. In the bare cluster, the oxygen charges roughly split into

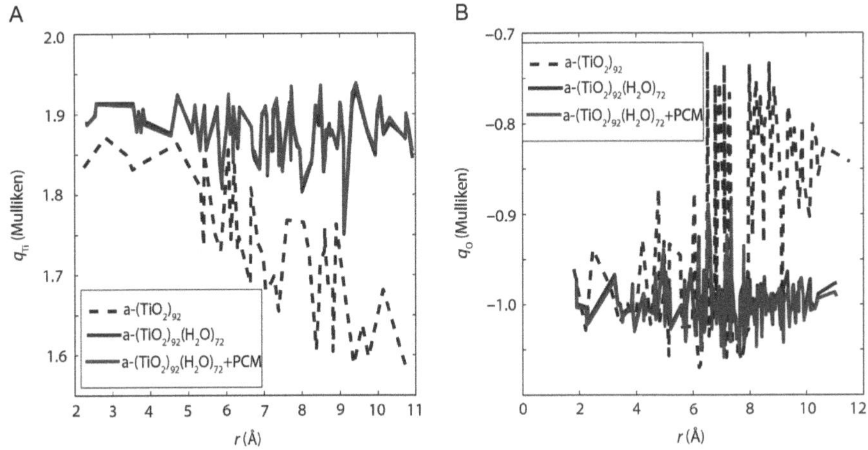

Figure 8.3 (A) Ti and (B) O Mulliken charge distributions for a-$(TiO_2)_{92}$ in vacuum, as well as for the a-$(TiO_2)_{92}(H_2O)_{72}$ clusters in vacuum and in PCM, all calculated at the B3LYP 6-311G//VDZ level of theory. The Mulliken charges are plotted as a function of distance from the center of the clusters. (For color version of this figure, the reader is referred to the online version of this chapter.)

two regions: one more ionic region from the middle of the cluster to ca. 6.5 Å, and one more covalent region with smaller negative Mulliken charges from 6.5 Å to the surfaces. It can furthermore be noted that the most negative ions are threefold coordinated O atoms, while the least negative oxygen ions have a coordination number of 2 and are located on the surface. The effect of passivation and PCM solvent agrees with the titanium ion case. The average oxygen charge becomes more negative, from −0.87 in the bare cluster to −0.95 in the saturated cluster. Also the charges become more homogenous, which makes surface and bulk ions undistinguishable in this respect. Addition of PCM brings no further changes to the charge distribution.

3.1.3 Electronic structures

The electronic structures for all $(TiO_2)_{46}$ and a-$(TiO_2)_{92}$ clusters have been analyzed in terms of molecular orbital (MO) pictures and projected density of state (PDOS) plots shown in Figs. 8.4 and 8.5, respectively. Also, the exact values of HOMO and LUMO energies, HOMO–LUMO (HL) gaps, and TD-DFT, the lowest excitation energy (BG) calculated in vacuum and in PCM, have been listed in Table 8.2. All electronic properties were calculated at the B3LYP 6-311G//VDZ level.

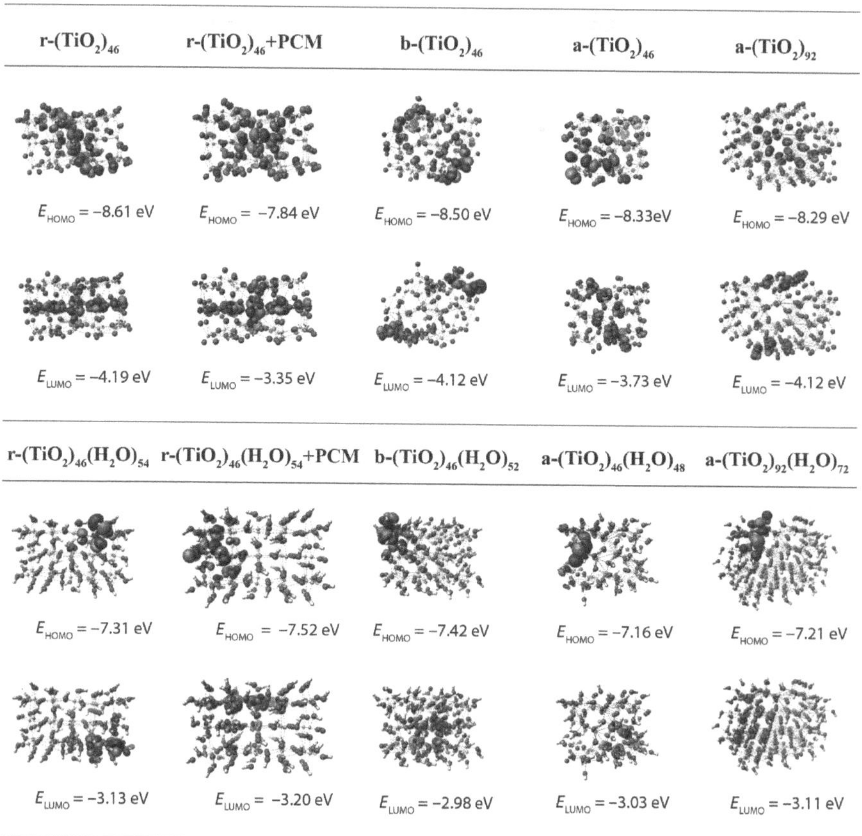

Figure 8.4 HOMO and LUMO orbitals with corresponding energies of bare and embedded clusters of the three polymorphic forms of TiO_2: anatase, brookite, and rutile (B3LYP 6-311G/VDZ). (For color version of this figure, the reader is referred to the online version of this chapter.)

The frontier MO of the bare and embedded clusters are compared in Fig. 8.4. HOMO and LUMO for the bare clusters are delocalized in the center of the cluster or symmetrically located on the surfaces. HOMO and LUMO for almost all saturated clusters are, in contrast, strongly localized in the surface, only the LUMOs of b-$(TiO_2)_{46}(H_2O)_{52}$ and a-$(TiO_2)_{92}(H_2O)_{72}$ are delocalized through the bulk of the clusters. An implicit solvent has also been applied for the bare and embedded r-$(TiO_2)_{46}$ clusters. The addition of the PCM solvent brings only small changes to the orbital appearances. In the case of the saturated cluster, the orbitals change location to other surfaces, and they become somewhat more delocalized inside the structure, similar to the corresponding bare clusters.

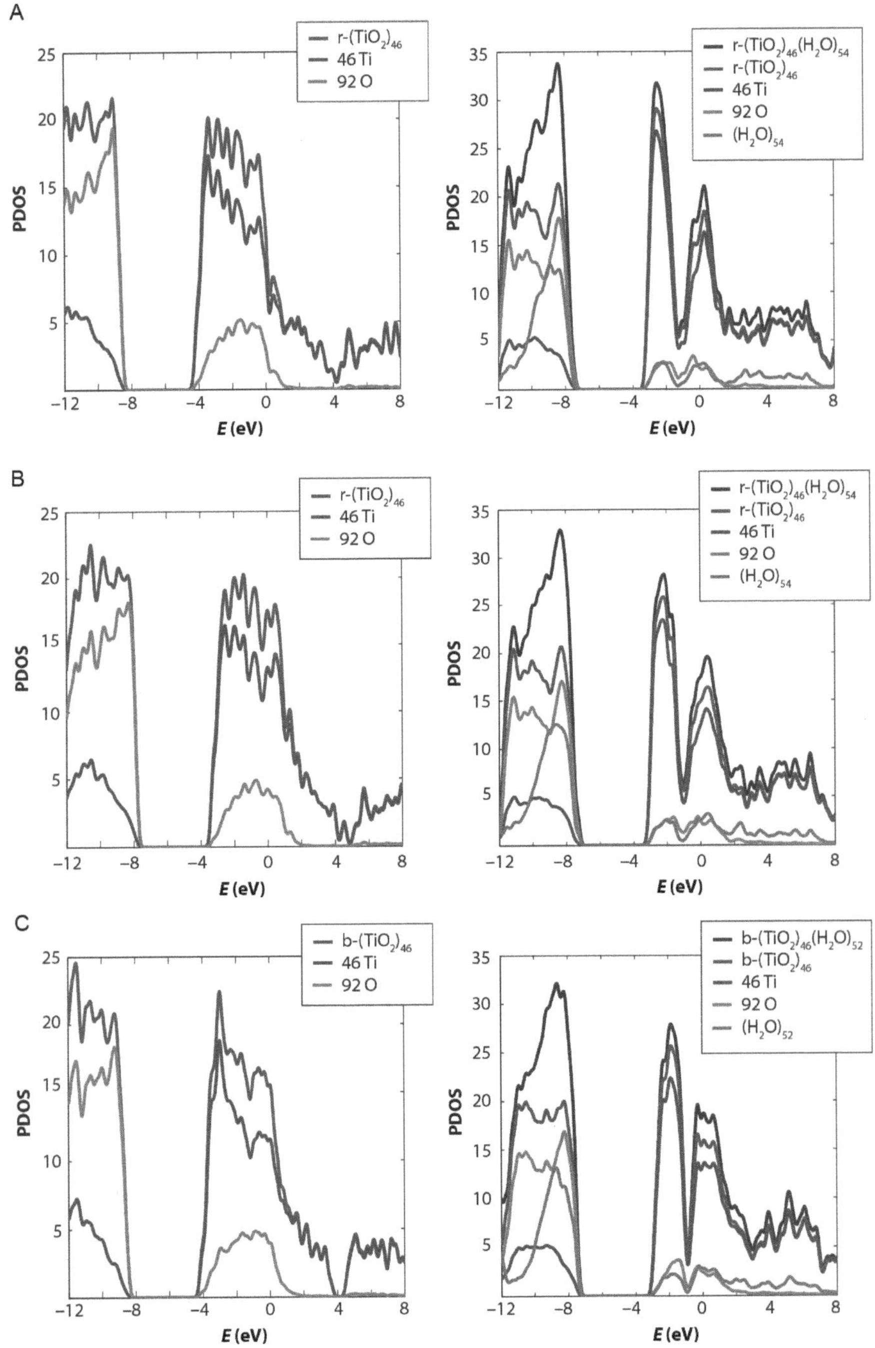
A
r-(TiO2)46
46 Ti
92 O
PDOS
E (eV)
r-(TiO2)46(H2O)54
r-(TiO2)46
46 Ti
92 O
(H2O)54
PDOS
E (eV)
B
r-(TiO2)46
46 Ti
92 O
PDOS
E (eV)
r-(TiO2)46(H2O)54
r-(TiO2)46
46 Ti
92 O
(H2O)54
PDOS
E (eV)
C
b-(TiO2)46
46 Ti
92 O
PDOS
E (eV)
b-(TiO2)46(H2O)52
b-(TiO2)46
46 Ti
92 O
(H2O)52
PDOS
E (eV)

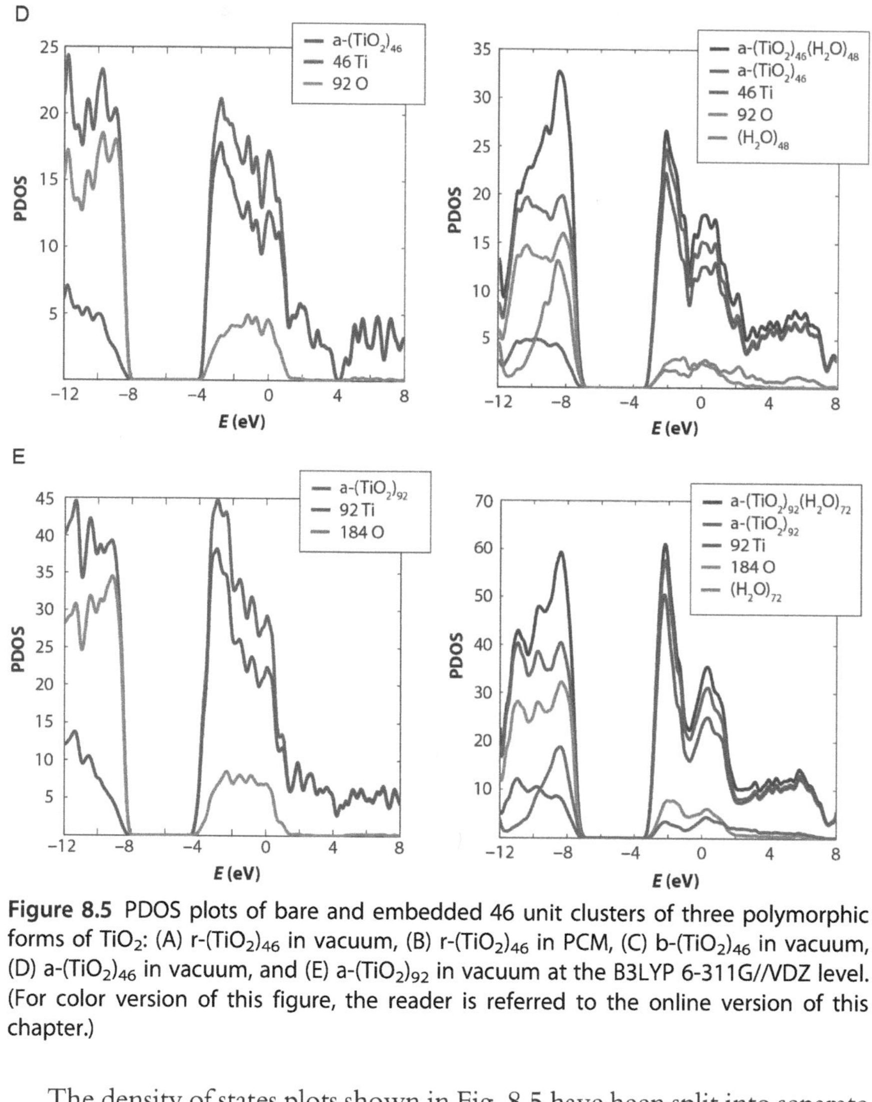

Figure 8.5 PDOS plots of bare and embedded 46 unit clusters of three polymorphic forms of TiO_2: (A) r-$(TiO_2)_{46}$ in vacuum, (B) r-$(TiO_2)_{46}$ in PCM, (C) b-$(TiO_2)_{46}$ in vacuum, (D) a-$(TiO_2)_{46}$ in vacuum, and (E) a-$(TiO_2)_{92}$ in vacuum at the B3LYP 6-311G//VDZ level. (For color version of this figure, the reader is referred to the online version of this chapter.)

The density of states plots shown in Fig. 8.5 have been split into separate contributions for cluster, titanium ions, oxygen ions, and dissociative water. As expected from bulk TiO_2, oxygen 2p levels make up the upper part of the titanium dioxide valence band, while the bottom of the conduction band is dominated by empty Ti 3d states. Similar band structures have been obtained for all investigated clusters. The valence and conduction band edges are both shifted up in energy by ca. 1 eV when embedding and/or PCM is added to the bare clusters, but otherwise the DOS contributions of the clusters calculated with explicit and/or implicit solvent look similar to those computed

Table 8.2 HOMO and LUMO energies, HOMO–LUMO (HL) gaps, and first excitation (BG) energies of bare and embedded a-$(TiO_2)_{46}$, b-$(TiO_2)_{46}$, and r-$(TiO_2)_{46}$ in vacuum and PCM at the 6-311G/VDZ level

	HOMO (eV)	LUMO (eV)	HL (eV)	BG (eV)
a-$(TiO_2)_{46}$	−8.33	−3.73	4.60	3.90
a-$(TiO_2)_{46}$/PCM	−7.58	−2.85	4.73	4.02
a-$(TiO_2)_{46}(H_2O)_{48}$	−7.16	−3.03	4.13	3.75
a-$(TiO_2)_{46}(H_2O)_{48}$/PCM	−7.62	−3.03	4.59	3.71
b-$(TiO_2)_{46}$	−8.50	−4.12	4.38	3.77
b-$(TiO_2)_{46}$/PCM	−7.64	−2.93	4.71	3.88
b-$(TiO_2)_{46}(H_2O)_{52}$	−7.42	−2.98	4.44	3.76
b-$(TiO_2)_{46}(H_2O)_{52}$/PCM	−7.68	−3.12	4.56	3.79
r-$(TiO_2)_{46}$	−8.61	−4.19	4.42	3.61
r-$(TiO_2)_{46}$/PCM	−7.84	−3.35	4.49	3.63
r-$(TiO_2)_{46}(H_2O)_{54}$	−7.31	−3.13	4.18	3.45
r-$(TiO_2)_{46}(H_2O)_{54}$/PCM	−7.52	−3.20	4.32	3.49

Energies given in eV.

without solution. One small but clearly noticeable difference for the PDOSs in water is that there is a clear split of the lowest (Ti 3d) part of the conduction band in all saturated clusters. It can also be noted that the explicitly added water molecules contribute significantly to the upper part of the valence band.

HOMO, LUMO, HL gaps, and the first TD-DFT excitation band gap (BG) energies for all $(TiO_2)_{46}$ clusters calculated in vacuum and PCM have been listed in Table 8.2. As already seen from the PDOS plots, the addition of implicit and explicit solvent significantly shifts the HOMO and LUMO energies to less negative values. The average shifts of the HOMO and LUMO energies are about 0.88 and 1.08 eV, respectively, after addition of implicit or explicit solvent. On the other hand, any further shift of valence and conduction bands for all water-saturated $(TiO_2)_{46}$ clusters when they are surrounded by implicit solvent is minor, on average 0.15 eV. This indicates that the two kinds of solvent models have a similar impact on the TiO_2 band structure. Although significant shifts are seen in both the HOMO and LUMO energy values when environmental treatments are added to the initially bare cluster, HLs and BGs are rather stable. The maximal change in HLs and BGs between clusters of the same kind calculated in different environments is 0.47 and 0.31 eV, respectively, so that the addition of implicit and/or explicit solvent does not strongly affect the calculated HL and BG values. All calculated BGs fall in the range of 3.45–4.02 eV and are, thus, in reasonable agreement, but they are all somewhat larger, compared to the experimental bulk band gaps (3.21, 3.12, and 3.0 eV for anatase, brookite, and rutile, respectively).[53]

3.2. Adsorption energies

Average adsorption energies, E_{ads}, have been calculated according to the equation:

$$-E_{ads} = \frac{E_{(MO_2)_n(H_2O)_m} - \left(E_{(MO_2)_n} + mE_{H_2O}\right)}{m}$$

where n is the number of MO_2 units, m is the number of H_2O molecules, and E_{ads} is defined so that positive adsorption energies correspond to stable surface binding. The average adsorption energies per H_2O molecule for a-$(TiO_2)_{46}(H_2O)_{48}$, b-$(TiO_2)_{46}(H_2O)_{52}$, r-$(TiO_2)_{46}(H_2O)_{54}$, and a-$(TiO_2)_{92}(H_2O)_{72}$ have been calculated in vacuum and PCM and are listed in Table 8.3. Interestingly, the differences in adsorption energy between the $(TiO_2)_{46}$ clusters of the different polymorphic forms are very small, at most 0.13 eV, and the addition of an implicit solvent decreases the adsorption energy by about 0.4 eV in all cases. It can also be noted that the calculated adsorption energies do not change noticeably when the cluster size is increased to 92 TiO_2 units. The adsorption energies obtained by periodic calculations, ranging between 0.60 and 1.01 eV, are significantly smaller compared to current results.[12,33] To a large extent, this difference is likely

Table 8.3 Average adsorption energies per H_2O molecule, E_{ads} (in eV), for a-$(TiO_2)_{46}(H_2O)_{48}$, a-$(TiO_2)_{92}(H_2O)_{72}$, b-$(TiO_2)_{46}(H_2O)_{52}$, and r-$(TiO_2)_{46}(H_2O)_{54}$, all calculated at the B3LYP 6-311G/VDZ level of theory

	E_{vacuum}	E_{PCM}
a-$(TiO_2)_{46}(H_2O)_{48}$	1.39	0.96
b-$(TiO_2)_{46}(H_2O)_{52}$	1.31	0.91
r-$(TiO_2)_{46}(H_2O)_{54}$	1.26	0.89
a-$(TiO_2)_{92}(H_2O)_{72}$	1.33	

to reflect a higher surface reactivity of the nanocrystals, with many surface defects and under-coordinated surface atoms.

3.3. Group IV oxide materials: SiO_2, GeO_2, and SnO_2

Metal dioxide clusters, $(MO_2)_n$, of rutile form have here been studied for group IV elements. The results for nano-SnO_2 clusters, which are used as catalyst or gas sensing materials in many modern applications due to their optoelectronic properties, have been combined with results for two other materials: GeO_2 and SiO_2. In contrast to SnO_2, the rutile forms are not the lowest energy polymorphs of SiO_2 and GeO_2, which commonly occur in hexagonal α-quartz structure. The investigation of the rutile-type SiO_2 (stishovite) is interesting as an example of a minority crystal form that can be realized synthetically, or can form naturally under unusual (high temperature and pressure) conditions such as in meteorite impact craters. In particular, it has a 6:3 coordination, which is uncommon with respect to the majority of tetrahedral SiO_2 polymorphs. GeO_2 has been included to facilitate a systematic study of similarities and differences in oxide nanoparticle properties for group IV elements of the periodic table.

3.3.1 Geometries

The rutile-TiO_2 clusters have been used to build corresponding bare and embedded group IV oxide nanoparticles, followed by complete structural reoptimization for the different nanomaterials at the B3LYP/LANL2DZ level of theory. The optimized structures are shown in Fig. 8.6. An analysis of the M—O bonds and M—O—M angle distribution for the bare and saturated 46 unit clusters of SnO_2, GeO_2, and SiO_2, as well as the $(SnO_2)_{92}$ clusters is shown in Fig. 8.7.

In the structures studied here, the crystals are built from MO_6 octahedral units characterized by a 6:3 coordination of the ions. The MO_6 octahedral units are, furthermore, characterized by two kinds of bonds: four equatorial and two axial, with significant possibilities for bond length variations in non-perfect crystal structures such as nanostuctured materials. Such bond length variations can, for example, be observed in the experimental rutile-SiO_2, which has a long axial (1.81 Å) and much shorter equatorial (1.76 Å) bonds (1.78 Å in average).[51,54] This is, in contrast, not seen in the optimized $(SiO_2)_{46}$ cluster (Fig. 8.7). The average bond distance in $(SiO_2)_{46}$ is, however, equal to 1.74 ± 0.10 Å, which indicates a shorter average Si—O bond length than in the experimental structure. This is probably due to the fact that after relaxation, several of the Si ions, which are initially

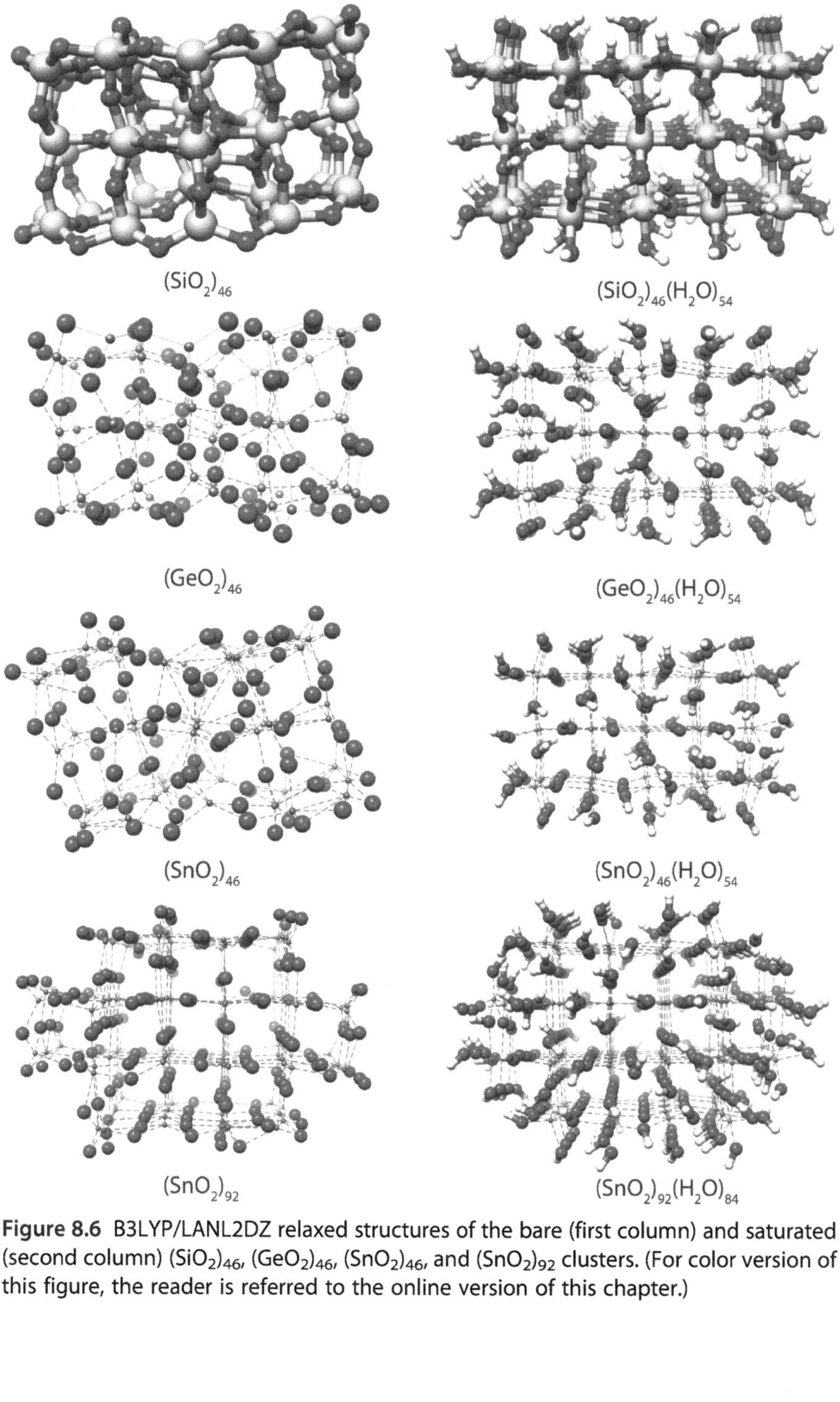

Figure 8.6 B3LYP/LANL2DZ relaxed structures of the bare (first column) and saturated (second column) $(SiO_2)_{46}$, $(GeO_2)_{46}$, $(SnO_2)_{46}$, and $(SnO_2)_{92}$ clusters. (For color version of this figure, the reader is referred to the online version of this chapter.)

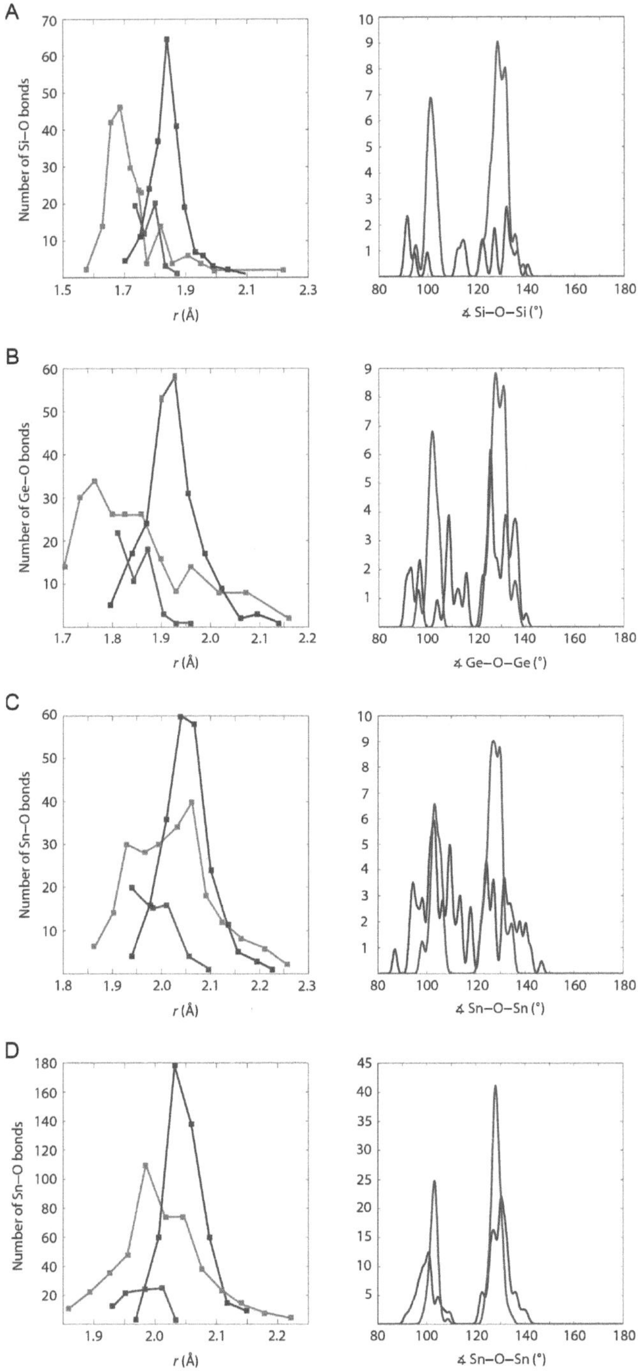

Figure 8.7 M—O bond distances (top) and M—O—M angle distributions (bottom) for the bare (gray line) and water-saturated (red line) (A) $(SiO_2)_{46}$, (B) $(GeO_2)_{46}$, (C) $(SnO_2)_{46}$, and (D) $(SnO_2)_{92}$ clusters optimized with B3LYP/LANL2DZ. The M—OH bonds in the water-saturated clusters are marked in blue. (For interpretation of the references to color in this figure legend, the reader is referred to the online version of this chapter.)

sixfold-coordinated, transform to fourfold coordinated Si ions, which are characterized by much shorter bond distances.[54] After saturating the cluster with water, the Si—O bonds become much longer with an average equal to 1.85 ± 0.06 Å, which is now significantly larger than the experimental bulk values. The number of both very short and very long bonds significantly decreases in the saturated $(SiO_2)_{46}$ cluster, which indicates a more uniform cluster structure characterized by a smaller degree of reconstruction (Fig. 8.7).

The GeO_6 units in GeO_2 are comparatively less distorted, with equatorial and axial bonds equal to 1.872 and 1.902 Å, respectively.[51] Similar to the $(SiO_2)_{46}$ case, the bare $(GeO_2)_{46}$ cluster has many under-coordinated ions with very sort bonds. The average calculated bond distance is again somewhat shorter (1.85 ± 0.10 Å) than the average experimental bulk value (1.88 Å), but significantly increases in the saturated cluster (1.93 ± 0.06 Å).

Much less variation in the experimental crystal structure has been found for SnO_2. All bonds in the SnO_6 unit have almost the same length of 2.05 Å.[51] In contrast to the other rutile-type clusters investigated here, the average bond length of the bare $(SnO_2)_{46}$ cluster equals 2.04 ± 0.10 Å, which is very close to the experimental values. Also, this increases only slightly after saturation of the cluster (2.06 ± 0.05 Å). Going up in calculated cluster size does not bring any significant change to the cluster bond distances. This is illustrated by the $(SnO_2)_{92}(H_2O)_{84}$ cluster where the calculated bond distances are 2.02 ± 0.08 and 2.05 ± 0.04 Å for the bare and saturated clusters, respectively.

Calculated average M—OH bond lengths are also listed for the group IV MO_2 clusters in Table 8.4. The average M—OH values range from 1.85 Å for the water-saturated $(SiO_2)_{46}$ cluster to 2.02 Å for the water-saturated $(SnO_2)_{46}$ cluster, that is, values that are rather similar to the average Ti—OH bond lengths discussed previously.

Table 8.4 Average M—O and M—OH bond lengths with mean deviation (in Å) for the bare and embedded IV-group element clusters for B3LYP/LANL2DZ optimized clusters

	$M—O_{bare}$	$M—O_{embedded}$	M—OH
$(SiO_2)_{46}$	1.74 ± 0.10	1.85 ± 0.06	1.85 ± 0.16
$(GeO_2)_{92}$	1.85 ± 0.10	1.93 ± 0.06	1.88 ± 0.05
$(SnO_2)_{46}$	2.04 ± 0.08	2.06 ± 0.05	2.02 ± 0.06
$(SnO_2)_{92}$	2.02 ± 0.07	2.02 ± 0.04	2.00 ± 0.04

The results of an M—O—M angle distribution analysis, which has been done for all calculated group IV clusters, are shown in Fig. 8.7. The rutile-type OM_3 units have one smaller and two larger characteristic angles, similar in all investigated rutile-type materials, equal to 98.6° and 130.7° in SiO_2, 99.7° and 130.1° in GeO_2, and 101.9° and 129.1° in SnO_2.[51]

The bare 46 unit clusters do not show any rutile angle pattern because of significant defects and under-coordinated ions in their structure. This behavior is, in fact, very similar to the small $(TiO_2)_{46}$ clusters. The M—O—M angle distribution changes completely after saturation of the clusters. The angle distributions of the hydrated clusters show two sharp maxima: one lower around 100° and one higher around 130° in all cases. The maxima are found to be much more pronounced than in the titanium dioxide clusters, indicating more ordered crystal structures of the investigated rutile-type group IV materials. When the size of the SnO_2 cluster is increased to 92 MO_2 units, the rutile angle pattern can be distinguished even for the bare cluster.

After relaxation, one molecular water molecule has formed in all investigated 46 units clusters. This suggests that it may be interesting to explore mixed molecular/dissociative adsorption patterns similar to TiO_2 also for the group IV oxide nanomaterials, but this is again outside the scope of the present investigation.

3.3.2 Charge distributions

A careful charge distribution analysis, similar to the one presented for nano-TiO_2 above, has been done for the bare and saturated $(SnO_2)_{92}$ clusters. The results of this analysis are plotted in Fig. 8.8. The charges for both clusters have, in particular, been plotted as functions of increasing distance from the center of the cluster for B3LYP/LANL2DZ calculations carried out in vacuum and with a PCM solvent model. The charge distribution analysis was restricted to this large cluster with significant amount of bulk, where surface and bulk ions can easy be distinguished. The smaller 46 unit rutile-type clusters, which contain mostly surface ions, have been excluded from the analysis for simplicity.

The charges of the bare $(SnO_2)_{92}$ cluster show a behavior similar to that of the charges of the bare $(TiO_2)_{92}$ cluster. The charge profile is not uniform and two regions, one more ionic in the middle of the cluster with fully coordinated ions, and one more covalent mostly located on the surfaces with ions with reduced coordination number can be distinguished. Unlike the $(TiO_2)_{92}$ case, however, the charge profile does not change dramatically

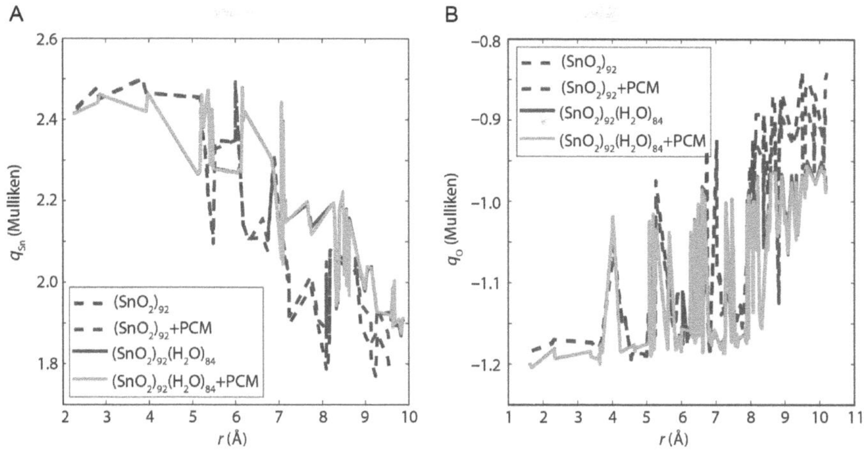

Figure 8.8 (A) Sn and (B) O charge distribution of bare and embedded $(SnO_2)_{92}$ clusters calculated in vacuum and in PCM at the B3LYP/LANL2DZ level of theory. The charges have been plotted with increasing distance from the center of the clusters to the edges. (For color version of this figure, the reader is referred to the online version of this chapter.)

after cluster saturation. Although the ionic and covalent parts can still be detected, the covalent character of the surface ions decreases slightly, which is especially visible for oxygen ions, which become more negative by ca. 0.1 unit of charge. The average atomic charges increase from 2.10 to 2.22 for tin and from −1.05 to −1.10 for oxygen. In both bare and passivated clusters, the addition of a PCM solvent does not bring any significant changes to the charge distribution.

3.3.3 Electronic structures

The electronic structures for all relaxed group IV oxide clusters have been analyzed for the B3LYP/LANL2DZ calculations with MO pictures and PDOS plots shown in Figs. 8.9 and 8.10, respectively. In addition, the calculated values of HOMO, LUMO, HL, and BG in vacuum and PCM are listed in Table 8.5.

Figure 8.9 shows the HOMO and LUMO for all investigated group IV oxide clusters. MOs for the $(SnO_2)_{46}$ cluster calculated in PCM have also been included for comparison. The HOMO and LUMO for the bare clusters are localized mostly on under-coordinated ions, which effectively constitute surface defects. Orbital delocalization can be seen in the case of the $(SnO_2)_{46}$ cluster calculated in PCM. The orbitals change location after

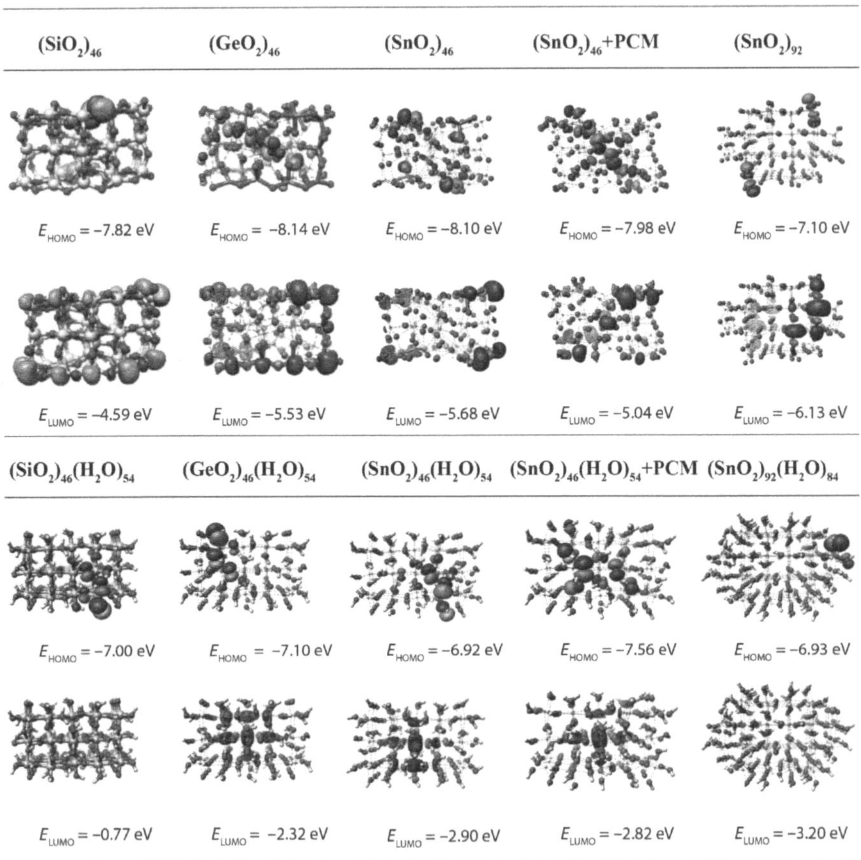

Figure 8.9 HOMO and LUMO orbitals with corresponding energies of bare (top) and water-saturated clusters (bottom) of the three rutile-type group IV materials: $(SiO_2)_{46}$, $(GeO_2)_{46}$, $(SnO_2)_{46}$, and $(SnO_2)_{92}$. All results have been obtained from B3LYP/LANL2DZ calculations. (For color version of this figure, the reader is referred to the online version of this chapter.)

cluster saturation but are still strongly localized on the surfaces as in the saturated titanium dioxide clusters.

Similar to titanium oxide, the valence band of the studied rutile-type oxides consist mostly of O 2p orbitals, but in contrast to TiO_2 the bottom of the conduction band is dominated by empty s-levels from the group IV elements, since their d orbitals are fully occupied.[55–57] The different contributions of particular cluster ions and water molecules to the valence and conduction bands of all investigated group IV rutile-type clusters are shown in the PDOS plots in Fig. 8.10. Although the valence band of all clusters is

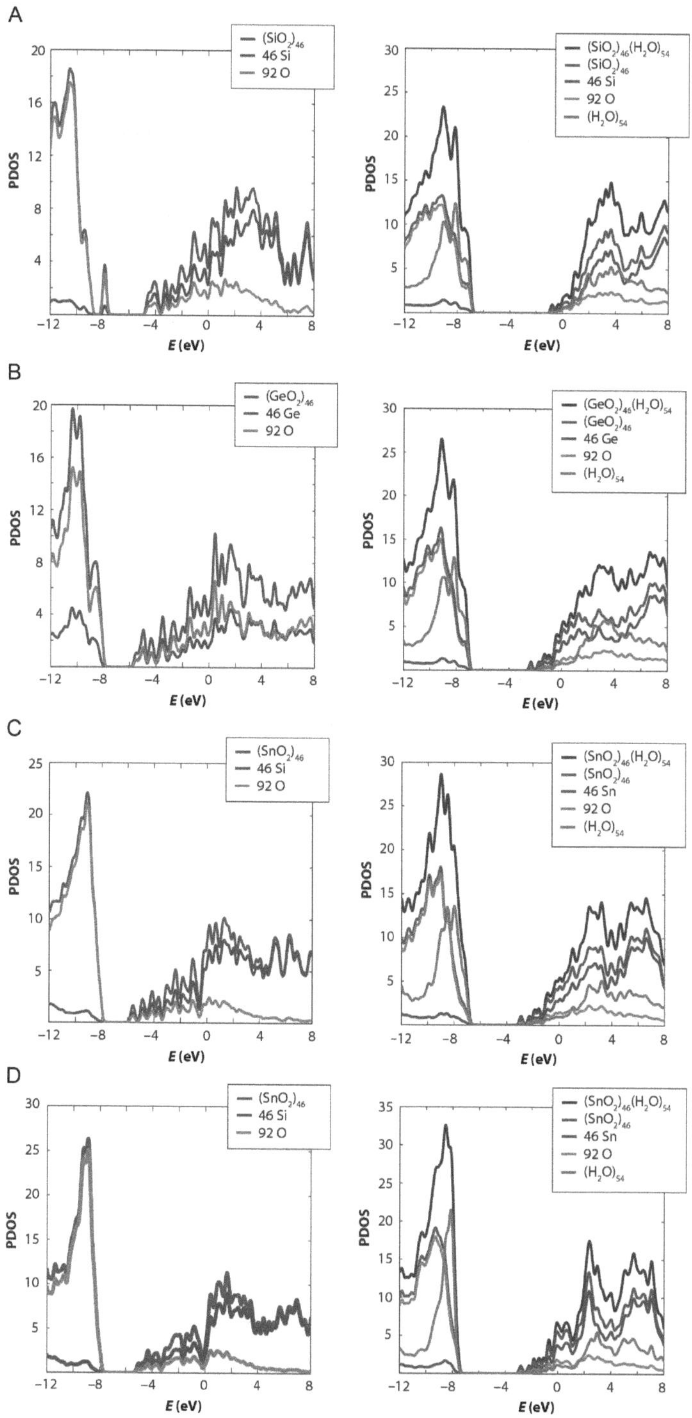

Figure 8.10—Cont'd

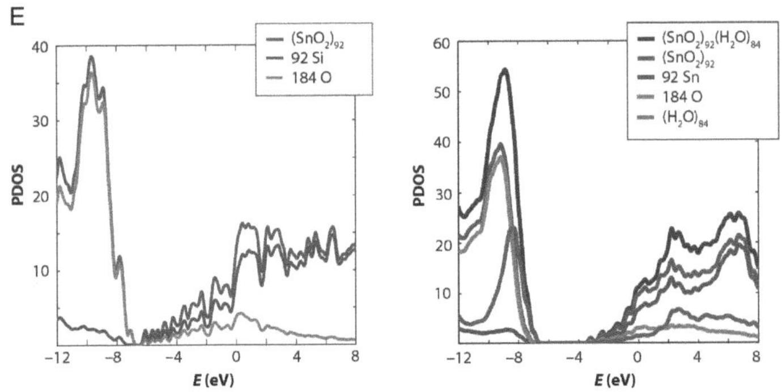

Figure 8.10 PDOS plots of bare and embedded clusters of three rutile-type materials: (A) $(SiO_2)_{46}$ in vacuum, (B) $(GeO_2)_{46}$ in vacuum, (C) $(SnO_2)_{46}$ in vacuum, (D) $(SnO_2)_{46}$ in PCM, and (E) $(SnO_2)_{92}$ in vacuum obtained at the B3LYP/LANL2DZ level of theory. (For color version of this figure, the reader is referred to the online version of this chapter.)

Table 8.5 HOMO and LUMO energies, HOMO–LUMO (HL) gaps, and first excitation energies (BG) for the bare and saturated $(SiO_2)_{46}$, $(GeO_2)_{46}$, and $(SnO_2)_{46}$ clusters calculated in vacuum and PCM at the B3LYP/LANL2DZ//B3LYP/LANL2DZ level of theory

	HOMO (eV)	LUMO (eV)	HL (eV)	BG (eV)
$(SiO_2)_{46}$	−7.82	−4.59	3.23	2.84
$(SiO_2)_{46}$/PCM	−8.22	−4.13	4.09	3.66
$(SiO_2)_{46}(H_2O)_{54}$	−7.00	−0.77	6.23	–
$(SiO_2)_{46}(H_2O)_{54}$/PCM	−7.49	−0.61	6.88	6.37
$(GeO_2)_{46}$	−8.14	−5.53	2.61	2.13
$(GeO_2)_{46}$/PCM	−7.92	−5.16	2.76	2.27
$(GeO_2)_{46}(H_2O)_{54}$	−7.10	−2.32	4.78	–
$(GeO_2)_{46}(H_2O)_{54}$/PCM	−7.61	−2.23	5.38	4.86
$(SnO_2)_{46}$	−8.10	−5.68	2.42	2.01
$(SnO_2)_{46}$/PCM	−7.98	−5.04	2.94	2.48
$(SnO_2)_{46}(H_2O)_{54}$	−6.92	−2.90	4.02	3.67
$(SnO_2)_{46}(H_2O)_{54}$/PCM	−7.56	−2.82	4.74	4.26

Energies are given in eV.

mainly constructed by orbitals from the oxygen ions, there is a mixture of oxygen and IV-group ions levels at the edge of the conduction band. It can also be noted that it appears as if the bands of the bare clusters are not fully developed for the comparatively small cluster size used in this study.

Especially, a lot of states with molecular character can be seen at the bottom of the conduction bands. Those levels, mostly located on the surface defects, significantly decrease the gaps between the valence and conduction bands.

The electronic structure changes dramatically with addition of embedding to the bare clusters. As for TiO_2, the addition of explicit water molecules causes the valence band edge to shift by about 1.0 eV for all investigated materials (see Table 8.5). Also the large contribution of O 2p-orbitals of water is visible, especially, at the valence band edge. The saturation of under-coordinated ions by hydroxide groups and protons decreases the number of surface defects, and thereby the number of molecular levels. The decreased number of molecular levels leads to a significant increase of the gap between the valence and conduction bands.

The exact bulk band gap is well known only for SnO_2 crystals where it is reported to be 3.6 eV.[58] For GeO_2 on the one hand, the direct-allowed band gap found experimentally is 5.35 eV, which has been supported by calculations giving a value of 5.52 eV.[59] On the other hand, another study showed the direct-forbidden transition at 4.68 eV.[60] The band gap calculated at Γ-point for tetragonal SiO_2 varies from 5.15 to 5.7 eV in different periodic calculations.[54,55,61]

Table 8.5 gives the calculated HOMO, LUMO, HL, and BG energies in vacuum and PCM for all rutile-type group IV $(MO_2)_{46}$ clusters. The calculated HL and BG energies are quite small for the bare clusters calculated in vacuum, but increase significantly after adding PCM. The water-saturated clusters give larger band gaps in significantly better agreement with experimental bulk values.

3.4. Adsorption energies

Average adsorption energies have been calculated for all investigated rutile-type clusters in the same manner as for the titanium dioxide case. The energies have been calculated in vacuum and PCM using the LANL2DZ basis set and with results shown in Table 8.6. The small tin dioxide and germanium dioxide clusters have similar adsorption energies close to 2 eV, while the silicon dioxide cluster has a calculated adsorption energy which is significantly lower by about 0.40 eV. The adsorption energy value increases only slightly by ca. 0.20 eV with increased size of the tin dioxide cluster. Previous periodic calculations of dissociative water adsorbed to SnO_2 surfaces reported significantly lower adsorption energies, between 1.39 and 1.77 eV, adsorption energies.[32–35,62] Again, the discrepancy between cluster and bulk results

Table 8.6 Average adsorption energies per H_2O molecule for the water-embedded $(SiO_2)_{46}$, $(GeO_2)_{46}$, $(SnO_2)_{46}$, and $(SnO_2)_{92}$ clusters calculated at the LANL2DZ//LANL2DZ level of theory

	E_{vacuum}	E_{PCM}
$(SiO_2)_{46}(H_2O)_{54}$	1.61	1.71
$(GeO_2)_{46}(H_2O)_{54}$	1.99	2.07
$(SnO_2)_{46}(H_2O)_{54}$	1.98	2.04
$(SnO_2)_{92}(H_2O)_{84}$	2.16	2.17

Energies are given in eV.

may to a large extent reflect differences in the nanoscale surface binding environment characterized to a higher degree by reactive surface sites.

4. CONCLUSIONS

Surface interactions between water and cluster models of three polymorphic forms of nanostructured TiO_2 and related rutile-type group IV materials have been investigated using DFT calculations. Complete cluster saturation has, in particular, been considered for dissociative adsorption of water molecules, employing a cluster modeling strategy that allows for unambiguously electroneutral clusters through the combination of stoichiometric clusters with equal numbers of OH^- and H^+ adsorbates.

The addition of water saturation of the bare TiO_2 and IV-group elements clusters has significant influence on the geometrical structures of the entire nanoclusters. The water terminated clusters show much less surface restructuring and bulk defects compared to the bare clusters in all investigated cases. The saturation with explicit and implicit solvent has almost no influence on the electronic structure of the titanium dioxide clusters, since the band structure of the bare TiO_2 clusters is already quite well described. Compared to TiO_2, the bare group IV rutile-type nanomaterials have a significantly less well-developed band structure characterized by very low band gaps compared to bulk. Introducing the water saturation to these nanocrystals has significant influence on the calculated band structure as the large number of molecular states near the bands edges deceases. This leads to significant increases of the calculated band gaps between the valence and conduction band in these materials. The surface saturation, therefore, leads to significantly more realistic representations of the electronic structures of the group IV nanoparticles compared to typical experimental conditions.

Taken together, the electronic structure results largely provide an *a posteriori* validation of several previous studies of nano-TiO_2 materials based on bare clusters, while at the same time showing that such an approach can fail to provide physically realistic models for other nanostructured materials such as the group IV oxides studied here for comparison.

Calculated adsorption energies in TiO_2 and the related rutile-type nanomaterials are found to be considerably larger compared to previously reported energies from periodic calculations of the corresponding materials. This is likely to largely reflect the fact that nanocrystalline materials have a lot of under-coordinated surface sites which are highly reactive.

In a broader perspective, it is clear that first principles atomistic calculations of explicitly nanostructured clusters of increasing size, such as the ones considered in this work, provide a promising avenue to explore complex nanoscale properties for a wide range of technologically important and/or scientifically interesting materials.

ACKNOWLEDGMENTS

The Swedish Research Council (VR), and the Knut and Alice Wallenberg (KAW) Foundation, as well as the Nanometer Structure Consortium in Lund, nmC@LU, are gratefully acknowledged for the financial support. The calculations have been performed using computing resources provided by the Swedish National Supercomputing Center (NSC) and the Lund University Supercomputing facility (LUNARC).

REFERENCES

1. Katusic, S.; Albers, P.; Kern, R.; Petrat, F.-M.; Sastrawan, R.; Hore, S.; et al. Production and Characterization of ITO-Pt Semiconductor Powder Containing Nanoscale Noble Metal Particles Catalytically Active in Dye-Sensitized Solar Cells. *Sol. Energy Mater. Sol. Cells* **2006**, *90*(13), 1983–1999.
2. Hagfeldt, A.; Graetzel, M. Light-Induced Redox Reactions in Nanocrystalline Systems. *Chem. Rev.* **1995**, *95*(1), 49–68.
3. Hagfeldt, A.; Grätzel, M. Molecular Photovoltaics. *Acc. Chem. Res.* **2000**, *33*(5), 269–277.
4. Graetzel, M.; Janssen, R. A. J.; Mitzi, D. B.; Sargent, E. H. Materials Interface Engineering for Solution-Processed Photovoltaics. *Nature* **2012**, *488*(7411), 304–312.
5. Tricoli, A.; Righettoni, M.; Pratsinis, S. E. Minimal Cross-Sensitivity to Humidity During Ethanol Detection by SnO_2–TiO_2 Solid Solutions. *Nanotechnology* **2009**, *20*(31), 315502.
6. McAleer, J. F.; Moseley, P. T.; Norris, J. O. W.; Williams, D. E.; Taylor, P.; Tofield, B. C. Tin Oxide Based Gas Sensors. *Mater. Chem. Phys.* **1987**, *17*(6), 577–583.
7. Bârsan, N.; Stetter, J. R.; Findlay, J. R.; Göpel, W. High-Performance Gas Sensing of CO: Comparative Tests for Semiconducting (SnO_2-Based) and for Amperometric Gas Sensors. *Anal. Chem.* **1999**, *71*(13), 2512–2517.
8. Batzill, M. Surface Science Studies of Gas Sensing Materials: SnO_2. *Sensors* **2006**, *6*(10), 1345–1366.

9. Persson, P.; Bergström, R.; Ojamäe, L.; Lunell, S. Quantum-Chemical Studies of Metal Oxides for Photoelectrochemical Applications. *Adv. Quantum Chem.* **2002**, *41*, 203–263, Available from: http://www.sciencedirect.com/science/article/pii/S0065327602410544.
10. Diebold, U. The Surface Science of Titanium Dioxide. *Surf. Sci. Rep.* **2003**, *48*(5–8), 53–229.
11. Zhao, Z.; Li, Z.; Zou, Z. Structure and Properties of Water on the Anatase TiO_2(101) Surface: From Single-Molecule Adsorption to Interface Formation. *J. Phys. Chem. C* **2012**, *116*(20), 11054–11061.
12. Lindan, P. J. D.; Harrison, N. M.; Gillan, M. J. Mixed Dissociative and Molecular Adsorption of Water on the Rutile (110) Surface. *Phys. Rev. Lett.* **1998**, *80*(4), 762–765.
13. He, Y.; Tilocca, A.; Dulub, O.; Selloni, A.; Diebold, U. Local Ordering and Electronic Signatures of Submonolayer Water on Anatase TiO_2(101). *Nat. Mater.* **2009**, *8*(7), 585–589.
14. Barnard, A. S.; Zapol, P.; Curtiss, L. A. Modeling the Morphology and Phase Stability of TiO_2 Nanocrystals in Water. *J. Chem. Theory Comput.* **2005**, *1*(1), 107–116.
15. Blagojevic, V.; Chen, Y.-R.; Steigerwald, M.; Brus, L.; Friesner, R. A. Quantum Chemical Investigation of Cluster Models for TiO_2 Nanoparticles with Water-Derived Ligand Passivation: Studies of Excess Electron States and Implications for Charge Transport in the Gratzel Cell. *J. Phys. Chem. C* **2009**, *113*(46), 19806–19811.
16. Sahoo, S. K.; Pal, S.; Sarkar, P.; Majumder, C. Size-Dependent Electronic Structure of Rutile TiO_2 Quantum Dots. *Chem. Phys. Lett.* **2011**, *516*(1–3), 68–71.
17. Iacomino, A.; Cantele, G.; Ninno, D.; Marri, I.; Ossicini, S. Structural, Electronic, and Surface Properties of Anatase TiO_2 Nanocrystals from First Principles. *Phys. Rev. B* **2008**, *78*(7), 075405.
18. Nunzi, F.; Mosconi, E.; Storchi, L.; Ronca, E.; Selloni, A.; Grätzel, M.; et al. Inherent Electronic Trap States in TiO_2 Nanocrystals: Effect of Saturation and Sintering. *Energy Environ. Sci.* **2013**, *6*(4), 1221–1229.
19. Cheng, H.; Selloni, A. Hydroxide Ions at the Water/Anatase TiO_2(101) Interface: Structure and Electronic States from First Principles Molecular Dynamics. *Langmuir* **2010**, *26*(13), 11518–11525.
20. Koparde, V. N.; Cummings, P. T. Molecular Dynamics Study of Water Adsorption on TiO_2 Nanoparticles. *J. Phys. Chem. C* **2007**, *111*(19), 6920–6926.
21. Alimohammadi, M.; Fichthorn, K. A. A Force Field for the Interaction of Water with TiO_2 Surfaces. *J. Phys. Chem. C* **2011**, *115*(49), 24206–24214.
22. Kim, S.-Y.; Kumar, N.; Persson, P.; Sofo, J.; van Duin, A. C. T.; Kubicki, J. D. Development of a ReaxFF Reactive Force Field for Titanium Dioxide/Water Systems. *Langmuir* **2013**, *29*(25), 7838–7846.
23. Prezhdo, O. V. Photoinduced Dynamics in Semiconductor Quantum Dots: Insights from Time-Domain Ab Initio Studies. *Acc. Chem. Res.* **2009**, *42*(12), 2005–2016.
24. Fantacci, S.; De Angelis, F. A Computational Approach to the Electronic and Optical Properties of Ru(II) and Ir(III) Polypyridyl Complexes: Applications to DSC, OLED and NLO. *Coord. Chem. Rev.* **2011**, *255*(21–22), 2704–2726.
25. Martsinovich, N.; Troisi, A. Theoretical Studies of Dye-Sensitized Solar Cells: From Electronic Structure to Elementary Processes. *Energy Environ. Sci.* **2011**, *4*(11), 4473–4495.
26. Labat, F.; Le Bahers, T.; Ciofini, I.; Adamo, C. First-Principles Modeling of Dye-Sensitized Solar Cells: Challenges and Perspectives. *Acc. Chem. Res.* **2012**, *45*(8), 1268–1277.
27. Persson P. *Multiscale Modelling of Interfacial Electron Transfer*. Sol Energy Convers. Dyn. Interfacial Electron Energy Transf. [Internet]; Royal Society of Chemistry; 2013;

Available from: http://pubs.rsc.org/en/content/chapter/bk9781849733878-00077/978-1-84973-387-8.
28. Gałyńska, M.; Persson, P. Emerging Polymorphism in Nanostructured TiO_2: Quantum Chemical Comparison of Anatase, Rutile, and Brookite Clusters. *Int. J. Quantum Chem.* **2013**, *113*, 2611–2620.
29. Persson, P.; Bergström, R.; Lunell, S. Quantum Chemical Study of Photoinjection Processes in Dye-Sensitized TiO_2 Nanoparticles. *J. Phys. Chem. B* **2000**, *104*(44), 10348–10351.
30. Persson, P.; Lundqvist, M. J. Calculated Structural and Electronic Interactions of the Ruthenium Dye N3 with a Titanium Dioxide Nanocrystal. *J. Phys. Chem. B* **2005**, *109*(24), 11918–11924.
31. Persson, P.; Lundqvist, M.; Ernstorfer, R.; Goddard, W.; Willig, F. Quantum Chemical Calculations of the Influence of Anchor-Cum-Spacer Groups on Femtosecond Electron Transfer Times in Dye-Sensitized Semiconductor Nanocrystals. *J. Chem. Theory Comput.* **2006**, *2*(2), 441–451.
32. Lindan, P. J. D. Water Chemistry at the SnO_2(1 1 0) Surface: The Role of Inter-Molecular Interactions and Surface Geometry. *Chem. Phys. Lett.* **2000**, *328*(4–6), 325–329.
33. Bandura, A. V.; Kubicki, J. D.; Sofo, J. O. Comparisons of Multilayer H_2O Adsorption onto the (110) Surfaces of α-TiO_2 and SnO_2 as Calculated with Density Functional Theory. *J. Phys. Chem. B* **2008**, *112*(37), 11616–11624.
34. Hahn, K. R.; Tricoli, A.; Santarossa, G.; Vargas, A.; Baiker, A. First Principles Analysis of H_2O Adsorption on the (110) Surfaces of SnO_2, TiO_2 and Their Solid Solutions. *Langmuir* **2012**, *28*(2), 1646–1656.
35. Goniakowski, J.; Gillan, M. J. The Adsorption of H_2O on TiO_2 and SnO_2(110) Studied by First-Principles Calculations. *Surf. Sci.* **1996**, *350*(1–3), 145–158.
36. Persson, P.; Gebhardt, J. C. M.; Lunell, S. The Smallest Possible Nanocrystals of Semiionic Oxides. *J. Phys. Chem. B* **2003**, *107*(15), 3336–3339.
37. Lundqvist, M. J.; Nilsing, M.; Persson, P.; Lunell, S. DFT Study of Bare and Dye-Sensitized TiO_2 Clusters and Nanocrystals. *Int. J. Quantum Chem.* **2006**, *106*(15), 3214–3234.
38. Becke, A. D. Density-Functional Thermochemistry. III. The Role of Exact Exchange. *J. Chem. Phys.* **1993**, *98*(7), 5648–5652.
39. Bouteiller, Y.; Mijoule, C.; Nizam, M.; Barthelat, J. C.; Daudey, J. P.; Pelissier, M.; et al. Extended Gaussian-Type Valence Basis Sets for Calculations Involving Non-Empirical Core Pseudopotentials. *Mol. Phys.* **1988**, *65*(2), 295–312.
40. Pal, S. K.; Sundström, V.; Galoppini, E.; Persson, P. Calculations of Interfacial Interactions in Pyrene-Ipa Rod Sensitized Nanostructured TiO_2. *Dalton Trans.* **2009**, *45*, 10021–10031.
41. Krishnan, R.; Binkley, J. S.; Seeger, R.; Pople, J. A. Self-Consistent Molecular Orbital Methods. XX. A Basis Set for Correlated Wave Functions. *J. Chem. Phys.* **1980**, *72*(1), 650–654.
42. Roos, B.; Veillard, A.; Vinot, G. Gaussian Basis Sets for Molecular Wavefunctions Containing Third-Row Atoms. *Theor. Chim. Acta* **1971**, *20*(1), 1–11.
43. Hay, P. J. Gaussian Basis Sets for Molecular Calculations. The Representation of 3d Orbitals in Transition-Metal Atoms. *J. Chem. Phys.* **1977**, *66*(10), 4377–4384.
44. Hay, P. J.; Wadt, W. R. Ab Initio Effective Core Potentials for Molecular Calculations. Potentials for the Transition Metal Atoms Sc to Hg. *J. Chem. Phys.* **1985**, *82*(1), 270–283.
45. Hay, P. J.; Wadt, W. R. Ab Initio Effective Core Potentials for Molecular Calculations. Potentials for K to Au Including the Outermost Core Orbitals. *J. Chem. Phys.* **1985**, *82*(1), 299–310.

46. Wadt, W. R.; Hay, P. J. Ab Initio Effective Core Potentials for Molecular Calculations. Potentials for Main Group Elements Na to Bi. *J. Chem. Phys.* **1985**, *82*(1), 284–298.
47. Frisch, M.; Trucks, G.; Schlegel, H.; Scuseria, G.; Robb, M.; Cheeseman, J.; et al. Gaussian 09, Revision B.01. In *Gaussian 09 Revision B01*, Gaussian Inc.: Wallingford, CT, 2009
48. Cangiani, G.; Baldereschi, A.; Posternak, M.; Krakauer, H. Born Charge Differences of TiO_2 Polytypes: Multipole Expansion of Wannier Charge Densities. *Phys. Rev. B* **2004**, *69*(12), 121101.
49. Posternak, M.; Baldereschi, A.; Walter, E. J.; Krakauer, H. Wannier Functions and Born Charge Tensors of Brookite TiO_2. *Phys. Rev. B* **2006**, *74*(12), 125113.
50. Wyckoff, R. W. G. *Crystal Structures 1*, 2nd ed.; Interscience Publishers: New York, 1963.
51. Baur, W. H.; Khan, A. A. Rutile-Type Compounds. IV. SiO_2, GeO_2 and a Comparison with Other Rutile-Type Structures. *Acta Crystallogr. B* **1971**, *27*(11), 2133–2139.
52. Meagher, E. P.; Lager, G. A. Polyhedral Thermal Expansion in the TiO_2 Polymorphs: Refinement of the Crystal Structures of Rutile and Brookite at High Temperature. *Can. Miner.* **1979**, *17*, 77–85.
53. Reyes-Coronado, D.; Rodríguez-Gattorno, G.; Espinosa-Pesqueira, M. E.; Cab, C.; de Coss, R.; Oskam, G. Phase-Pure TiO_2 Nanoparticles: Anatase, Brookite and Rutile. *Nanotechnology* **2008**, *19*(14), 145605.
54. Xu, Y.; Ching, W. Y. Electronic and Optical Properties of All Polymorphic Forms of Silicon Dioxide. *Phys. Rev. B* **1991**, *44*(20), 11048–11059.
55. Alvarez, J. R.; Rez, P. Electronic Structure of Stishovite. *Solid State Commun.* **1998**, *108*(1), 37–42.
56. Jacquemin, J. L.; Bordure, G. Band Structure and Optical Properties of Intrinsic Tetragonal Dioxides of Groups-IV Elements. *J. Phys. Chem. Solids* **1975**, *36*(10), 1081–1087.
57. Robertson, J. Electronic-Structure of SnO_2, GeO_2, PbO_2, TeO_2 and MgF_2. *J. Phys. C: Solid State Phys.* **1979**, *12*(22), 4767–4776.
58. Agekyan, V. T. Spectroscopic Properties of Semiconductor Crystals with Direct Forbidden Energy Gap. *Phys. Status Solidi* **1977**, *43*(1), 11–42.
59. Arlinghaus, F. J.; Albers, W. A., Jr. Electronic Energy Bands and Optical Transitions in Tetragonal Germanium Dioxide. *J. Phys. Chem. Solids* **1971**, *32*(7), 1455–1462.
60. Stapelbroek, M.; Evans, B. D. Exciton Structure in the U.V.-Absorption Edge of Tetragonal GeO_2. *Solid State Commun.* **1978**, *25*(11), 959–962.
61. Cohen, R. E. Bonding and Elasticity of Stishovite SiO_2 at High Pressure: Linearized Augmented Plane Wave Calculations. *Am. Miner.* **1991**, *76*(5–6), 733–742.
62. Evarestov, R. A.; Bandura, A. V.; Proskurov, E. V. Plain DFT and Hybrid HF-DFT LCAO Calculations of SnO_2 (110) and (100) Bare and Hydroxylated Surfaces. *Phys. Status Solidi B* **2006**, *243*(8), 1823–1834.

INDEX

Note: Page numbers followed by "*f*" indicate figures and "*t*" indicate tables.

H

R

S

T

U

Kolibri 360 GmbH
Ferdinand-Jühlke-Straße 7
99095 Erfurt, Deutschland
produktsicherheit@kolibri360.de